Biotechnology of Orchids

Biotechnology of Orchids

Jayaram Reddy
Professor
St. Joseph's Post Graduate and Research Centre
Langford Road, Bangalore, India

I.K. International Publishing House Pvt. Ltd.
NEW DELHI • BANGALORE • MUMBAI

Published by
I.K. International Publishing House Pvt. Ltd.
S-25, Green Park Extension
Uphaar Cinema Market
New Delhi - 110 016 (India)
E-mail : ik_in@vsnl.net

ISBN 81-88237-67-1

Published by Krishan Makhijani for I.K. International Publishing House Pvt. Ltd., S-25, Green Park Extension, Uphaar Cinema Market, New Delhi-110 016 and Printed by Rekha Printers Pvt. Ltd., Okhla Industrial Area, Phase II, New Delhi-110 020.

PREFACE

Biotechnology encompasses experimental techniques that enable scientists to manipulate and use biological systems in basic research and in the development of agricultural, medicinal, and commercial products. The history of biotechnology has often been running back through the discovery of recombinant DNA techniques. However biotechnology was an established industry before the impact of molecular genetics. It was associated with the most important drugs ever made, the antibiotics and with hopes for world-shaking products such as starvation curing single-cell protein, and energy-producing gasohol. Biotechnology is as old as fermentation, but revolutionary advances during the early 1970s gave scientists unprecedented ability to isolate genes and manipulate the genetic structure of organisms. The scope and pace of biotechnological innovation expanded dramatically during this period and continue to grow at a rapid pace. Biotechnology has led to new kinds of partnerships between university scientists, government organizations and private industry, blurring the line between basic and applied research in biology. What made biotechnology seem a potential counterpart to information technology in a third industrial revolution were the recent advances in molecular biology. They seemed to offer the possibilities of making industrially synthesized proteins identical to those produced in the normal human body. There was also the hope that industrially useful materials such as human body parts and tissues, hormones, vaccines, interferons, interleukins, plastics or ready coloured fibres might be made in the laboratory environment. The attractions of a potentially new agricultural regime might be as great as the implications for medicine. Agricultural benefits sometimes overshadowed the medical implications. However, furore over the agricultural use of genetically modified organisms meant a growing distinction between agricultural and medical aspects of biotechnology. Practically plant biotechnology research centered around two main aspects; one being plant transformation and the other is the successful regeneration and rapid multiplication of the genetically modified (GM) plants.

Since the first reports of transgenic plants in 1983, the transformation of plants with foreign genes has been a key feature in the development of modern plant biology. Transgenic plants, apart from their value for crop improvement, provide an opportunity for the analysis of function and regulation of plant genes. In addition, studies on the transformation process have played a key role in the development of plant molecular biology. Different methods have been developed to introduce the foreign genes into plants and animals. A common feature is that the transforming DNA has to bypass different membrane barriers; it has first to enter the plant cell by penetrating the plant cell wall and the plasma membrane; then it must pass the nuclear and cytoplasmic barriers. Finally, the newly introduced DNA segment has to establish itself in the plant genome. Once this is achieved, transformation events can be selected for and, if regeneration competence overlaps with transformation competence, transgenic plants can be regenerated

Any plant transformation experiment relies at some point or the other on tissue culture. There are some exceptions to this generalization but the ability to regenerate plants from isolated cells or tissues *in vitro* underpins most plant transformation systems. Most methods of plant transformation applied to GM crops require that a whole plant is regenerated from isolated plant cells or tissue which has been genetically transformed. This regeneration is conducted *in vitro* so that the environment and growth medium can be manipulated to ensure high frequency regeneration. Two concepts, plasticity and totipotency, are central to understand plant cell culture and regeneration. Plasticity allows one type of tissue or organ to be initiated from another type. In this way, whole plants can be subsequently regenerated. The regeneration of whole organisms depends upon the concept that all plant cells can, given the correct stimuli, express the total genetic potential of the parent plant. The maintenance of genetic potential is called 'totipotency'.

Orchids, comprise a unique group of plants and the flowers are a most enchanting and exquisite creation of nature. Taxonomically, they represent the most highly evolved family. Orchids exhibit an incredible range of diversity in size, shape and colour. They are known for their long lasting and bewitchingly beautiful flowers which fetch a very high price in the multi billion dollar floriculture market.

Majority of the cultivated orchids are native of tropical countries and occur in their greatest diversity in humid tropical forest of South and Central America, Mexico, India, Ceylon, Burma, South China, Thailand, Malaysia, Philippines, New Guinea and Australia. It is estimated that about 1,300 species (140 genera) of orchids are found in India with Himalayas as their homeland and the others being scattered in Eastern and Western Ghats. Majority of the orchids being marketed in the world today are Indian in origin. Orchids account for 2.7% of the global cut flower production. India though strategically well ahead does not even account for 1 percent of the world orchid trade. Ironically Taiwan has monopolized the world orchid trade. India with a rich orchid germplasm has much to gain by the utilization of biotechnology to improve orchid cultivation in order to stake a claim in the global floriculture trade. The floriculture sector is gaining importance throughout the world. Flower production is burgeoning in many countries. There has also been the broad expansion in the international trade. Sales are largely centred on markets with strong purchasing powers, like Germany, England, the USA and Japan. The Netherlands is the major player in this sector, both in production and distribution.

In flowering plants the shape and colour of the flower determine ornamental value. Biotechnology is thus indispensable for the production of flowers of International acceptance. Hence the application of the techniques of biotechnology such as tissue culture and transformation will have profound impact on the Indian floriculture prospects.

This book is written with a passion for orchids, to assist researchers in plant biotechnology and to those who contribute to the growth of the Indian floriculture. The book is largely based on the experience gained by the author during the research conducted on orchids for Ph. D degree. The main areas considered in this book are orchids and biotechnology. Though the book is entitled **"Biotechnology of Orchids"** it deals with plant biotechnology in general. An attempt has been made to collect all the necessary information regarding orchids and plant biotechnology. About 650 research papers and books have been referred and cited in the book and the protocols developed for more than 260 orchid cultivars are reviewed, which are quite useful to researchers. In the initial chapters the information about orchids is written. History, uniqueness, habit, habitat, orchid seed and their germination, distribution, evolutionary status, hybridization, registration of hybrids, taxonomy, methods of cultivation and orchid-mycorrhizal association are discussed to the extent necessary for a scientist. The status of world floriculture industry and India's position is also explained with a critical note.

The next focus of the book is on plant biotechnology. History, prospects, potentials and techniques are discussed in detail. Since the present trend in plant biotechnology is to produce genetically modified plants by transgenesis, importance has been given to the techniques related to this process. Methods of transgenesis, transgenic crops, and the related controversies are treated in separate chapters. An attempt is made to clear the confusion caused by the press media and NGOs regarding GMOs. Different methods by which foreign genes can be introduced into an organism are described with specific examples. These techniques are discussed for plant systems in general and also for orchids in particular. Protoplast isolation, fusion and culture techniques are also reviewed for plants in general and orchids with specific examples. Cytology is always important in biology and it is noticed that most biotechnology books neglect this section; hence a chapter dedicated to cytological techniques has been introduced, which also includes the methods for the preparation of different types of stains used in cytology. Tissue culture technique forms the backbone of biotechnology and hence various steps of the technique are described in detail for orchids which of course is not much different for other plants. Compositions of the nutrient media are not given in the appendix as they are easily available and nowadays media are available in the "ready mix" forms. The last section carries protocols for the micropropagaion of four different orchid hybrids and also includes original results obtained during morphogenetic studies carried out by the author.

This book can help the readers to prepare for examination/interviews, conduct basic plant research, participate in plant genetic improvement programmes, and enter the plant production industry in any role requiring the knowledge of plant biotechnology, or even start a plant production industry. The contents and design of the chapters in the book will meet the needs of students, researchers, trainees, professional scientists, and research readers as well as be of great value to orchid lovers and floriculturists who may or may not have had the knowledge of biotechnology.

I am grateful to my guides, mentors, students, colleagues, and all those who have been directly or indirectly helpful in making this book a reality. I thank my wife, my son and all others for patience, encouragement and support.

Jayaram Reddy
Professor, St. Joseph's
Post Graduate and Research Centre, Langford Road, Bangalore, India

CONTENTS

1

INTRODUCTION

General Account

Orchids are enchanting and exquisite creations of nature. "It is interesting to look at one of the magnificent exotic species, or, indeed, at one of our humblest forms, and observe how profoundly it has been modified, as compared with all ordinary flowers". This is a statement made by Charles Darwin about orchids and quoted in the first volume of *Orchid Review*. Jacob Breynius, a German botanist, in the 17th century collected information about plants unfamiliar to Europeans and compiled *Exortium Aliarum Que Minus Cognitum Plantarum*, in which he described the extraordinary diversity of orchids. They take any form, be it little birds, lizards, insects, etc. They resemble man, woman, sometimes they look like an austere sinister fighter and sometimes like a clown. They represent the image of a lazy tortoise, a melancholy toad or an ever chattering agile monkey, and so on (cf. Black, 1977).

Orchids, the most beautiful flowers in god's creation, comprise a unique group of plants. Taxonomically, they represent the most highly evolved family among monocotyledons. Orchids exhibit an incredible range of flower with a vast diversity in size, shape and colour. They are the most pampered of all plants and occupy a top position among all the flowering plants. They have been of great value in cut flower production and as potted plants. They are known for their long-lasting and bewitchingly beautiful flowers, which fetch a very high price in the international market. Orchids create immense interest because, orchid flowers are irregular, and extremely variable in size and shape with a sparkling texture. They may grow solitarily or in spikes, emitting the fragrance of lemons, cloves or fresh lavender oil. A few are, however highly odorous. Orchids are highly diverse in montane tropics (Fig. 1.1).

Orchids have extended a strange fascination over many races of people. The early Greeks revered their testiculate root tubers as a symbol of virility.

Fig.1.1 Hyperdiverse orchids of montane tropics; *Telipogon* spp of Equador.

The Chinese called many of their orchids, "The plants of King's Fragrance", as long back as the time of Confucius (Hunt, 1987). Indians have known orchids even from the Vedic period. In the Vedic scriptures there is a mention of plants under the name "VANDA", which has been adapted as a generic name for one of the most beautiful group of orchids. Most of the orchids are perennial herbs with simple leaves. Although their specialised flower structure conforms to a standard plan, the vegetative parts show great variations, a large number of them being epiphytes, or terrestrials and a few being saprophytes and leafless in nature. A majority of the cultivated orchids are native to tropical countries and they occur in their greatest diversity in humid tropical forests of South and Central America, Mexico, India, Srilanka, Burma, South China, Thailand, Malaysia, Philippines, New Guinea and Australia. **Brazilian Cattleyas**, **Mexican Laelias** and **Indian Dendrobiums**, **Cymbidiums** and **Vandas** have played a major role in the development of modern orchid industry of the world.

Most of the known species of orchids blend unusually varied combinations of attractive colours and bizarrely shaped floral parts, with an almost overpowering perfume. Most of the orchids grown today are artificial hybrids produced by the plant breeder. Orchids are unique in their floral structure, their pollination mechanisms, and their association with a fungal partner, their miniscule seeds and their excessive promiscuity which can lead to features from many species being combined in one plant. In a greenhouse, orchids can be a frustrating challenge to the grower's skill, and in the wild, they are a reward at the end of a day's exploration. Today orchids are often sold at exorbitant prices that are many times their intrinsic value, and they are much higher priced than any other plants because of the extraordinary beauty of their flowers, variety of fragrance and brilliance of colour. Now floriculture is a multibillion dollar industry with orchids holding a very prominent place. There is a big demand for this fascinating group of plants. In order to meet this demand, development of advanced micropropagation techniques has become necessary.

Theophrastus, a Greek philosopher, who is also called the father of botany (370-285 BC), gave the name "***Orchids***" to the group of bizarre plants, on the basis of the resemblance of paired underground tubers of these plants to masculine anatomy (the testes). This resemblance was also responsible for the mistaken belief that orchids possess aphrodisiac properties, and that eating underground tubers might provoke "Venus" and they may beget male children.

HISTORY OF ORCHID CULTIVATION

Orchids have been known to horticulturists and herbalists for the past 400 years (Arditti 1966a). But the first reference to orchids is found in eastern literature in the writings of Confucius, a Chinese philosopher (551 - 479 BC). The first reference to this particular species occurs in contemporary Chinese literature in the early Christian era. For example, the Chinese minister of state, Ki-Han (in a work written in about 300 AD) mentions two orchids, now known as *Cymbidium ensifolium* and *Dendrobium moniliformae* (cf. Black, 1977). Several other books were also written from this time upto the middle of the 18th century. Most of the orchids described, were cultivated and admired for their aesthetic value, the delicacy and refinement of their flowers, and were almost invariably fragrant, for according to the old Chinese proverb *"A flower without fragrance is like a person without virtue"*. What is probably the first book on the cultivation of orchids was written in Chinese in about 1000 AD. It lists species and varieties of orchids and also gives cultural instructions (cf. Black, 1977). It is clear from this information that the cultivation of orchids in China and Japan was an occupation of the aristocracy.

The evolution of orchid culture from a hobby to commercialisation was very slow, as most of the orchids collected from different regions died because they could not adopt to local conditions. The methods of propagation were not known and most of the orchids which flowered under utmost care were found to be self sterile. Further, it was difficult to germinate the seeds, as they lacked functional endosperms. The seeds sown in the nursery beds required a long period of germination and any disturbance to the soil or physical environment destroyed the whole population.

According to available literature, orchids have been known to Indians, since about 500 BC. They are described by Charaka (300 BC), Shushruta (250 BC), Waghbhata Shiva (200 BC) and others in their Ayurvedic texts, which were the *materia medica* of those days. There are unmistakable descriptions of many common orchids like *Habenaria grandifloraformis* called by the name *"Sthala Padmini"* which means terrestrial lotus. *Eulophia nuda* was called *"Malakanda"*, as its tubers look like a beaded necklace (cf. Bopaiah,1986). The Chinese word for orchid is *'ran'* or *'lan'* (cf. Black, 1977) and in Sanskrit, they are called *'Rasanas'* (cf. Diwakar,1987).

In the west, however, the interest in orchids was mainly medicinal. They were first mentioned in *History of Plants and the Causes of Plants* by

Theophrastus, who gave them the name *"Orchids"*, the Greek word for testicles, from the shape of the paired underground bulbs of an orchid, *Orchis morio* (cf. Hunt, 1987). The first Western reference to an exotic species is found in *Aztec Herbal of 1552, the Bandianus MS* in which *Vanilla* is illustrated. This was also the first tropical orchid named. Tropical orchids were first described by HA Reede tot Draakenstein, who was appointed governor of the Dutch island of Malabar during the second half of the 17th century (cf. Black, 1977).

The botanical interest in orchids progressed gradually, but gained increasing momentum as more knowledge became available, and the improvement in microscopy made detailed observations possible. By the start of 18th century, orchid collection was firmly established as a fashionable horticultural practice in many parts of the world. The colonising Europeans, in the tropics, picked orchids from the rain forest trees and drew them or pressed them or even tried to send them home for domestication.

Orchidomania

In 1800 itself, the trade of exotic plants was a competitive business. There was a *cactus fever*, a *fern fever* and at the close of the nineteenth century, a *Vogue of orchid*.

In the later half of the 19th century, a sort of *'Orchidomania'* gripped many of the Europeans. In fact from the beginning of the 19th century, the number of tropical orchids finding their way back to England gradually increased. As more and more specimens were seen by the public, the beauty and strangeness of the flowers created great excitement and a growing demand to own specimens of these rare and exotic blooms brought into being many nurseries, whose names are now classics in the world of orchids. It has been recorded that a major commercial orchid nursery in England imported over a million plants of *Odontoglossum crispum*, from South Africa, in a single shipment.

Orchid Hybrids and Hybridisation

There is long list of orchid hybrids, both natural and artificial. A number of natural hybrids, both intergeneric and interspecific, have been reported in different genera like *Odontoglossum, Phalaenopsis, Cattleya, Laelia, Miltonia and Oncidium. Cattleya guatemalensis,* collected from Guatemala in 1861 by Skinner, was found to be a natural hybrid between *Epidendrum aurantiacum* and *Cattleya skinneri*. It was later named as *Epicattleya*. Similarly,

Phalaenopsis intermedia is a natural hybrid between *P. aphrodite* and *P. rosea*.

Hybridisation was considered to be an impossibility upto the middle of the 19th century. It was John Dominy, a relatively untutored Scot Gardner (who was described as the devoted and faithful servant and friend of James Veitch), who first achieved this success. In 1853, Dominy crossed *Calanthe masuca* and *C. fuscata* and this cross, or *Orchidaceous mule* as the Victorians termed it, produced its first flowers in 1856, at Veitch Nursery (Withner, 1959). From then on, it became possible to breed selectively for colour, size, shape and any other desired characteristics. Since 1856, when the first orchid hybrid *Calanthe dominyi* flowered, a large number of artificial hybrids have been produced both at intergeneric and interspecific levels. To date, more than 150,000 hybrids have been registered with an average of 10,000 or more every year. The success and ease with which such a large number of hybrids are being produced every year depends upon the fact that most of the orchid genera and species have no generic barriers and they cross freely with each other. It must be mentioned that most of the orchid genera are still in the process of evolution and most of the orchid groups are in reality only artificial constructs. The other two factors which have played a major role in the development of orchid hybrids are polyploidy and introgressive hybridisation. In some genera, like *Cymbidium, Paphiopedilum, Phalaenopsis, Cattleya, Laelia,* and *Sophronitis*, Polyploidy has shown an extraordinary effect, and coupled with intergeneric compatibility, has culminated in the formation of hybrid groups which show, both greater size and hybrid vigour as compared to the parental species. The important genera which have given maximum number of man-made hybrids are *Cattleya, Cymbidium, Paphiopedilum, Vanda, Dendrobium*, etc.

It was only in 1821, that Conrad Loddiges and sons started growing orchid plants commercially at their nursery in Hackney. They were followed by John Dominy of M/s Veitch & Sons who produced the first orchid hybrid between two Calanthe species in 1852. This was followed by a large number of man-made hybrids produced by hobbyists and small growers. In 1913, Sun Kee Nursery, Singapore, started the first cut-flower production of *Arachnis* type hybrids.

In addition to working on the creation of new hybrids, there was a continued search for new species in their natural habitat. Fierce competition developed between rival orchid establishments, and collectors were sent all over the world, often at a great personal risk, to track down rare orchids.

Owing to the difficulty orchid growers experienced in reproducing new stock from seed, an enormous number of plants were sent home. Later on regulations were brought into effect prohibiting the massive export of orchid in most countries to which they are native. The giant among nineteenth century orchid hunters was Frederick Sander, whose teams of collectors were sent to all parts of the world, and whose name is commemorated in countless species and hybrids (cf. Black, 1977). Orchids are sometimes capable of fooling taxonomists by being closely similar (Fig. 1.2).

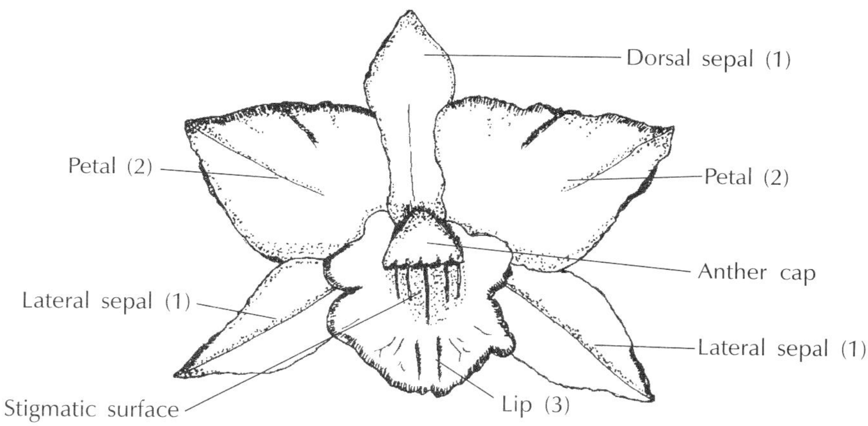

Fig. 1.2 *Dendrobium type*

The huge volume of literature published about orchids upto the beginning of the century has been, in great part, written by scientists. A different form of work on orchids was inevitable, hence many books have been written by cultivators of these plants, giving instructions on the different requirements of great number of species. Towards the end of the 18th century when orchids were first imported into Northern Europe in significant quantities, little was known of their needs in captivity. They were introduced by seamen or non-botanists and were put into stores or hot houses heated by a bricked fire underneath the house, and to counteract the heat, large quantities of water was used, as a result, not surprisingly, most plants died soon after their arrival. It was only the discovery of cast iron and the use of four and six inch iron pipes, filled with water and heated by a boiler, that long-term cultivation of greenhouse plants was possible. Even then cooler growing orchids did not thrive for many years. However, it was later realised that although the orchids brought home were tropical in origin, not all of them were used to a hot, steamy, Amazonic atmosphere, and when the growers ceased to be afraid of allowing fresh air into their orchid houses, success quickly followed. The book written by Watson and Bean in 1893, *"Orchids, Their Culture and Management with Descriptions of all the Kinds in General Cultivation"* gives some idea about their passage in to European homes. The most popular and ambitious of the publications was *"The Orchid Grower's Manual"* by Williams, published in 1894.

All these books helped to create a growing interest in orchids and orchid culture among a wider public. Periodicals such as The *"Gardeners' Chronicle"* contained illustrated articles about orchids. *"The Orchid Review"* was started in 1893 by R.A. Rolfe. It is still a vigorous publication. The other important publications were the *"Sander's List of Orchid Hybrids"*, and *"The Orchid Stud Book"*. It was started by Sander in collaboration with C.C. Hurst. The Book by James Veitch, *"Hortus Veitchii"* contains a list of orchid hybrids and important species covering all branches of horticulture. *"Sander's Orchid Guide"* started in 1901 gives cultural details and instructions in great detail. The *Horticultural Society*, which became the *Royal Horticultural Society* in 1861, is one of the institutions which have influenced public interest in orchids.

Interest in orchids, particularly in the raising of hybrids, continued to increase, and was greatly helped by the discovery of the asymbiotic method of germinating orchid seeds, and the technique of producing orchids by meristem culture, developed by Morel (1965). It led to an increase in the number of orchid societies; before 1939, there was only one orchid society in England—The *Manchester and North of England Society*. Now there are several hundred orchid societies. In USA, every city has its own orchid society. These societies provide all necessary information regarding orchids through various ways. The American Orchid Society was formed in 1921 by amateur, private, and commercial growers. The monthly bulletin of this society, *American Orchid Society Bulletin,* is full of information gained from practical experience. In addition, information is also available on the internet.

Orchid Review is the oldest journal. Besides, *American Orchid Society Bulletin*, there are several popular journals such as *Australian Orchid Review, Die Orchidee, Florida Orchidist, New Zealand Orchid Review, Orchidian, Orchid Advocate, Orchidata, Orchid Digest, Orchid Society of Great Britain Journal, Orcuidea Orcuideologea South African Ochid Society Journal*, etc. (cf. Hunt, 1987). In India the only major orchid journal is the *Journal of*

Orchid Society of India. Now, *Zoo Outreach Organisation* has been publishing *Zoo Print Journal*, devoted to the orchids of the Western Ghats.

Orchid Breeding and Registration of Hybrids

Ever since the success achieved by Dominy in 1853, in producing hybrids breeders have worked so assiduously that now there are several thousand registered hybrids. There are much more hybrids than the known species. Some of these hybrids are naturally formed. Orchids being notoriously promiscuous, a great majority of the orchids are produced by man. Since orchid hybrids are invariably fertile and equally promiscuous, it is possible to combine the characters of many genera in one plant species. For example, *Schafferar* **Martha Schaffer** involves species from five different genera, *Aspasia, Brassia, Cochioda, Miltonia* and *Odontoglossum* (cf. Hunt, 1987).

More hybrids are recorded in orchids than in any other plant family. The history of orchid hybridisation has been discussed by Arditti (1984). It is estimated that from 1883 to 1974, approximately 106,410 orchid hybrids have been derived in different parts of the world (Rao, 1977). There are 450 hybrid genera derived from about 120 natural genera, and over 70,000 hybrid grexes, in single or hybrid genera, derived from over 1100 natural species. A single genus (hybrid or natural) may contain upto 1300 distinct grexes. A single grex may be related upto 35 species in its parentage. In addition, a single hybrid genus may have been derived from upto 8 natural genera (Hunt, 1986).

International Registration Authority of *Orchid hybrids*, constituted by Royal Horticultural Society is the official body to register orchid hybrids (Rao, 1977). Details of the rules and registration procedures of all the hybrids produced and reported so far can be found in the *"Handbook on Orchid Nomenclature and Registration"*, published by the *International Orchid Commission* (Hunt, 1987). Naming of orchid hybrids is regulated by a set of rules and regulations (Cribb *et. al.*,1985), for instance in India, Singh (1986) registered two new orchid hybrids, *Dendrobium* **IIHR 38** and *Vanda* **IIHR 164.**

Morphology

It is paradoxical that though orchids exhibit such exquisite floral features, their stems and leaves are relatively dull. Orchids can be broadly classified as terrestrial and epiphytic species. Terrestrial or ground orchids comprise about half of the known species. Ground orchids may arise from a creeping underground rhizome or a root-like stem, or underground tubers, stem above the ground level is of varying lengths, normally bearing leaves and inflorescence at the tip. The leaves of terrestrial orchids vary from orbicular and heart shaped to cylindrical and awl shaped, but the majority of species possess, what is loosely termed as, lanceolate or spear shaped leaves. In a few cases, leaves are hairy and can sometimes be lobed or otherwise divided, but usually they are entire, shiny and waxy in texture.

Epiphytism is not a unique, but an extensive feature of this family. The majority of the tropical orchids are epiphytes, living high on the branches of tropical rain forest trees. To obtain its necessary moisture and food, an epiphyte develops aerial roots which dangle down and absorb the necessities, from the atmosphere and the humus which collects in the tissues of the bark of tree. Sometimes, the stems enlarge to form bulb-like structures called pseudobulbs. It is a unique feature of orchids. About 95 percent of orchids are autotrophs, but the remaining species are saprophytes. *Neottia nudus-avis* (Bird's nest orchid) and *Corallorhiza trifida* (Coral root orchid) are saprophytes (cf. Hunt, 1987).

Orchid Flowers

Although the flowers of orchids appear to be very complicated, they are very simple in structure. Like many of their relatives, such as the members of Liliaceae, orchids have an outer floral whorl of three sepals and an inner whorl of three petals. However, this similarity is superficial as all orchids have their middle petal transformed into a unique structure called *labellum* or *lip*. This is morphologically different from the other two petals and sepals, being unusually larger, differently coloured and ornamented with hair, crests and cushion like *callus*. Often the lip is extended backwards to form a sector containing *spur*, and in most species, the front of the lip is lobed and may be considerably convoluted. In the orchid flower 6 non essential parts are divided into threes, i.e. 3 sepals and 3 petals, collectively known as tepals due to their resemblance in texture and colour, similarity to each other and subsimilarity to sepals. It is, however, the third petal which is different and distinctive and is called labellum or the lip. The lip which is highly polymorphous is responsible for the different names of the Orchids, like 'Frog orchid', 'Dove orchid', 'Spider orchid', 'Lady slipper orchid'; etc. The reproductive organs of the orchids, the stamens and pistil, are condensed and form a consolidated complex body, the column, the male and female part are separated by a flap or projection of a tissue called rostellum (Figs. 1.3, 1.4, 1.5, and 1.6).

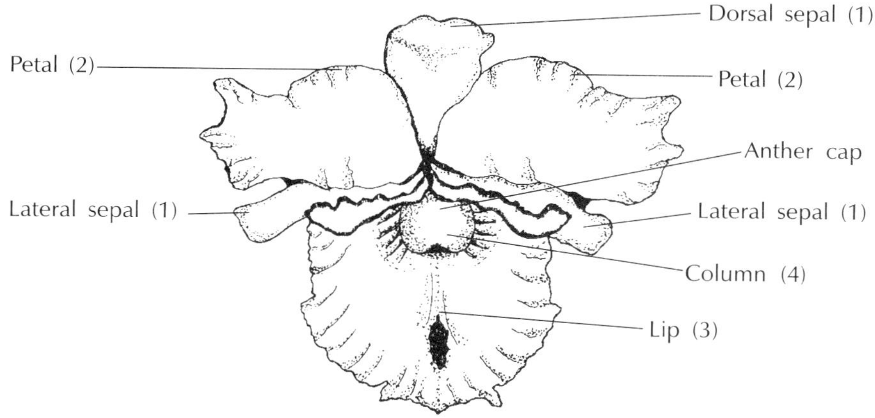

Fig. 1.3 *Cattleya type*

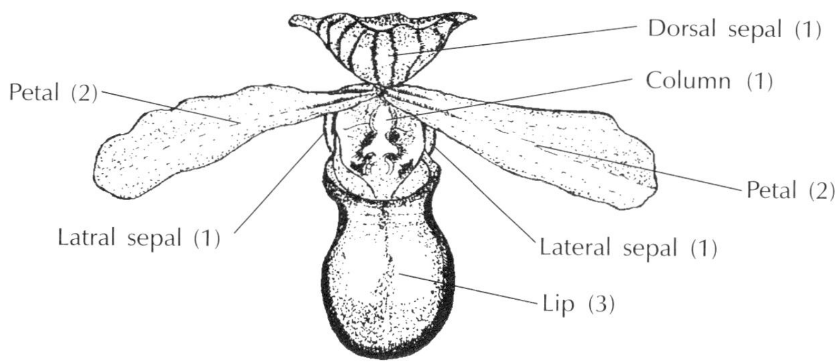

Fig. 1.4 *Cypripedium type*

Fig. 1.5 Orchid floral parts.

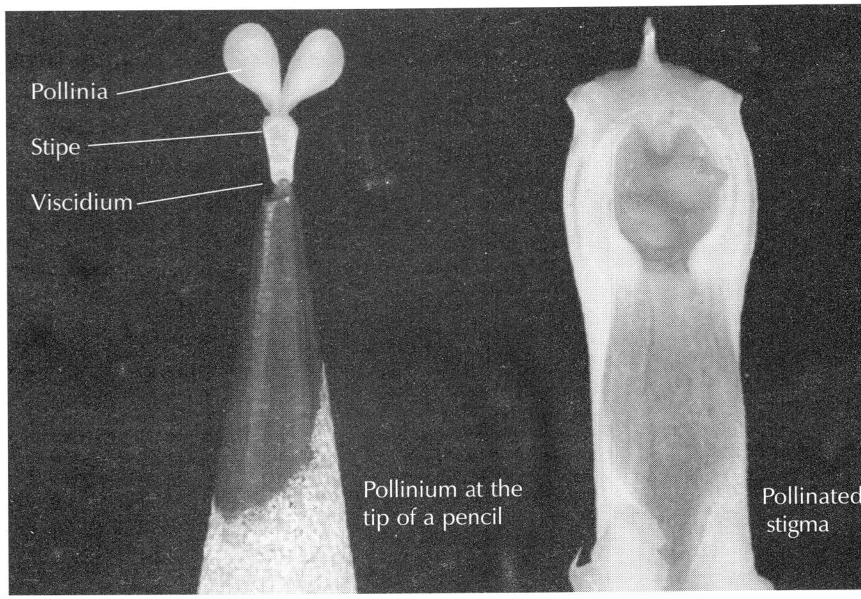

Fig. 1.6 Pollinia of *Aspasia epidendroides*

Fig. 1.7 Pollination of *Disa chrysostachya* by Sunbirds in South Africa
(Reported by S.D. Johnson and M. Brown, 2004)

a. A sun bird sitting on the inflorescence, b. Inflorescence of Disa, c. Inflorescence of Disa; a close up, d. Parts of the flower, e. Transfer of pollinaria through sun bird's feet

Pollination

Orchids have other unique features, related more directly to reproduction. The stamens and pistil are combined into a single structure called *column*. The pollen is variously agglutinated into two, four, six or eight stalked ovoid or spherical masses called *pollinia*. Below the *pollinia* is a sterile flap of tissue called *rostellum*, which prevents the *pollinia* from reaching the stigmatic surface under the *rostellum*, thus preventing self-pollination.

The *lip, spur, column, rostellum* and *pollinia* are all involved in a complex pollination mechanism that is another hallmark of the family, Orchidaceae. Although some orchids are self-pollinated, most of them have evolved various ways to attract a pollinating agent, and ensure transfer of the pollen by fly, wasp, bee, butterfly, moth, humming bird or bat. The flower's shape, colour and ornamentation, especially of the lip, are the most important aspects in attracting pollinators. Floral odour and nectar at the base of the spur are also important factors in attracting the insects (Phatak, 1983). Flowers are designed specifically to suit a particular type of pollinator (Figs. 1.7, 1.8, 1.9 and 1.10).

Fig.1.8 Orchids designed for butterfly pollination

Fig.1.9 Active trap mechanisms for pollination. Flowers of Porroglossum spp have sensitive, motile lips (similar to Venus Fly Trap). Insect landing on lip triggers lip to close, lifting insect into contact with column.

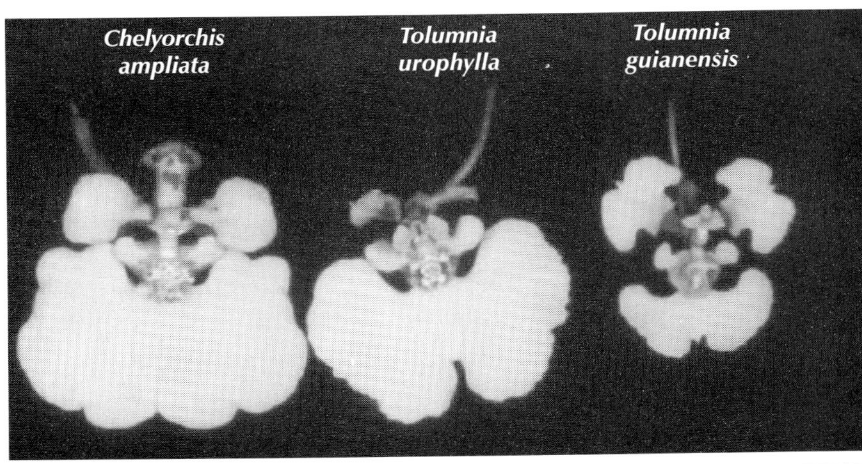

Fig. 1.10 Orchids that are good at fooling pollinators are also good at fooling orchid taxonomists. Flowers look similar but are from different species.

Orchid Seeds and Germination

Orchid seeds are very small, only 1-2 mm x 0.5-1 mm. About 1,300-4,000,000 seeds are produced in a capsule (Pierik, 1987). In *Cycnoches egertonianum*, it is computed that over 4,000,000 seeds can occur in a single capsule. Each seed has a few embryonic cells and a transparent seed coat or testa, which is net-like and species-specific. Most orchid seeds are hardly differentiated, if at all, at the distal end of the embryo, there is a potential growing point, which is barely recognisable. (Figs. 1.11, 1.12, 1.13, 1.14 and 1.15)

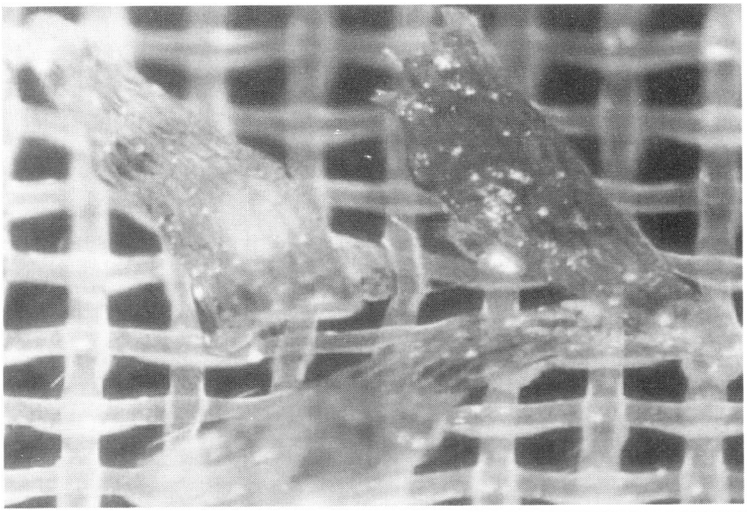

Fig. 1.11 Seeds of *Cypripedium reginae*

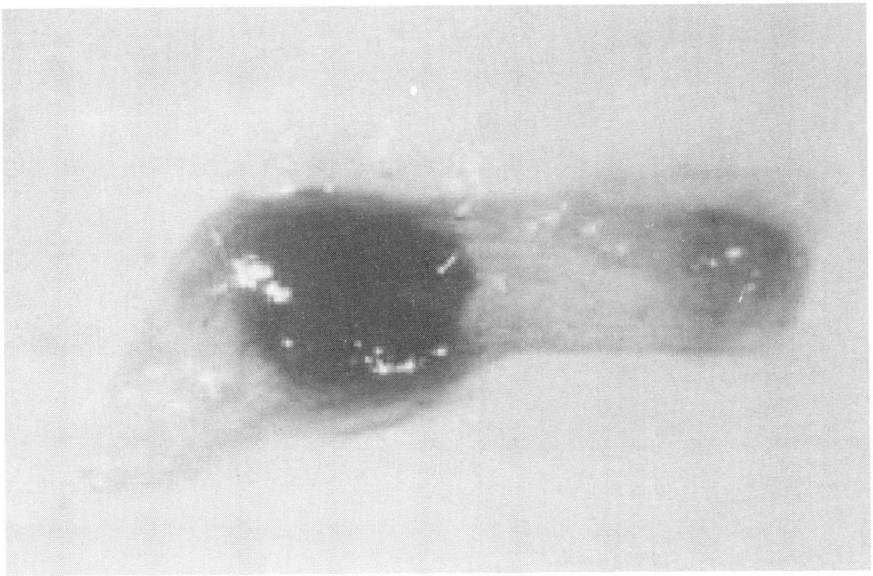

Fig.1.12 Seed of *Cypripedium reginae showing germination*

Fig.1.13 *Anoectochilus formosanus* seeds under stereo microscope

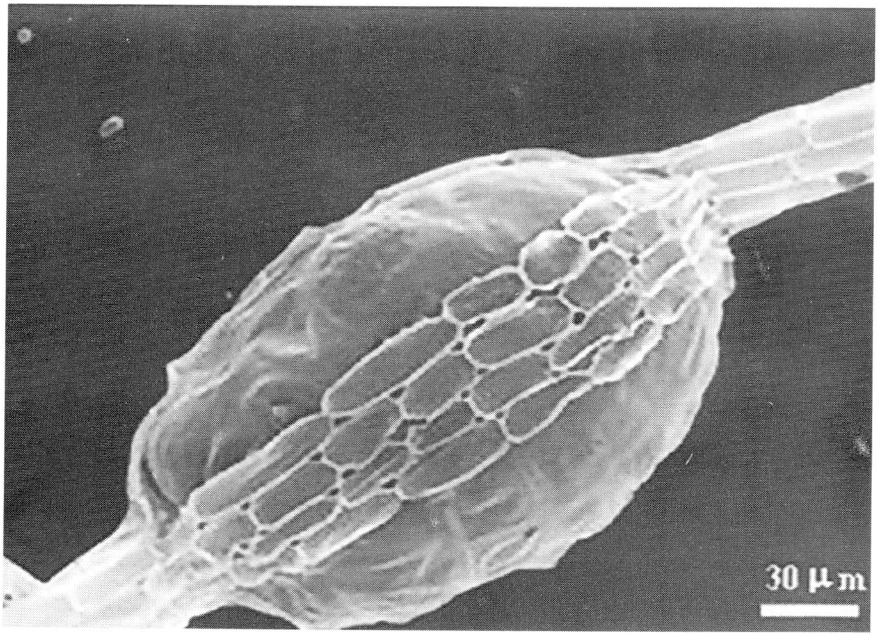

Fig.1.14 Electron micrograph of *Anoectochilus formosanus* seed

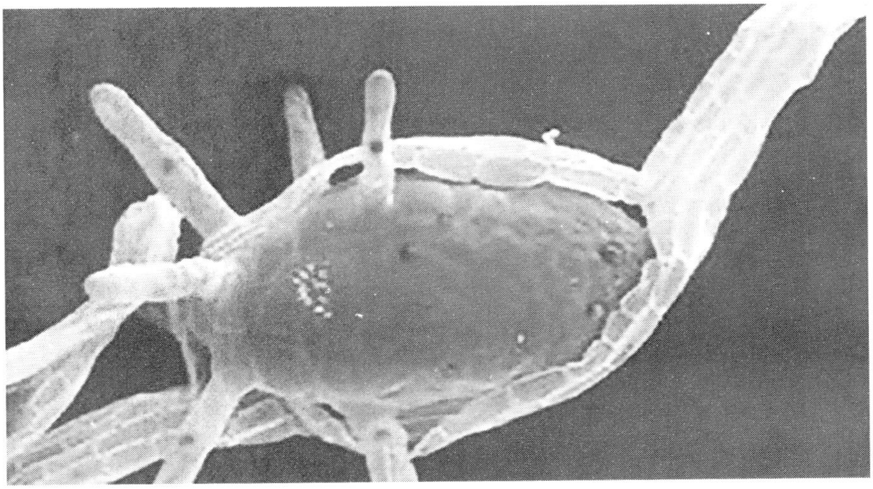

Fig.1.15 *Anoectochilus formosanus* seed forming a protocorm after germination

According to Arditti (Pierik, 1987), orchid seeds imbibe water via the *testa*, and swell. After cell division, the embryo emerges out of the seed coat. A protocorm is formed from the clump of cells and on this a shoot meristem can be distinguished. The protocorm becomes green and at the same time, leaves and roots are produced. Growth is very slow and an early identifiable seedling may take upto a year to appear.

While working on the influence of carbohydrates on green plants, Lewis Knudson (1916) started experimenting with the germination of orchid seeds. On the basis of the analysis of orchid syrup which contained starch, protein, sugars and minerals, he formulated a medium and successfully germinated seeds of *Cattleya, Laelia, Epidendrum*, and concluded that fungus was not necessary for orchid seed germination. This was a breakthrough in orchid seed germination and was followed by others who tried to improve the original medium of Knudson in order to germinate other orchid species. When the seeds were inoculated on a nutrient medium *in vitro* not only is the percentage of germination improved upto 100% in some cases, but it also took less time for further development. It may be mentioned here that while growing an orchid seed differentiates both biochemically as well as morphologically.

Taxonomy

Carl Linnaeus adopted the name Orchid in his *Species Plantarum* published in 1973. John Lindley British botanist, in 1830 introduced the name

'Orchidaceae'. A systematic study of orchids was published by Lindley in his book *The genera and species of orchidaceous plants,* in which he classified the family. Lindley's classification was later modified and adopted in the *Genera Plantarum* by Bentham(1881). Since then, many taxonomists have classified orchids. Pioneering taxonomic work on Indian orchids was done by J.D. Hooker (1888), and it had been compiled in the *Flora of British India*. In 1835, he published another important work, *A century of Indian Orchids,* in the Annals of Royal Botanical Gardens, Calcutta. According to Dressler (1981), the family, Orchidaceace includes about 725 genera and 19,192 species. According to Abraham (1991), there are over 25,000 species. Number of orchids being reported every year is very high (Fig; 1.16).

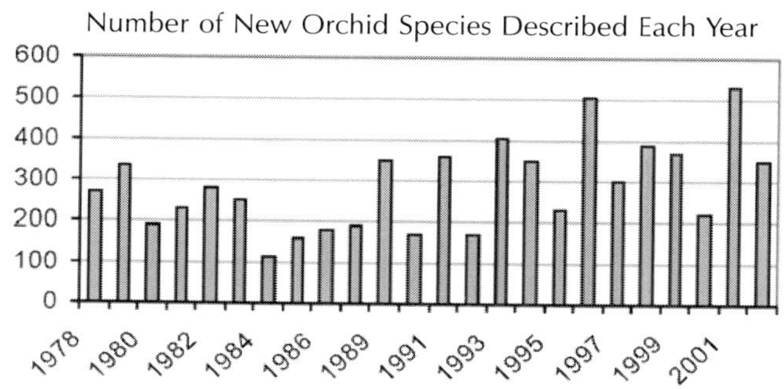

Fig.1.16 Number of new orchids reported in different periods

The family is placed under the division monocotyledonae and has been divided into three subfamilies, Apostasioideae, Cypripedioideae, and Orchidioideae. Orchidioideae is further subdivided into four tribes Orchideae, Neottieae, Epidendreae and Vandeae. Each tribe is further divided into subtribes. Each subtribe contains one to many genera. Each genus may contain 1-150 species. Many species are further divided into subspecies and varieties. When two different species are hybridised (many times the same two species are hybridised) each of the resulting offspring, is called 'grex'. Each 'grex' can contain thousands of different forms, each of which, if grown to flowering size, is called a cultivated variety or cultivar (cv).

The most important characters in orchid classification have always been stamen characters. These include stamen number, anther orientation, pollinium number, and pollinium orientation. Even though they have been used for nearly 200 years, few details are known about them. A study of the development of these features gives additional information and a new perspective on homology.

In a study of the development of pollinium number, which in orchids is usually 2, 4 or 8, it was found that a particular number can be achieved in different ways. In the earliest stages, anther primordia has a single mass of sporogenous tissue in each anther theca, which is then subdivided into the appropriate number of pollinia. This information helps to make more detailed statements about homology and characters, and therefore increases the amount of information in the cladistic analyses.

Distribution of Orchids

Orchids are found throughout the world, except in the Antarctic. The tropics have over 90% of the orchids. According to Jain and Malhotra (1984), there are about 132 genera and 924 species in India. But, according to Abraham (1991), there are over 1200 species described from India, itself. According to Devi and Deka (2001), there are about 700-800 genera and 25,000-35,0000 species in the world.

India has a very large variety of orchids, and hilly regions have one or the other orchid flowering almost throughout the year. The diversity is so large that there are large-flowered, terrestrial, epiphytic and also saprophytic orchids. In general, terrestrial orchids are more common in North-Western India, epiphytic orchids in North-Eastern India and small flowered orchids in the Western Ghats. The largest terrestrial genus is *Habenaria* (100 species) and the largest epiphytic genus is *Dendrobium* (70 species). Most of the *Paphiopedilum (lady's slipper)* species are restricted to NE Himalayas except for *P. druryi* which has been reported from Kerala but is now almost extinct from its original habitat. Orchids form 9% of our flora and are the largest botanical family of higher plants in India. It is estimated that about 1,300 species (140 genera) of orchids are found in the Himalayas of India as their main home and others scattered in the Eastern and Western Ghats. The following is the distribution of orchid species in different regions of India.

North-Western Himalayas	200 species
North-Eastern India	800 species
Western Ghats	300 species

North-Eastern India owing to its peculiar gradient and varied climatic conditions contains the largest group of temperate and sub-tropical orchids.

More than 46% of the known orchids in the Western Ghats are endemic to this region (Anand Rao, 1998). Fischer (1928), while completing the work of Gamble on the *Flora of Presidency of Madras,* has described 191 species belonging to 60 genera. Many of them occur in Karnataka. Sharma *et al.,* (1984) included 173 species in 51 genera from Karnataka. As many as 62 species in 32 genera are reported from Kodagu district in Karnataka alone (Keshavamurthy and Yoganarsimhan, 1990). Among these 26 species in 10 genera i.e., 42% are endemic. This includes the highly endangered species like *Xenikophyton, Acanthephippium bicolor* and *Trichoglottis tenera* (Anand Rao, 1998).

Formerly, the list of plants banned or restricted for export from India included a few orchids, but now they include all orchids growing wild. The Convention of International Trade in Endangered Species of Fauna and Flora (CITES), ratified by India, places all species of Orchidaceae under Appendix II, meaning thereby, that their trade will be only through export permits. Steps have also been taken to conserve Indian native species by establishing orchidaria, sanctuaries and germplasm conservation centres. Botanical Survey of India has established two orchidaria, one at Shillong, Meghalaya; and the other at Yercaud, Tamil Nadu; to conserve rare and endangered species. The ICAR research complex at Shillong, the Indian Institute of Horticultural Research at Hessaraghatta, Bangalore, and the Indian Botanical Gardens at Calcutta, maintain collections of orchids in their orchidaria. Some states have also established orchid sanctuaries, e.g., Sikkim at Singtom and Deorali, and Arunachal Pradesh at Tapi.

Some orchids endemic to India are so ornamental and so much in demand that their natural populations have been over exploited. Some species in the genera like, *Arundina, Cymbidium, Coelogyne, Dendrobium, Paphiopedilum, Renanthera, and Vanda* are almost extinct. The provisional list of 150 endangered plants of India includes many orchids like *Acanthephippium sylhetense, Anoectochilus sikkimensis, Aphyllorchis montana, Arachnanthe clarkei, Arundina graminifolio, Cymbidium macrorhizon, Dendrobium densiflorum, Didiciea cunninghamii, Eria crassicaulis, Galeola lindleyana, Gastrodia exilis, Paphiopedilum fairrieanum, P. druryi, Pleione humilis, Renanthera imschootiana, Vanda coerulea, V. pumila and V. roxburghii.*

CULTIVATION OF ORCHIDS AND UTILISATION OF ORCHID WEALTH

In the past, explorers employed by the royalty and wealthy Europeans were sent to far-off places to gather orchids. The orchids collected were then grown in greenhouses. Initially horticulturists struggled to develop suitable techniques to grow them but gradually the methods were standardised.

In the world, the major developments in the cultivation of orchids have been due to modern scientific technology, which has been suitably used in case of orchid seed germination and meristem culture. Today, orchids are grown on assembly-line method in extensive glasshouses in a controlled environment, with the sale of orchid flowers running to millions of dollars. The modern methods of propagation have brought orchid cultivation at par with other commercial crops.

Propagation of orchids can be done by sexual and asexual methods. Orchids like *Aerides, Arachnis, Epidendrum, Renanthera, Phalaenopsis, Vanda* and *Dendrobium* can be propagated by cutting. Orchid cuttings are usually bigger and should possess one or more roots. Cuttings are usually potted in propagation beds or directly in pots after treating the cut end with fungicides to prevent rotting. Cuttings of genera, like *Aerides, Arachnis, Vanda,* etc., are very hardy and can be directly potted in pots, whereas those of *Dendrobium* and *Phalaenopsis* need special care to root and should be potted in propagation beds. The propagation of orchids through cuttings is getting popular again and some nurseries like to propagate their orchids through cuttings, so as to get uniform plants. The percentage of variation through this method is almost nil as compared to *in vitro* propagation through tissue culture. Further, some orchids like *Anaectochilus* respond more to vegetative propagation through cutting than to any other method. Most of the sympodial orchids, like *Coelogyne, Cattleya, Dendrobium* and *Cymbidium*, are propogated through this method. The method involved, consists of dividing large clumps into smaller units. However, care should be taken not to divide the plants unless there are 8-10 pseudo-bulbs. *Dendrobiums* which are very fast growing can be divided every year.

In some monopodial orchids, like *Ascocenda* and *Phalaenopsis*, Keikis or off-shoots emerge frequently on the main stem. This usually happens when the apex has lost its effectiveness in suppressing axillary buds. In most of the commercial orchid nurseries, topping of the stem is commonly practised to induce Keikis formation.

Most of the *Dendrobiums* produce aerial shoots or bulbs on old back bulbs, devoid of leaves. They usually arise on the upper part of the back bulbs and grow out slowly. These aerial shoots take 90-120 days to develop roots. At this stage, they are detached along with a portion of back bulb and potted as independent plants.

Vanda and other monopodial orchids can also be multiplied by air-layering or morcotage. A cut is made through the stem, 20 to 30 cm below the apex, and moist *Sphagnum* moss is wrapped around the cut portion. The rooting media is kept moist and once the roots are formed, the layer is removed from the mother plant and potted in small-sized pots.

The pseudo-bulbs or thickened stems are very useful devices for the storage of food and water and the function like bulbs.

In addition to the epiphytic orchids, ground orchids or terrestrial orchids grow like ordinary plants with their roots in the soil. While most of the temperate zone orchids are terrestrial and tropical orchids are epiphytes. In their natural surroundings orchids are protected from the tropical sun by the shades of trees. However, under controlled condition, orchids can be grown in specially designed orchidaria or orchid houses, running North and South made from materials like split bamboo, glass, shadenets, etc. A central tank filled with water or by using artificial fogging nozzles help in increasing humidity.

However, it must be clear that all types of orchids cannot be grown under one roof. While tropical orchids enjoy warm humid, atmosphere, and burst into activity during the rainy season, temperate orchids should be grown in cool houses. Orchids dislike sudden changes in temperature, however, a difference of 10°C to 20°C between day and night temperature is beneficial. The most suitable range is 18°C to 30°C. Proper ventilation is a must as it provides fresh air and also helps in reducing the temperature. In fact, some orchids can also grow in the open sun. Various terete leaved species of *Vanda, Aranda, Arachnis, Renanthera, Kegawara, Mokara,* etc. can be grown in open trenches, filled with brick pieces and charcoal, as is done in Srilanka, Singapore and Thailand. Indirect sunlight is ideal for orchids. Seedlings require less sunlight than adult plants. Very poor light tends to produce weak plants and retards flowering. A plant which has been grown in shade should be gradually shifted to sunlight conditions. The optimum requirement of light varies from species to species. Whereas, *Cypripedium* and *Phalaenopsis* require only 200-300 foot candles, genera like *Vanda* and *Aranda* thrive best under 800 foot-candles. A majority of orchids are day neutral and are not influenced by the length of the day. But in *Cattleya,* both short-day and long-day plants are seen. Usually in the orchid growing areas of Thailand, Malaysia and Singapore, growers use shadenets with 35% to 85% shade percentage to grow orchids of different genera. Warm humid atmosphere is most essential for the growth of most of the tropical orchids which do not have a well established root system. It is a good idea to have a water tank or pool in the centre of the orchidaria to maintain humidity, which should not be less than 30% at night and 80% during the day time, The plants should be watered 2-3 times a day and should not be allowed to dry up during hot climate. Active growing plants require more water. Similarly, plants in baskets require more water than those in pots. Care should be taken to water the plants with a fine spray by using standard nozzles, and not to hit the plants with powerful jets of water. Plants which are freshly potted should be watered very sparingly till the new roots appear, and watering should be gradually increased. Orchids should be potted in small spots according to the size of the plants. As a rule, orchids should be underpotted to get more flowers. Any kind of pot which can hold the medium and provide aeration is suitable. Mostly people prefer plastic pots, which retain moisture longer than mud pots. Vandaceous and Sarcenthene orchids can be grown in teak-wood baskets. Orchid plants should not be disturbed frequently and repotting is done only when absolutely necessary. Orchids like *Cymbidium*, react favourably when repotted after 2-3 years, whereas Vandeceous orchids and *Paphiopedilum* should not be disturbed unless very necessary.

Terrestrial orchids, like *Spathoglottis, Phaius* and *Calanthe*, should be grown in 20-25 cm pots. For *Paphiopedilum,* a mixture of 2 parts leaf mould, 2 parts loam soil and 1 part each of brick pieces and charcoal is recommended. Experiments conducted at Botanical Survey of India, Calcutta, have revealed that in case of *Rhychostylis gigantea*, chunks of hardwood charcoal, alone as potting substance, were superior than other potting media. Tree fern fibre also performed better than the other media, while coconut husk and over-burnt brick as planting substance had adverse effects on the growth and flowering of *R. gigentea* plants. Some other media tried for growing orchids, are gravel jelly, fir bark, tree fern fibre and polyurethene foam. In nature, orchids obtain their supply of inorganic nutrients, like calcium, magnesium, iron, potassium, nitrogen and traces of manganese, boron, copper, zinc, etc., from the tree on which they are growing, from the atmosphere, and decaying litter and the droppings of birds. However, under controlled conditions they have to be supplied with all these major and minor nutrients. Taking into consideration the special need of different orchids, a large number of fertiliser mixtures, both solid and liquid, are available in the market. Liquid fertilisers are much more quickly absorbed and can be applied more frequently. As the orchids slowly grow, slow release fertilisers like Osmocote can be used to get very good results. Even foliar sprays, containing necessory nutrients are beneficial. Usage of fertilisers should also depend on the stage of growth. During vegetative growth, large quantities of nitrogen are required, while during flowering, nitrogen should be reduced and the amount of phosphate increased.

Epiphytic orchids are more commonly cultivated due to their natural potential for vegetative propagation which gives a strong impetus for their cultivation. Most of the cultivated orchids, however, are found to be self-sterile. The seeds sown in nursery beds require long periods of germination and any disturbance in the soil or physical environment destroy the whole population. Nowadays the natural environmental conditions of cultivated orchids are well known and the successful growth of the most species is no longer a matter of trial and error. But many orchids are very difficult to be cultivated even today. One important rule in orchid cultivation is that the plants must be appropriate for the conditions that the grower can guarantee to maintain, even during the darkest and coldest days of winter, and even when normal food supplies are limited.

Orchids are usually grown in flower pots, preferably made up of fireclay. However, many people use only plastic ones filled with compost. Epiphytic orchids are grown in a potting medium consisting of the chopped stem of royal fern and several different media in various combinations. The terrestrial orchids require ordinary potting compost with organic, fibrous material or chopped biomass.

Even though enthusiasts have achieved considerable success with window sill and collar culture. Any type of greenhouse, except the cheaper plastic kind, can be used if it can be divided into 2 or 3 sections, to provide the correct conditions for warm, intermediate and cool plants. Ideally heating should involve a controllable degree of warm air circulation, so as to avoid local over-heating or under-heating or excessive condensation. Electric fan heaters comprise the ideal system. Adequate shading ventilating system, watering and air humidification damping down is to be maintained to avoid irreparable damage to orchids.

FUNGAL, *VIRAL* AND BACTERIAL DISEASES OF ORCHIDS

Orhids, like any other plants, are infected by pathogens. The following is a list of common fungal and bacterial diseases reported in orchids.

Disease	***Pathogen***
Leaf spot	*Colletotrichum and Gleosporium*
Leaf blight	*Pythium*
Collar blocth	*Penicillium thomii*
Collar rot	*Sclerotium*
Orchid wilt	*Sclerotium rolfsli*

More than 32 viral diseases are known to occur in orchids. In some cases, the same virus has been known to produce more than one disease in different species, the most common being *Cymbidium mosaic virus* .

The most commonly reported insect pests on orchids are thrips, aphids, spidermites, soft scale mealy bugs, orchid weevil, snail and slugs. These insect pests harm the plants in many ways. They feed on tender young shoots, suck the sap and damage the young buds and shoots, and also act as carriers of different diseases.

2

ORCHIDS AND FLORICULTURE

GLOBAL FLORICULTURE SCENARIO

Floriculture today is a multibillion dollar industry. It is a potential money spinner. According to a report of the International Labour Organisation (ILO) - 'The world cut flower industry: Trends and prospects' by Gijsbert van Liemt, China and India dominate in terms of area under cultivation, but their yield per hectare is low. "In India and China, purchasing power is too low for sizeable imports to take place and quality is too low for sizeable exports to develop". Germany is the typical example of a second group of countries, with sizeable markets where imports satisfy most demands (up to 70 per cent). Roses are the main traded products and Germany is the largest import market, followed by the United States. A third group, which includes Columbia and Kenya, has a small home market, but a large volume of exports of Flowers. Lastly, The Netherlands is a member of a group with a large home market combined with a large export share. It is the world's leading exporter. The developed countries have a competitive advantage in their being close to the market and to quality research and extension services. They have room for innovation and boast of high quality physical infrastructure. However, high labour costs are a competitive disadvantage. Though developing countries like India benefit from low cost of manual labour, but skilled labour and technicians can be expensive. They also have abundant light, good climatic conditions and low land costs. However, airfreight adds significantly to the total cost of exports.

Quality is the catchword as far as exports are concerned. Quality benchmarking of flowers has many aspects. For instance, flowers should be free from diseases and they should be undamaged – elements, which can be judged on visual inspection. Other quality aspects, however, are more difficult to judge. For instance, once cut, it is hard to see whether flowers have been correctly handled. Yet, this is an important determinant of vase life and whether or not the bud will open. This is why reputation is so important and why growers who have consistently delivered high quality flowers fetch higher prices than little-known or irregular suppliers.

FLORICULTURE IN INDIA

After an initial bloom in floriculture in India, a decline seems to have set in. Many of the ambitious floriculture ventures have not lived up to the expectations. It is essential that an evaluation be carried out to find out as to what ails the Indian floriculture industry.

There was a rapid growth of floricultural projects during 1970 to 1980, with most of these producing roses for export. This led to near saturation of the market and a decline in prices of the floricultural produce. Most of the producers in the flower business concentrated on the production, without having any link to the market needs. Shipments were dumped at the auction markets, without paying any regard to the appropriate time, volume or quality ratio of the different flower varieties. No promotional efforts were made to make the sale of flowers popular in competition with other comparable goods available in the market for the same end, especially the gift and decoration segments.

A majority of Indian flowers blooming in the fields are certainly at par in quality, to the ones grown in Europe. However, the problem begins the moment they are cut and commence their journey to the market. Improper handling after harvest, unavailability of optimum temperature and moisture conditions during storage and transportation and under or over-packaging leading to damage and wilting of flowers, are a few factors which have an adverse impact on the quality of Indian flowers reaching the consumer-both at domestic and international markets.

Though Indain flower industry suffers from logistical bottlenecks and inadequate post-harvest infrastructure, government support is focused mostly on the growers. Delhi is India's apex flower market and it is the silent hub of the nation's enormous flower traffic. The flowers moving in and out of Delhi, in all directions of the globe are as diverse as the various stages of handling them pass through. However, roses from Pune decorate coffee tables in a plush German café, the perennial struggle of flower growers, suppliers and distributors never makes it to the desktops. The space around the Coffee Home near Hanuman Mandir in Connaught Place in Delhi, hosts what is arguably the largest flower market in Asia, with 99 per cent cut flowers and one per cent marigolds travelling up from

Kolkata. Cut flowers are typically used as 'vase-decoration' flowers, e.g., Gladioli, Tulips, etc. Marigold and Indian rose petals are used for *pooja* and other ceremonial purposes. There are two other flower markets in the capital – one at Fatehpuri in Chandni Chowk, which specialises in *pooja* flowers and the other at Mehrauli, where one primarily finds marigolds with only 20 per cent of the total flowers being cut flowers. Similarly, each and every city and town in India have a small or a big flower market catering to the local needs. Indians buy flowers mostly for *pooja* purpose. Wedding and festival seasons are the peak times for the boom in flower trade. Unfortunately, the preference of buyers is limited to a few flower varieties like jasmine, marigold, etc. and he does not pay much importance to the quality of the flowers. In this situation orchids are unknown to many and they are available only in few markets.

But flower trade is not hassle-free in India. Corruption, preferential treatment and other unfair practices are common. Still, a licence is better than not having one. It helps to have one while dealing with airfreight (imports/ exports), inter-state transportation and knocking on the government's door.

Floriculture in Taiwan

Taiwan's agricultural industry, as a whole, is going through a rocky stretch. However, in the light of the country's predicted admittance to the World Trade Organization (WTO), the island's floriculture industry is thriving. The government has targeted the industry as one of the seven agricultural sectors to be strategically developed in the future. With its excellent topography and climate, Taiwan has an outstanding environment for growing flowers. The Department of Agriculture and Forestry has budgeted US $1.09 million to promote the floriculture industry over four years. It also intends to establish a databank of export flowers, strengthen flower marketing and sales systems, and develop new overseas markets. The size of the land utilised in floral production is indicative of the importance of the industry in Taiwan and its growth over the last 15 years. According to statistics provided by the Council of Agriculture (COA), 9,314 hectares of land, mainly in Changhua and Nantou regions in central Taiwan, are being used for floral production in 1994, as compared to only 1,672 hectares in 1981. The number of farmers devoted to the production of flowers, potted plants, seeds and sprouts, and cut-flowers has also increased considerably in recent years. Currently more than 20,000 farmers are engaged in floriculture. Collectively, they churned out some US $282 million worth of floral products in 1994. Cut-flowers are the vanguard of Taiwan's floral exports. Of the approximately US $370 million flowers grown last year, Chrysanthemum and gladiolus claimed the lion's share with 30 million and eight million flowers each, being shipped overseas. Other major exports include *Anthurium*, *Oncidium*, Golden Spider, Lily, Calla, Tuberose, and Butterfly orchid. Due to convenient transportation between Taiwan and Japan, which not only reduces costs and delivery time, but also guarantees the quality of flowers, Japan has become the island's largest cut-flower export market, followed by South Korea and the United States. In 1994, Taiwan exported a total of US $27.2 million worth of floral products, while importing US $13.4 million worth.

But the very nature of Taiwan's floriculture industry means that a number of changes will be necessary in order to remain competitive. The industry is largely composed of individual farmers operating on a relatively small scale. Rapid price changes and complicated laws regarding land use and ownership are viewed as barriers to further increases in exports. Possible solutions to the present bottleneck are the establishments of special flower growing zones for export purposes, and the establishment of an information system capable of providing information to local growers about access to foreign markets. The successful production and marketing of *butterfly orchids*, by the state-owned Taiwan Sugar Corporation (TSC), is an example of a successful government enterprise. Bolstered by cheap land costs, rich human resources, and abundant financial support, the TSC moved into floriculture in 1987 and later it initiated the mass propagation of *butterfly orchids*. These orchids sold primarily to Japan, the U.S., and European countries, have won worldwide acclaim.

Floriculture in the Netherlands

The floriculture sector is gaining importance throughout the world. Flower production is burgeoning in many countries like India. There has also been broad expansion in international trade. Sales are largely centred on markets like Europe, the USA and Japan with strong purchasing power. The Netherlands is a major player in this sector, both in production and distribution. Dutch flower auctions play an intermediary role between the Dutch growers and the wholesale trade, foreign flowers and plants are increasingly sold at auctions in the Netherlands.

The cultivation of flowers and plants accounts for only around 4% of horticultural land use in the Netherlands. However, floriculture under glass is extremely intensive. Thus the floriculture sector generates over EUR 3000,000,000 - half of the total production value of Dutch horticulture.

The rapid expansion in the market for flowers and plants has induced many vegetable growers to switch to floriculture. In fact, they often opt to grow flowers for the cut flower market. Cut flower growers in turn are moving into the production of pot plants. This trend has led to a further increase in the share of the Dutch floriculture sector in the total production value of Dutch horticulture.

At least 70% of Dutch production is exported. The Netherlands dominates the market in the cultivation of starting material. Dutch cuttings and young plants are finding their way into all the flower-growing countries in the world. Dutch producers of starting material, and cut flower growers have often contributed to the development of the sector in these countries. A substantial proportion of world horticultural production finds its way to the customer through auctions in the Netherlands. Tons of Asiatic, African, and South American flowers, and other plants are flown into Amsterdam airport every day, vast majority of which are destined for actions. In this way, the Dutch flower auctions serve as a hub for around half of the worlds production of flowers and plants. Prices set here are often adopted as guide prices for international trade. The Dutch wholesalers who re-export the imported flowers and plants often provide added value. They have bouquets made up to the specifications of the customers and provide the flowers with labels. The supply of foreign flowers and plants is extremely important for Dutch wholesalers and exporters. They are well aware that customers want to make a selection from a complete range. However, there are certain periods when Dutch growers cannot supply particular products, but foreign growers often can.

The main centres of production are situated in the west of the Netherlands, in the area around the airport and in the greenhouse district, Westland, between the cities of The Hague and Rotterdam. Flowers and plants are largely produced by family firms, in which the owner plays an active role. As a rule, the growers do not sell their products themselves, but are affiliated to one of the flower auctions. The auctions concentrate supply and demand in the same location to ensure problem free sales. This leaves the growers entirely free to concentrate on production. This has led to far-reaching specialisation in the nurseries, themselves. This specialisation relates largely to the selection of crops. By growing only one crop, or sometimes even just one variety or cultivar, growers can optimise their production.

The production of cut flowers under glass is the largest sector in floriculture. The major cut flower crops are roses, tulips, chrysanthemums, freesias and gerberas.

In flowering plants, the shape and colour of the flowers determine ornamental value. Among the important flowering pot plants are *Kalanchoë, Phalaenopsis, Dendranthema* (potted chrysanthemum), *Spatyphillum*, Gerberas, Roses *Begonia*, *Cyclamen, Saint-Paulia, Primula,* and *Hortensia*. These are all largely cultivated on growing tables. Taller flowering plants, such as *Hortensia* and larger-sized *Euphorbia pulcherima,* are grown on concrete floors or anti-rooting cloth. A proportion of pot plants are grown hydroponically, particularly for interior plants in offices and similar uses. The plants stand in pots of fired clay granules, which can absorb a lot of water. One great advantage of this system is that the plants get a large water buffer, which need only be topped up once a month. Young plants for hydroponic cultivation are raised on phenol foam or rooted directly into the clay granules. Some are raised in potting compost. Before the plants are transplanted, the roots of these plants are rinsed clean of any compost which could be transferred to the clay granules.

Auctioning of flowers in the Netherlands is a well known activity in the world. **The auction clock** is still the primary means of trade at the auctions. The flowers and plants are displayed in front of the auction clock. This system has recently been linked to computer sales, where the buyer sees the product for sale on the screen. The auction clock works on the price reduction principle, whereby the price goes downwards and the product is sold to the first buyer to respond. Despite the advent of other sales methods, the price set by the auction clock is the determining factor for the pricing of flowers and plants at home and abroad. The auctioneers also act as intermediaries between wholesalers and growers for immediately available lots. Advance sales of floriculture products by this method are also growing in importance. Electronic sales have recently been introduced, in which the cultivated product no longer passes physically through the auction; instead the auction just offers the electronic infrastructure of the sale. This system enables growers and wholesalers to respond to product supply and demand.

The Netherlands' central role in the distribution of flowers and plants is also manifest in the fact that many foreign producers choose to sell their products through the Dutch flower auctions and wholesale trade, despite the high transport costs. All the major imported flowers are supplied all year-round. The supply only drops off in the summer months when there is a larger demand.

The main imports in these auctions are roses, carnations, *Solidago, Hypericum* and orchids. Orchids are mainly imported from Taiwan and to some extent from the other Asian nations with India accounting for a very small percentage. A lot of cut foliage, such as fern fronds, is also imported. However, foreign suppliers must be affiliated in order to supply to the auction. Imported flowers must also meet the same quality criteria as that of Dutch products. Almost three-quarters of all Dutch production of flowers and plants are exported. The main export markets in Europe are Germany, France and Great Britain. Other important European markets are Italy, Belgium, Switzerland and Austria. The Eastern European markets are also expanding. Dutch exporters also serve remote destinations, such as the United States, Japan and the Middle East. In the destination countries, most Dutch floriculture products are sold to florists, garden centers, markets and street traders, sometimes through the local wholesale trade. The majority of Dutch flowers and plants are sold within Europe. For this reason, most of the products are transported by road. The lorries are 'temperature controlled', to maintain optimum product quality. All flowers and plants arrive at the auction in one of the two standard rolling containers, Danish containers and auction containers. These containers are designed for easy loading onto lorries. Air transport is also gaining in importance, particularly for cut flower exports. This is mainly because Dutch exporters have access to remote markets, like Russia, Japan and the United States. Many imported plants also arrive in the Netherlands by air, from countries such as Israel, Kenya, Colombia, Taiwan, Israel and Ecuador.

The Dutch Floriculture Environmental Project (MPS) has been running since 1995, with the aim of reducing dependency on chemical crop protection products. It is a certification system in which participants register environmental data, such as energy consumption, use of pesticides and fertilisers and the sorting of waste. The aim of the system is to minimise environmental pollution in horticulture.

Growers participating in the programme often use integrated crop protection, in which natural predators (parasitic wasps, predatory bugs and mites, galflies, nematodes and bacterial preparations) are used to control insect pests. The use of chemical crop protection products is kept to a minimum, to avoid disturbing the biological equilibrium between useful and harmful insects. The registration data is used to assign businesses to one of the three environmental grades: A, B or C. Currently 3600 nurseries are taking part in the project, 3300 in the Netherlands, 154 in Belgium and 90 in Israel. The other participants are based in Denmark, and East and Southern Africa. Those with the MPS-A certifications are the most environmentally aware. Over two thirds of the participating nurseries have this qualification. The programme is not only concerned with environmental considerations, but social factors such as safety, health and working conditions which also play an important role. Wholesalers and retailers also participate in the programme.

Since 2001, applied plant research in the Netherlands has been brought together under the Practical Plant Research Organisation (PPO). The glasshouse horticulture strand of this organisation works for the cut flower, pot plant and glasshouse vegetable sectors.The PPO has research establishments in Aalsmeer, Naaldwijk, Horst andKlazienaveen, where work is carried out in modern, well-equipped glasshouse complexes. In addition, the PPO has laboratories, cold stores and climatic test chambers, and different areas of expertise to support research and facilities for light measurement, substrate and nutrient research. The PPO's clients are the Ministry of Agriculture, Nature Management and Fisheries, and the Commodity Board for Horticulture, which represent the private sector.

Dutch floriculture is a very good example for a very well developed organization of controlled and regulated trade of agricultural products. More than anybody India, being an agriculture based country, should learn a lot from the system followed by Holland.

ROLE OF ORCHIDS IN INDIAN FLORICULTURE

Man's wish to surround himself with ornamental plants, even in his indoor habitat, dates from earliest times. But it is in the modern times, that an enthusiasm for new ornamental plants began to develop on the part of botanists, flower growers, missionaries and traders. England played a prominent role in the commercialisation of exotic plants. George III and his mother founded the celebrated Kew Botanical Gardens (Prasad and Kumar, 1998).

In 1800 itself, the trade of exotic plants was a competitive business. Thus, over the years, a fascination for ornamental plants sustained the practice of floriculture and horticulture, at large. The term horticulture is derived from two Latin words, *hortus*, meaning a garden and *cultura*, meaning cultivation. Initially, horticulture referred to the culture of crops grown in gardens. At present, fruits, vegetables, flowers, and other ornamentals are grown not only in home gardens, but also on a commercial scale. Floriculture is a branch of horticulture dealing exclusively with the cultivation of flowering plants. It has been the endeavour of scientists all over the world to develop better varieties of plants in terms of quality and to improve the procedures for production. To this endeavour, the application of biotechnology has actually given an edge. Thus, the present day floriculture may be defined as the science and technology involved in the production, processing and merchandising of flowers and flowering plants.

Horticulture has emerged as a major venture in the world scene, by the propagation of ornamental by tissue culture. Over the past few decades, the flower business in the world has taken root and blossomed into a US $20 billion industry (Chopra, 1995). Flowers are fast emerging as potential money spinners for many third world countries. The annual consumption of cut flowers in the world is worth US $13 billion (Prasad and Kumar, 1998). The main importers of cut flowers are USA, Germany, France, UK, Switzerland, Sweden, Norway, the Netherlands, Denmark, Belgium, Italy, Austria and Japan. Germany is the largest importer, followed by the USA, France and UK. Among the exporters of cut flowers, the Netherlands dominates. It accounts for 70% of the world's cut flower trade. Columbia accounts for 12%, Israel's share is 6% in the world cut flower trade and India accounts for just 0.1% (Koshoo, 1995). Thailand which is the world's sixth largest exporter of cut flowers, earns US $30 million a year from orchid exports, and Singapore earns US $16 million a year. Out of the production of tropical orchids through tissue culture in Asian countries, Thailand accounts for 33% (Gavinlertvatana and Prutpongse, 1994). In the Netherlands, floriculture has expanded so much that practically no more land is available for expansion (Prasad and Kumar, 1998). It supplies more than 17,000 tons of flowers per year to Germany alone. There are 11,000 growers, 2500 wholesalers, 10,000 propagation companies and nine flower auction centres in the Netherlands. They control the worlds export and auctioning of flowers. As the global market is expanding rapidly, the Dutch companies are looking for other countries like India for procuring floricultural products. It is likely that Singapore may become an official auction centre for flowers in Asia. Various aspects of commercialisation of orchids are discussed by Pemphishey (1997).

The world trade in flowers was expected to touch US $38 billion by the end of the year 2000. The demand for flowers in West European countries has risen from US $12 billion in 1991 to US $18 billion 2000. The consumption of cut flowers in the USA was expected to rise from US $5 billion in 1991 to US $11 billion in 2000 (Prasad and Kumar, 1998). Though 75% of the area in Israel is desert, a daily Rose flight takes off from Tel Aviv to the Netherlands, carrying flowers worth US $135 million per year. Kenya, the youngest member of floriculture business, has the world's largest carnation plantation. Columbia, the second largest exporter of flowers meet 60% of the demands of the USA. A small country like Thailand has monopolised the world orchid trade (Chopra, 1995).

A majority of the cultivated orchids in India are native of tropical climates and are found in abundance in India in the states of Assam, Meghalaya, West Bengal, Karnataka and Kerala. Kalimpong, Shillong, Trivandrum, Bangalore, Yercaud and almost all the coastal areas of India, which are the places most suitable for the cultivation of orchids.

In this context, one should strengthen the fields of agriculture and horticulture to achieve healthy economic growth, through a number of efforts have been put, there is not much of a success in the Indian Floriculture. Floriculture has the potential for generating employment within the country and earning valuable foreign exchange.

According to an estimate, the area under floriculture in India was around 34,000 hectares in 1989 including 24,000 hectares under traditional cultivation (Chopra, 1995). It is now estimated that about 70,000 hectares of land is covered under floriculture. Climatically controlled greenhouses are available in about 500 hectares for growing flowers of export quality, more than two-thirds of the area is occupied by traditional flowers, while in the remaining area modern cut flowers are grown (Ghosh, 1998). Floriculture has by far a great annual growth potential of 25-30% which is 25 to 30 times more than that of cereals or any other agricultural produce (Prasad and Kumar, 1998). Since most of the flowers in the international cut flower industry are being grown in our country, we are at an advantageous position. India is well placed to meet the international demand of cut flowers which, peaks during the winter months. Being a tropical country, we are a treasure house of ornamental plants, especially orchids, as there are about 1200 orchid species growing wild in India. However, the country is yet to reach the full potential of production and commercial exploitation in this area. There was a target to achieve an export of flowers worth Rs.200 crores by 2000 (Khoshoo, 1995). But unfortunately it could not be achieved due to various reasons.

In the recent past, a number of Export Oriented Floriculture Units (EOU) have been established in India. By December 1996, more than 200 EOUs were approved including 157 units having foreign collaboration and direct foreign investment. The total investment in this sector was approximately Rs. 1000 millions (Ghosh, 1998). At present, there are 230 EOUs in India with an average unit size between 3-4 hectares. Of the 230 units only 52 are under production, 44 are under implantation and 134 units have been officially approved. More than 50% of the EOUs are in the South Zone, including Bangalore, Hyderabad and Chennai. The total area of units under production and implementation is 323 hectares (Apte and Bhargava, 1998). At least 30 biotech companies in India are engaged in micropropagation of orchids and other commercial crops (Ravishankar and Venkataraman,1997).

Apart from the need to meet international demands, there is a strong local market for flowers in India. The use of flowers has a central place in Indian culture. There is no reliable study of the turnover of floriculture trade in India. However, some empirical estimates exist at the local level. For instance, in Delhi during the peak season the sale of cut flowers touches Rs. 80,000 per day in the winters and Rs. 50,000 in the summers. Roughly 10% of the sales are accounted for consumption in hotels, 20% in the offices and embassies and the remaining 70% through retail shops (Khoshoo, 1995). The annual sales in Delhi, on a cumulative basis, amount to a total of Rs. 100 crores (Deshpande, 1992).

In India growing of orchids commercially is not organised and it is still in the hands of hobbyist and a few dealers who mainly depend on wild collections from forests to meet a large part of their foreign and local demands. As a result, some of the orchid-growing areas are without any knowledge about orchids and very rare species are facing the danger of depletion now. Deforestation through burning of forests and felling of trees for timber, has been the major cause for the depletion of Indian orchid wealth. A large number of orchid species which were present in plenty in Indian forests are now on the verge of extinction and some of them have become so rare that a large number of botanical teams are unable to trace them. To cite an example, *Paphiopedilum druryi* which was once found in plenty in Agastaya Hills in South India is now difficult to locate. Steps should therefore be taken to conserve this natural wealth.

Floriculture is a highly specialised and remunerative venture, which has tremendous potential for export besides local consumption. Indian floriculture industry is definitely poised to bloom (Thomas, 1996). According to Govil and Gupta (1994), the total annual production of ornamental plants by tissue culture is 67,00,000, as compared to all other plants produced by tissue culture, which is estimated to be 37,60,000. According to Agricultural and Processed Food Products Export Development Authority (APEDA). The industry has grown at an annual rate of 10-15 percent during the last two decades. However, commercially it is still developing in India. Due to the availability of area and the natural advantages of harvesting rich solar energy, India is in a favorite position. Moreover, it has the advantages of a diverse climate and soil, unlike the Westren countries which have severe climatic condition. Israel, where desert conditions prevail in 75% of the area, is the third largest exporter of horticultural products, next only to Columbia and the Netherlands (Bezalel, 1995).

India's trade of cut flowers is negligible, when compared even to some smaller countries in Asia like Singapore, Japan, and Taiwan, etc. Efforts are being made to improve the floriculture industry in India through the establishment of infrastructure, congenial Government policies and strengthening international collaborations. An Indo-Dutch co-operation started around 1992 with the implementation of the Eight Five Year Plan (1992-1997) in India. This plan has played a crucial role in the development of an export floriculture industry. The Government allocated an amount of Rs. 1000 crores in the eighth plan, as against Rs. 24 crores in the seventh plan, an unprecedented boost to floriculture. With the implementation of economic reforms and new seed development policies in India, the Government of the Netherlands, together with the Financial Bank for Developing Countries at Hague, ordered a study by a commercial company in 1992, which published the report entitled, "Opportunities for Floriculture in India". This lead to a series of dialogues resulting in the building up of an export floriculture industry in India. The Government of India kept short term goals enthusiastically high, to be achieved within a period of 5 years (1992-1997). The Government's target for export was Rs. 100 crores by 1996-97 and Rs. 200 crores by 2000. At present, there are mainly three Zones of importance in the high tech floriculture sector in India, North Zone with Gurgaon and areas around New Delhi, South zone with Karnataka, Tamilnadu and Andhra Pradesh, and West Zone including Pune and some areas of Maharastra.

Since orchids account for 2.7% of the global cut flower production, India with its vast orchid germplasm has a lot to gain by the use of tissue culture technology in the area of orchid cultivation (Singh, 1993).

3

MYCORRHIZAL ASSOCIATION IN ORCHIDS

Introduction

Mycorrhizae (mycorrhizas) are highly evolved, mutualistic associations between soil fungi and plant roots. The partners in this association are members of the fungus kingdom (Basidiomycetes, Ascomycetes and Zygomycetes) and most vascular plants (Harley and Smith 1983, Kendrick 1992, Brundrett 1991). In the mycorrhizal literature, the term symbiosis is often used to describe these highly interdependent mutualistic relationships where the host plant receives mineral nutrients while the fungus obtains photosynthetically derived carbon compounds (Harley and Smith 1983). Mycorrhizal associations involve three-way interactions between host plants, mutualistic fungi and soil factors.

There are six kinds of mycorrhizae recognized in the scientific literature. The most common is vesicular-arbuscular mycorrhizae (VAM). This is an endomycorrhiza, which means the fungus enters the cells of the root. VAM are now divided into two types, AM and VM. Both are found on grasses, most crop plants, many trees, shrubs, flowers, and in fact about 80% of the world's plant species. The fungi are nondescript soil fungi that are not evident without a microscope.

The next most common kind is ectomycorrhiza (ECM), which means that the fungus enters the root but not the root cells. There is only one kind of ECM, so this name is unambiguous. The host plants are *Pinus*, firs, spruce, oaks, and several other kinds of plants, mostly forest trees. The fungi involved are generally mushrooms or truffles. A related category is ectendomycorrhiza, in which the fungus enters the root cells of *Manzanita, madrone*, and a few other plants and form arbutoid mycorrhizas. These look like ECM and have similar fungi, but are technically endomycorrhizas. A separate but apparently related category is monotropoid, found on certain plants without chlorophyll. While these mycorrhizae traditionally have been called "saprophytic," it turns out that they share a mycorrhizal fungus with a nearby tree, and they are in effect parasites of the tree by way of the mycorrhizal fungus. Blueberries and related plants have another type of mycorrhizae called ericoid mycorrhizas.

Orchid mycorrhizas are unique in that they are required for orchid seed germination. It is well known that the orchid seeds are microscopic and non-endospermous. They never in nature germinate without the mycorrhizal association. Some kinds of orchids never photosynthesize, but instead parasitize the mycorrhizal fungi.

The relative importance of different mycorrhizal associations for plant species has been discussed using data for Angiosperms from the mycorrhizal literature (Trappe 1987). Many Pteridophytes are also known to have VAM associations, while Gymnosperms may have ECM or VAM associations. Most information about the mycorrhizal status of plants is from the northern hemisphere in temperate regions, especially for crop plants and weeds (Trappe 1987). Field surveys have found that plants with mycorrhizal associations predominate in most natural ecosystems in Australia and throughout the world (Brundrett 1991). Members of the fungus kingdom obtain nutrition from many sources, including decomposition of organic substrates, predation and parasitism, and involvement in mutualistic associations (Christensen 1989, Kendrick 1992). Many soil fungi are saprobes with the enzymatic ability to digest organic substrates of varying degrees of complexity, but some subsist on very low levels of organic or inorganic substrates (Wainwright 1988). Mycorrhizal fungi are a major component of the soil microflora in many ecosystems, but usually have limited saprophytic abilities (Harley and Smith 1983).

Mycorrhizal fungi are considered to have many important roles in natural and managed ecosystems. Very little is known about the taxonomic or functional diversity of mycorrhizal fungi in Australia (Bougher 1995). Indeed, there is a paucity of knowledge about all Australian fungi, with only about 10% of these organisms having been named, while an estimated 80% of them remain undiscovered (Pascoe 1991). Same is the case with almost all mycorrhizae of world.

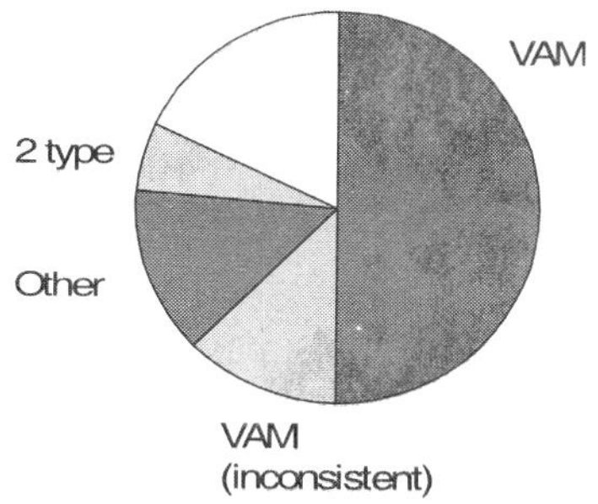

Proportion of angiosperm species:
18% are not found to have mycorrhizas 50% reported to have VAM 12% reported to have VAM in some cases, but not in others 20% had another type of association (ECM, orchid, ericoid, etc.)
Data compiled by Trappe (1987) from a dataset representing 3% of Angiosperm species.

Fig. 3.1 Types of mycorrhizae in different angiosperms.

Types of mycorrhizal associations

1. Vesicular-arbuscular mycorrhizas

They are also called arbuscular mycorrhizas (VAM or AM). They have associations where Zygomycetes fungi in the order Glomales produce arbuscules, and vesicles within root cells. Spores are formed in soil or in roots. These associations are defined by the presence of arbuscules. Fungi within roots spread by linear hyphae or coiled hyphae VAM associations produced by Glomalean fungi.

Fig. 3.2 Mature arbuscule of *Glomus mosseae* with numerous fine branch hyphae.

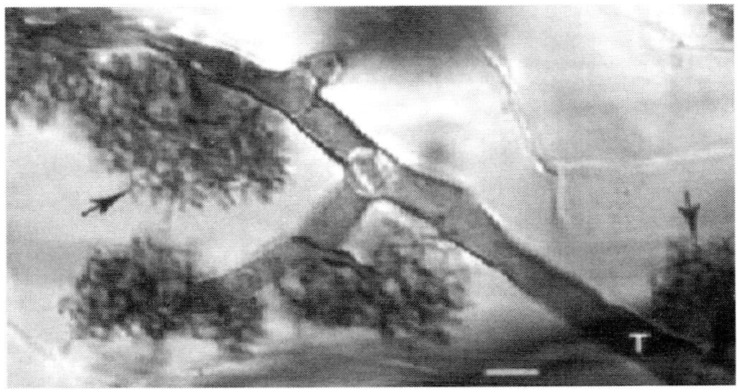

Fig. 3.3 Arbuscule of *Gigaspora margarita* with trunk hypha (T) and tufts of fine branch hyphae. (→)

These associations are known as arbuscular mycorrhizas, VA mycorrhizas, endomycorrizas, or endotrophic mycorrhizas and are generally abbreviated as VAM. There is disagreement about whether arbuscular or mycorrhizas or vesicular-arbuscular mycorrhizas, but the latter term(VAM) is the most appropriate name, because some fungi do not produce vesicles, arbuscules are not always present in roots and the role of arbuscules in nutrient exchange has not been confirmed (Smith 1995, Walker 1995). VAM associations involve fungi in the Zygomycetes order Glomales and the roots of a wide diversity of plants. Arbuscules are intricately branched haustoria that formed within a root cortex cell. They were named by Gallaud (1905), because they look like little trees. Arbuscules are formed by repeated dichotomous branching and reductions in hyphal width, starting from an initial trunk hypha (5-10 μm in diameter) and ending in a proliferation of fine branch hyphae (< 1 μm diameter).

Arbuscules start to form approximately 2 days after root penetration. They grow inside individual cells of the root cortex, but remain outside their cytoplasm, due to invagination of the plasma membrane. Arbuscules are considered the major site of exchange between the fungus and host. This assumption is based on the large surface area of the arbuscular interface, but has not been confirmed (Smith 1995). Arbuscule formation follows hyphal growth, progressing outwards from the entry point. Arbuscules are short-lived and begin to collapse after a few days, but hyphae and vesicles can remain in roots for months or years.

Vesicles develop to accumulate storage products in many VAM associations. Vesicles are initiated soon after the first arbuscules, but continue to develop when the arbuscules senesce. Vesicles are hyphal swellings in the root cortex that contain lipids and cytoplasm. These may be inter- or intracellular. Vesicles can develop thick walls in older roots and may function as propagules (Biermann and Linderman 1983). Some fungi produce vesicles which are similar in structure to the spores they produced in soil, but in other cases they are different.

2. Ectomycorrhizas (ECM)

Ectomycorrhizas (ECM) are associations where Basidiomycetes and other fungi form short, swollen lateral roots covered by mantle hyphae. These roots have Hartig net hyphae around cells in the epidermis or cortex.

3. Ectendo-, arbutoid- and monotropoid mycorrhizas (EECM)

This kind of associations are similar to ectomycorrhizas. These associations are seen in Monotropa, Manzanita. They also have same fungi as that of ECM. Technically EECM are endomycorrhizas.

4. Ericoid mycorrhizas

This group of fungi have hyphal coils in outer cells of the narrow "hair roots" of plants in the plant order Ericales. These associations also occur in thicker roots of the members of the family Epacridaceae.

5. Australian lily mycorrhizal association

The Australian lily *Thysanotus* has a unique mycorrhizal association where fungus hyphae only grow under epidermal cells.

6. Orchid mycorrhizas

The Orchidaceae is one of the most advanced plant families with many adaptations enabling long-term survival. All orchids that have been studied form endogenous mycorrhizal structures reminiscent of arbuscular mycorrhizae (Smith and Read, 1997). However unlike AM, which belong to the Glomales of the Zygomycotina , all orchid mycorrhizal fungi are classified in the form genus, *Rhizoctina* (Arditti, 1992). The *Rhizoctina* species associated with orchids include opportunistic soil pathogens and free living saprophytes as well as fungi with poorly known trophic niches (Taylor and Burns, 1999). This group includes anamorphs of *Tulasnella, Ceratobasidium* and *Thanatephorus. Rhizoctina* can be classified on the basis of number of nuclei in the young cells. Most orchids demonstrate considerable variation in mycorrhizal associations even among related orchid species.The phylogeny of the orchid mycorrhizae has been studied on the basis of DNA analysis.

Since the discovery that fungi are involved in the germination of orchid seeds (Bernard,1909) specificity in orchid mycorrhizae has been controversial (Harley and Smith, 1983). However, data from an extensive study of European orchids suggests that the degree of specificity is variable among species (Otero *et al.*, 2002). Non-photosynthetic orchids are considered to have specific mycorrhizal associations in the sense that they can involve only ectomycorrhizal fungi rather than *Rhizoctina*-like fungi. However, the amount of variation shown by these ectomycorrhizae is comparable to that of *Rhizoctina*-like fungi of an orchid, *Tolumnia variegata*. Unlike photosynthetic plants non-photosynthetic plants are highly specialized towards their mycorrhizal associations. Non-photosynthetic orchids, *Corallorhiza maculata* and *C. mertensiana* were investigated by amplifying fungal internal transcribed spacer (ITS) fragments from widespread samples of mycorrhiza (Taylor and Burns, 1999). The family orchidaceae contains a disproportionate number of myco-heterotrophic species which have lost the capacity to photosynthesize. Some orchids that have been studied acquire sugar from associated fungi, rather than providing it to the fungi as generally occurs in nature. This demonstrates the evolutionary propensity of orchids for the loss of photosynthesis. As the plant-fungal exchange upon which mutualistic mycorrhizal interactions are based has been short-circuited by orchids, the entire family appears to consist of mycorrhizal "cheaters". Many non-photosynthetic orchids have abandoned interactions with *Rhizoctina* species in favour of other Basidiomycetes species obviously due to the greater carbon availability. Fungal associations in several of these myco-heterotrophic orchids have been described as specific and confirmed by a number of studies (Taylor and Burns, 1997). Orchid mycorrhiza are found to consist of coils of hyphae within roots or stems of plants in the family Orchidaceae. Young orchid seedlings and some adult plants which lack chlorophyll are entirely dependent on mycorrhizal fungi for their survival. Orchid mycorrhizas are found in all of 144 taxa examined (Ramsay *et al.*, 1986).

Adaptations in orchids

Orchids are known to be the highly advanced group of plants. According to Engler and Prantl's system of classification this group is the most evolved one among monocots. This credit has been achieved by the number of adaptations seen in orchids. The significant adaptive mechanism of the orchids is that the roots are, especially in the tropical species have a multiple epidermis of dead cells, called velamen, that protects the root cortex from excessive drying and aids in water uptake. Another example can be seen in orchid flowers which have evolved to have a close relationship with insect pollinators such that sympatric species (species growing in close proximity that have the potential to interbreed and form hybrids) can each be pollinated by a separate insect species , maintaining species isolation (necessary to prevent interbreeding of species). A reliance on mycorrhizal interactions is a third adaptive mechanism that has allowed orchids to persist in less than ideal habitats and has led to their worldwide distribution. Along with root velamen, the symbiotic relationship of primarily terrestrial orchids with soil fungi has allowed these plants to adapt to habitats as diverse as arctic tundra and rainforest canopies. These mechanisms have aided the colonization of the epiphytic realm, which led to large and rapid species radiation. Mycorrhizal interactions also allow temperate orchids to persist for many years in a dormant state as the plant feeds on its fungal symbiont. In the most derived form of this interaction, some orchids have become wholly parasitic on fungi and do not produce chlorophyll. Another aspect of this mycorrhizal interaction is the development of microspermy, which means small seed and refers to the tiny, non-endospermous underdeveloped orchid seed. Microspermy is believed to have aided in the development of the all or nothing event associated with orchid pollination. Pollen in the Orchidaceae is delivered *en masse* as a pollinium, a structure in which the pollen within an anther is fused into a mass (Arditti, 1992). If an insect fails to deliver this mass of pollen, or no insect pollinators are attracted to the flower, the pollen is not delivered and there is no fruit set. If the pollinia are deposited on the stigma, all the pollen is delivered at once and can potentially fertilize every ovule. One pollination event leads to the production of hundreds of thousands of seeds. Microspermy is energetically inexpensive and is considered to be an adaptation. At pollination ovules are not developed and only three placental ridges can be seen in the ovary. After pollination, ovule development proceeds so that fertilization can take place. Double fertilization is undertaken as in other angiosperms, however the endosperm fails to develop. Even after fertilization, the zygote may rest for some time before division and embryo development. At maturity, the seeds contain little more than a rudimentary embryo and no endosperm (Arditti, 1992). This helps to ensure the survival of the parent plant since little energy is channeled to both the unpollinated ovary and the thousands of fertilized ovules. However, the seeds have no stored nutrients for subsequent growth and cannot survive without a mycorrhizal interaction. The adaptive mechanism that has allowed orchids to colonize diverse habitats and to persist in a semi-dormant state can also prevent seed germination and seedling growth. The orchid seed consists of a testa surrounding a tiny embryo in the globular stage with no defined cellular organization suggestive of embryonic tissue found in other angiosperms (Arditti, 1992). The chalazal end of the embryo develops into the shoot apex and roots arise from other parts. Upon germination, the embryo swells to form a protocorm that will develop rhizoids (Arditti, 1992). Hence, the tiny plantlets developed *in vitro* are called protocorm like bodies(PLBs). After a few days (or several weeks) the protocorm develops a shoot apex with leaf primordium and a root will form. During the early stages of growth prior to shoot apex formation, cell division takes place in the chalazal end of the embryo while cells enlarge but show little or no division at the micropylar end. This is combined with DNA synthesis; however, micropylar cells do not divide following DNA replication suggesting endo-reduplication (Raghaven and Goh, 1994). Levels of mRNA also increase especially at the chalazal end of the embryo. These results indicate changes in gene transcription in association with germination though changes in gene transcription are not related to fungal infection.

Orchid-mycorrhizal interactions

The mycorrhizal fungi colonise the soil and the bark of trees but also penetrate the living tissue of orchids. The fungi are of great importance for young orchid seedlings (protocorms). Lacking leaves and roots the protocorms are dependent on their fungi for the uptake of carbohydrates, minerals and water. However leafy adult orchids still benefit from the association and are often heavily infected. The degree of dependency varies within the genera and species. Epiphytic orchids are in general regarded as less dependent than terrestrial orchids. In protocorms, the fungus stays restricted to the basal part, not penetrating the cell division zone (meristem) and as roots develop they become infected from the soil. In *Caladenia* and *Pterostylis* and some other genera native to Australia, where roots are sparse or absent the fungus colonises the thickened stem just below the soil surface (stem collar). Masses of fungi are present around the stem and the roots. The outer region of the root (epidermis) is only sparsely colonised, the inner root region (vascular system) is fungus-free and most of the fungus is crowded in the cortex. The distribution of the fungus seems to be well organised and it becomes obvious that the plant controls the growth of the fungus within its tissue. Bernard discovered that the relationship between orchid and fungi is specific and only certain fungi are able to stimulate the growth of seedlings. For example, the neo-tropical epiphytic orchid *Cattleya* will form a mycorrhiza with a fungus originating from *Cattleya* but when combined with a fungus from *Phaleonopsis* or *Odontoglossum*, either the fungus infects and kills the orchid or the fungus infects the orchid but becomes killed by reactions of the host plant cells and seedlings do not develop further.

The orchid mycorrhizal relationship is fairly unique in the plant kingdom. However, it shares aspects in common with other types of mycorrhizae. Some orchid mycorrhizal fungi are pathogens of other plants suggesting that the orchid is able to circumvent or utilize possible virulence pathways. Furthermore, all orchids are mycotrophic at some stage in their lives, which again suggests that the orchid is able to control the infection process. These two points make the orchid an excellent model to study plant-fungal interactions. It is possible that orchids have evolved to utilize pathogen virulence as a means to establish symbiosis. By studying the orchid mycorrhizal process we can see how it is similar and how it differs from pathogenesis. It will then be possible to develop new strategies to overcome pathogenic interactions in other plant species. By understanding how.orchids and fungi are able to form a partnership in which the orchid is the dominant partner, it will be possible to analyze the evolution of plant fungal interactions and gain greater insight into symbiotic and pathogenic interactions. Probably the research in this area will help in finding solutions for many of the problems caused by fungi by causing myriads of fungal infections and there by causing enormous loss to the agriculture, horticulture and floriculture sectors across the world.

All orchids have an obligate relationship with mycorrhizal symbionts during germination. The relationship of orchids with fungi is relatively unique in the plant kingdom of the greater than 80 percent of plant species that form mycorrhizal relationships, orchids along with members of the order Ericales do not form endomycorrhizal relationships with genera of the Zygomycota. The main group of fungi inhabiting orchid roots is Basidiomycetes, though Ascomycetes have also been found most of the symbionts being *Rhizoctina*-like fungi. Some of the Basidiomycetes with which orchids form a relationship are pathogenic on other crops, e.g., *Rhizoctonia solani.* Even within the Orchidaceae, symbiotic fungi of one orchid species may be pathogenic on another. Furthermore, most studies on orchid mycorrhizal interactions have concentrated on terrestrial orchids from temperate regions, whereas the majority of orchid species are epiphytic in tropical regions.

Most orchids are able to control the infection and growth of endomycorrhizal fungi. Orchid mycorrhizal fungi are found intracellularly in cells of the cortex and they are confined to roots . Infection is limited to suspensor cells of the embryo and epidermal hairs and is highly restricted compared to fungal infection of non-orchid species. This suggests that orchids control the infection process, and that fungal symbionts are adapted to this control. Within the cells, the mycorrhizae form dense coils of mycelium called pelotons which are thought to be adaptations to the host cell (Hadley, 1982). Within the cell, the pelotons are surrounded by a membrane and interfacial matrix material (Peterson *et al.,* 1997). The membrane lacks adenylate cyclase activity but is otherwise similar to the plasma membrane. The orientation of microtubules and cell wall microfibrils is altered during infection and may be necessary to alteration in the cytoplasm and synthesis of the membrane surrounding the pelotons (Peterson *et al.,* 1997). Other factors of the orchid-mycorrhizal interaction are also important to the knowledge of symbiotic seed germination. Hadley (1982) reviews several studies that show how terrestrial plants grown with a symbiont have greater growth compared to plants grown asymbiotically. Anderson (1991) shows that *Spiranthes magnicamporum* had greater growth when grown with a fungal symbiont than without. Hadley (1982) also reviews the process of nutrient exchange between plant and fungus. It has been shown that vitamins, amino acids and sugars are translocated from the fungus to the orchid. However, it is not known whether these compounds are translocated across a living interface or released upon digestion of the fungus (Hadley, 1982). It has been studied that starch, accumulated during asymbiotic culture, is rapidly broken down upon infection and seeds germinated in the presence of an endophyte do not accumulate starch. Arditti *et al.,* (1990) state that orchids do digest their fungal symbiont and Hadley (1982) notes that, though digestion is the ultimate end for endomycorrhizal fungi, it is not known whether the enzymes for digestion are produced by the plant or the fungus. The activity of peroxidase, ascorbic acid oxidase, polyphenol oxidase and catalase found to be increased upon mycorrhizal infection and that this was in part due to digestion of the fungus. It has even been suggested that mycotrophy, the digestion of mycorrhizal fungi within infected seedlings can increase the water content of infected seedlings. Hadley (1982) pointed out that orchids control the level of infection and that phytoalexins appear to mediate this control. The level of this control may relate to the orchid-fungus specificity of some species and the pathogenecity of some fungi in other orchid species. Also, it implies a moderate defense response by the orchid. A compound called orhcinol has been found in protocorms of *Orchis*. This was surprising as orchinol is a phytoalexin and believed to inhibit fungal growth. Fungi readily infect protocorms so orchinol would not be expected to be present; however, its presence suggests expression of defense genes albeit at a level that does not interfere with infection. Mycorrhizae are not found in mature pseudobulbs or tubers (Hadley, 1982) supports the fact that the expression of defense genes at some point during mycorrhizal infection. It is hypothesized that an increase in oxidative enzyme activity upon mycorrhizal infection was similar to the oxidative activity of plants resistant to pathogens again indicates a defense response on the part of the orchid. Inferences can be made about some of the events that lead to mycorrhizal formation in orchids from looking at other types of mycorrhiza. In endomycorrhiza, flavonoids and phenolic compounds have been suggested as signal molecules released by the plant that induce spore germination, pre-infection hyphal growth, and branching of *Glomus* species. These compounds are possibly effective in directing growth of ectomycorrhizal mycelium to a suitable host. Once fungal mycelium contacts the host root, specific cell-cell interactions lead to the formation of mycorrhiza. These cell-cell interactions are host specific such that interaction of a fungus with a non-host will not lead to formation of structures characteristic of mycorhizal infection. The plant fungus interface represents one of the earliest stages of mycorrhizal formation and as such would reveal much about the physiological processes involved with mycorrhizal formation. In ectomycorrhiza, several wall associated proteins show changes in expression when the fungus interacts with the host root. Glycoproteins consisting of N-linked mannose residues seem to be involved in recognition and early adhesion events. Another group of proteins, the hydrophobins, which make the fungal cell wall hydrophobic, also seem to be involved in fungal adhesion to the host. The most interesting group is the symbiosis-regulated acidic proteins (SRAPs). These proteins are up-regulated after adhesion during Hartig net and mantle formation. They also contain an RGD (Arg – Gly – Asp) motif which has been shown to interact with fibronectin- and integrin-like proteins. Since fibronectin and integrin are associated with the cytoskeleton it is possible that SRAPs can alter cell dynamics. Membrane proteins that show up-regulation in response to mycorrhizal association are plant and fungal ATPases. These membrane proteins are involved in transfer of phosphate and other nutrients. The presence of ATPase in both fungus and plant provides evidence that nutrient transfer between the two organisms is critical to successful symbiosis. Carbon is one such nutrient and its metabolism is dynamic in developing mycorrhiza. In the plant host, sucrose levels in the roots rapidly lower and are concomitant with an increase in fungal sugars. The depletion of sucrose is due to an increase in sucrose synthase and invertase activity in cell wall and cytoplasm of *Trifolium repens* root cells. One fungal sugar that accumulates in mycelium of mycorrhizal fungi is trehalose. In cultures of *Amanita muscaria* grown in glucose, trehalose accumulates early and rapidly with a simultaneous decrease of glucose in the medium.

Still a lot of work needs to be done to understand the interaction of fungus and plant if a true relationship is to be obtained. It seems obvious that the fungus elicits changes within the plant and, since the orchid seed germinates prior to infection, plants must stimulate the fungus. At the heart of these issues are changes in the proteome of both fungus and plant which are mediated through gene expression. A more comprehensive knowledge of fungal plant interactions can be gained through analysis of the genes expressed by both plant and fungus during symbiotic germination. Peterson *et al.,* (1997) suggested that the orchid-fungus relationship "may prove very useful in determining the molecular events that underlie the pronounced structural changes that occur in both symbionts during mycorrhiza formation."

Plants interact with a variety of organisms throughout their lives and these interactions can result in parasitism, symbiosis, helotism or they may have no effect on the plant. Of the symbiotic interactions, the mycorrhizal interaction is by far the most widespread. Some have stated that the formation of this symbiosis allowed plants to colonize land. This would then suggest that symbiosis predates parasitism/pathogenesis, or that pathogenic interactions evolved from symbiotic ones or symbiosis and pathogenesis share, at some level, commonalities is not a new concept. Indeed, the similarities can be seen during initial contact, establishment of infection and in the formation of new structures. The symbiotic interaction even initiates defense responses though these are curtailed at a later point. With this parallel between pathogenic and symbiotic pathways, it is possible that with an increased understanding of plant-fungal symbiosis we can better understand pathogenesis. It has been suggested that a model system would serve well to promote the understanding of plant symbiotic interactions. A model that could be helpful in unraveling this sort of a mysterious relationship could be orchid, though the genome size is relatively large, difficulties in seed germination and relatively long life span preclude it from use as a general model. Though there are problems associated with the use of orchids as a model for studies on mycorrhizal symbiosis, there are some aspects not found in other plants, which make orchids essential for such studies. First, all orchids are mycotrophic at some stage of their lives, that is, the orchid is completely dependent on the fungus for its carbon. This would suggest that the orchid is capable of subverting fungal defense or of using symbiotic responses to its own advantage. Some orchids are mycoparasites, that need a fungal partner for the entire life of the plant. This display of reverse parasitism suggests again that orchids have the potential to utilize symbiotic interactions for their own purposes and are able to subvert fungal defenses. Second, some fungi associated with orchids are pathogens on other plants. This suggests the similarity of pathogenesis and symbiosis and that orchids can use fungal pathogenesis to establish symbiosis. The orchid is able to overcome the pathogenic response which makes them good candidates to look for novel resistance mechanisms. Third, the orchid mycorrhizal symbiosis is considered to be a more derived symbiosis and as such shares aspects in common with symbiosis in several other derived plant families (particularly mycoparasitic members of monocot families — Triuraceae, Burmanniaceae, and the dicot family — Gentianaceae). All the plants in this group form intracellular fungal structures that are similar to the orchid pelotons. These plants also exhibit varying degrees of mycoparasitism and these adaptations have arisen several times in different plant families suggests that the mechanisms to predate fungi are universal. It may also mean that the ability to overcome fungal defense or even subvert fungal virulence is latent in all plants. Interestingly, more primitive plants (mosses, ferns and liverworts) display a similar form of mycotrophy in the gametophytic stage. Even lichens are fully capable of subverting the fungal partner in what has been described as "a spectacular display of parasitism". It is important to note, however, that orchids are not immune to fungal attack, but have evolved to curtail certain pathogens for their own use.

Molecular Biology

Liang and Pardee first described, differential display using the polymerase chain reaction (DD-PCR). Since then it has become the most widely used method for identifying genes expressed at some developmental stage or in response to some stimuli. The method as described by Liang and Pardee uses pairs of primers to randomly amplify mRNAs at any given time. Since the same primer pairs are used for separate cells grown under different conditions, the overall banding patterns should be similar; however, because growing conditions are different, some cDNA bands will be present or absent in one cell extract only. These bands represent genes activated or silenced by the treatment. Though the sensitivity of PCR does allow for false positives, screening methods are available to minimize them. Recently, amplified fragment length polymorphisms has been used for differential display analysis (AFLP-DDPCR). The greater stringency afforded by this method helps eliminate unnecessary bands. The primers used in DD-PCR consist of an oligo dT and a 10 nucleotide upstream primer. The oligo dT primer has 2 additional bases at its 3′ end which serve to anchor the primer to the 5′ end of the mRNA's poly-A tract. The upstream primer is random, though some sequences function better than others. The fragment sizes generated by these primers range from 100 to 500 nucleotides and are easily separated on a sequencing gel by electrophoresis. From one primer pair, between 50 and 100 bands can be generated. Combinations of primer pairs will thus provide a suitable subset of the mRNAs expressed in the cell. In AFLP-DDPCR, the cDNA is restricted with a frequent cutter and a rare cutter. Adapters are then ligated to the ends of the cDNA. Primers designed to these adapters are then used in PCR. In theory, only fragments cut once by each enzyme should be amplified via PCR. The stringency produced by this method is believed to reduce false positives. It is also designed to produce fragments that are more likely to come from coding regions of genes. Seed research is one area where DD-PCR is being used. The study with *Botrytis cinerea* and *Lycopersicon esculentum* led to the identification of fungal genes activated during the infection process. Differential display has also been used successfully in orchids. Yu and Goh (2000) identified several genes including one belonging to the MADS box family associated with transition to flowering in a hybrid *Dendrobium*. Orchid molecular biology is a young and burgeoning field. Molecular techniques have been used to deduce systematic relationships and mechanisms of post-pollination ovule development in orchid. Some research is now being directed toward creating transgenic orchids that facilitate studies on molecular physiology. Many plant physiological studies use transgenic plants to uncover physiological processes. For example, T-DNA (mediated by *Agrobacterium tumefaciens)* has been used to identify genes involved in cytokinin signaling. Also, antisense constructs allow researchers to knock-out the function of a particular gene and screen for altered phenotype. However, these studies cannot be undertaken without methods to introduce genes into plants. Experiments have shown that several methods could be used for gene transfer into orchids with microparticle bombardment being the most effective (discussed in a separate chapter). As exciting as these advancements are, their use may be limited by poor seed germination preventing the formation of homozygous transgenics.

Fig. 3.4 T.S of stem of *Pterostylis vittata* (orchid) showing the intercellular and intracellular fungal hyphae

Fig. 3.5 *Pterostylis vittata.*

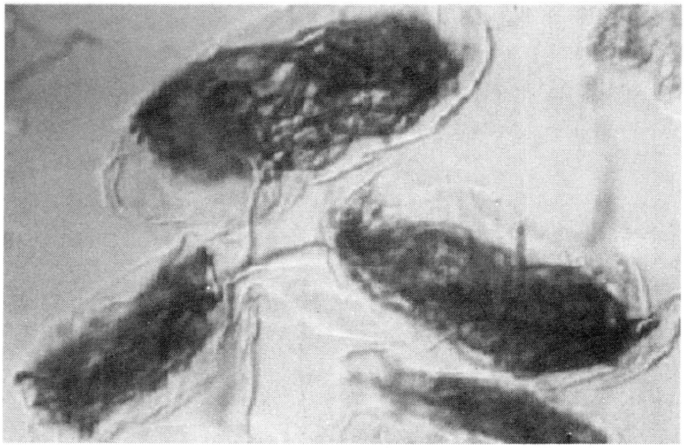

Fig. 3.6 T.S of stem of *Thelymitra crinita* (orchid) showing the intercellular and intracellular hyphae.

Fig. 3.7 Flower of *Thelymitra crinita.*

Benefits of mycorrhizae to orchids and other plants

1. Increased plant nutrient supply by extending the volume of soil accessible to plants. Orchid mycorrhizas are different from other types of mycorrhizas in the nature of the nutrient exchange. After

establishment of a mycorrhiza, organic carbon and other nutrients are passed from the fungus to the seed. Because asymbiotic germination requires sugars, amino acids and vitamins, it is assumed that these are also obtained from the fungus. The fungus continues to supply the protocorm with all its organic energy until such times as the plant starts to photosynthesise. Even then, it appears that the fungus does not gain significant supplies of carbon from the photosynthetic partner.

2. Increased plant nutrient supply by acquiring nutrient forms that would not normally be available to plants. The interaction between plant and fungus is highly regulated by the plant. The plant releases orchinol, a phytoalexin that causes the peletons to collapse. The degree of colonisation changes over the season, indicating that the orchid is controlling uptake of nutrients while preventing parasitism by the fungus.

4. Hyphae help in the germination of seeds. Specifically in case of orchids the germination of seeds cannot occur without the symbiotic association of mycorrhizae.

5. Some ECM and ericoid fungi have the capacity to breakdown phenolic compounds in soils which can interfere with nutrient uptake (Bending and Read 1997).

6. Root colonisation by ECM and VAM fungi can provide protection from parasitic fungi and nematodes (Duchesne *et al.,* 1989, Grandmaison *et al.,* 1993, Newsham *et al.,* 1995, Little and Maun 1996, Cordier *et al.,* 1998, Morin *et al.,* 1999).

7. Non-nutritional benefits to plants due to changes in water relations, phytohormone levels, carbon assimilation, etc. have been reported, but are difficult to interpret (Brundrett 1991, Smith and Read 1997).

8. Mycorrhizal benefits can include greater yield, nutrient accumulation, and/or reproductive success (Lewis and Koide 1990, Stanley *et al.,*1993).

9. Mycorrhizas can cause growth form changes to root architecture, vascular tissue, etc. (Daniels Hetrick *et al.,* 1988, Miller *et al.,* 1997).

10. Suppression of competing non-host plants, by mycorrhizal fungi has been observed (Allen *et al.,* 1989).

11. Significant amounts of carbon transfer through ECM fungus mycelia connecting different plant species has been measured (Simard

et al.,1997). This could reduce competition between plants and contribute to the stability and diversity of ecosystems.

12. Networks of hyphae supported by dominant trees may help seedlings become established or contribute to the growth of shaded understorey plants (Hogberg *et al.*, 1999, Horton *et al.*, 1999).
13. Nutrient transfer from dead to living plants can occur (Eason *et al.*, 1991).

4

INTRODUCTION TO BIOTECHNOLOGY

Overview

Biotechnology is now a byword with all starting from a farmer to the President of India. The branch has ramified so much that it has its extensions in all disciplines of science. In fact it is rather difficult to carry out research in biotechnology without the aide of experts from other disciplines. Thorough knowledge in physics, chemistry, mathematics and even computer science is a must for getting a proper hold on biotechnology. The land available for cultivation on the earth is limited on the other hand the population is growing unabated, especially in India. At this juncture the role of biotechnology becomes almost indispensable. The branch is perceived as the only solution to the problems of population explosion. It can be used to improve the crop plants in order to achieve higher production and better quality. Besides this the branch has vast potentials. Many of which have to be explored yet. The important achievements of biotechnology are listed in Table 12.1.

The term biotechnology was coined by Karl Ereky in 1917 initially to describe a process of large scale production of products from living organisms and their systems. Ereky's definition of biotechnology is, "the branch of science dealing with the synthesis of products from raw materials using living organisms".

But over the years biotechnology has grown beyond leaps and bounds During 20th century it has emerged as a new discipline of biology integrating with technology and it has been defined variously. According to Smith (1996), "Biotechnology is the formation of new combinations of heritable material by the insertion of nucleic acid molecules produced by whatever means outside the cell, into any virus, bacterial plasmid or other vector system so as to allow their incorporation into a host organism in which they do not naturally occur but in which they are capable of continued propagation".

This is more an advanced definition oriented towards genetic engineering but actually biotechnology has larger perspectives covering all areas of biology. It is not a "sudden sprout" but has a long history with many of the methods having been developed from 19th century itself. Thus it can broadly defined as "Use of techniques for improving the utility of the beneficial features and products of living organisms, tissues, cells and their systems in an economical way for the welfare of mankind".

US National Science Federation defines biotechnology as "the controlled use of biological agents such as microorganisms or cellular components for beneficial use". According to International Union of Pure and Applied Chemistry, biotechnology is "the application of biochemistry, biology, microbiology and chemical engineering to industrial process and products and on environment". Though there are many definitions, the meaning principally remains the same.

Probably the production of fat (ghee) use of which has been cited in the Vedic literature (5000-7000 BC) is one of the first examples for the, application of biotechnology. 'Soma' (fermented microbial beverage) was traditionally offered to God as a part of prayer. Bread was being produced by Egyptians in 4000 BC. In 6000 BC itself summarians and Babylonians in the present Iraq were making and using beer. These are some of the ancient examples for the use of biotechnology in the past and the biotechnology in the present. Before the starting of this discipline also it existed but at that time living organism and their system were used to make the products for the welfare of man without really understanding the mechanism. Fermentation had been used even in the Vedic and Babylonians times but they never knew what happens during the formation of either curds or beer. Biotechnology at present is different and it is basically production of useful substances with the help of living organisms only after understanding the underlying mechanisms. As the knowledge about the biomechanism and their genetic basis is growing. It is becoming more and more easier for producing biological products, which is always increasing. Designer drugs, vaccines, interferons, antibodies, etc., are some of the products which are becoming easily available. The technology is also being used for improving the plants and animals and it is possible because the science has the capacity to change, alter and translocate the genetic material.

Modern biotechnology embraced all advanced methods of genetic manipulation. One of the first steps for the foundation of biotechnology is the establishment of Institute of Biotechnology by the Leeds City Council

(UK) in 1920. Later in 1978 European Federation of biotechnology was setup. Agriculture and industries were brought closer to each other to work in tandem by biotechnology in the 20th century. Pharmaceutical companies too embraced biotechnology during the period of world wars during which the antibiotic, penicillin was invented.

After the discoveries of DNA double helix, polymerases, restriction endoncleases and a host of other events biotechnology grew from strength to strength.

The major milestones in the 20th century which can be called the attributes of biotechnology are human genome project (HGP) and cloning of sheap (Dolly).

The primary report of human genome was published in 2001. The final report was released in 2005. A France based company "Clonaid" claimed to have produced the first cloned human baby girl "Eve". They call themselves as the members of "Human Cloning Soceity". But birth of "Dolly" in 1996 is the first cloned mammal (sheap). In 2002 a designer baby was born to cure the genetic disease of her elder sister.

Besides genetic engineering there are some advanced areas such as treatment and utilization of biomaterials. Plant protoplast, cell, tissue and organ culture, nitrogen fixation, biofuels, enzymology, fermentation, and process engineering are some of the important areas now.

Biotechnology is sometimes can be called a bioindustry as scientists and industrialists are joining hands in the mass production of hormones, vaccines, antibodies, proteins enzymes, chemicals, designer drugs, biopesticides, antibiotics, tissues, organs, designer crops, designer plants and designer animals. Several major multinational companies have put their hands in this business. Elly Lilly, Merck, Monsanto, National Pituitary Agency, Genetech, Biogen are of the major companies.

Biotechnology has not suddenly cropped up. It is coming of age. It is a technology that was initiated several decades ago. The tremendous growth is because of the accumulation of knowledge in the areas of chemistry, physics, biology mathematics and computer science. Multidisciplinary strategies have been developed and advanced substantially over the years. A novel spectrum of investigation occured through the true interdisciplinary synthesis. This has led to the evolution of biotechnology which is an outcome of integrated efforts of biology with technology, the base ofcourse is in biology. The branches of biology such as cell biology, molecular biology, genetics, biochemistry, microbiology and immunology have been integrated together along with engineering

Biotechnology has also given rise to new fields like, bioinformatics, biophysics, biostatistics, bioengineering, biomimetics and biometry.

Agriculture has been the major beneficiary of biotechnology by the development of methods for producing advanced and improved plants. Similarly Health care has also been getting benefits especially in the areas of medical diagnostics and pharmaceuticals. Currently there are more than 40 biotechnology derived drugs and vaccines approved by the FDA in US, for medical use and more than 500 drugs are waiting to reach the market. The US authorities have approved nearly 130 gene therapy protocols. Nearly 600 diagnostic kits have also been developed.

APPLICATIONS OF GENETIC ENGINEERING

At present the important area of interest in biotechnology is genetic engineering. It focuses on isolation, alteration and translocation of genetic material. The techniques developed by molecular biologists are helpful in carefully isolating and inserting a gene from one organism into another totally unrelated organism.

This technology was initially used by Baltimore in 1976 for successfully transferring HGH (human growth hormone) gene into rabbit. Similarly insulin gene was cloned into *E. coli* for the production of the hormone. This was an achievement of Saran Narang and others from Canada. Elly Lilly, an Ameican pharma company was permitted to synthesize humilin (human insulin) industrially. USFDA (US food and Drug Administration) approved the product in 1982. Geneting engineering has been successfully used for the production of transgenic organisims. The first transgenic plant released for human consumption was Flavr-Savr variety of tomato, which carried a translocated gene coding for "longer shelf life". This transgenic tomato was released to the market in 1994.

(*Helicoverpa zea*) can reduce the amount of pesticide applications on a specific crop. According to the company growers using this transgenic crop need an average of 2.5 fewer applications pesticides per acre than conventional. According to an EPA estimate the cotton farmers cut down their expenditure by US $1.6 million in one year (1999) in the United Kingdom on pesticides. Mycogen, a biotech company developed a genetically modified peanut variety producing high levels of oleic oil which result in longer shelf life of nuts, candy and peanut butter. Mycogen has also developed a sunflower variety, which produces sunflower oil with improved temperature stability.

The Messenger technology developed by EDEN bioscience has been used to stimulate growth and defence pathway inherent within each plant without altering the plants DNA. Messenger treatments promote healthier plants and increased disease resistance and deterrence of insects and nematodes. Messenger product is available for cotton, citrus, apples, straw berries, rice, tomatoes, peppers, cucurbit, vegetables, cane berries, grass seed, potatoes and many other crops.

Bt cotton developed by Monsanto created many ripples in both scientific and political communities. Bt cotton is a transgenic cotton plant carrying a gene coding for an endotoxin against the bollworms (*Helicoverpa zea*) feeding on cotton plants. The controversy was raised because the *btgene* isolated from a bacterium Bacillus *thuringiensis* naturally kills the worms of pathogenic insects on cotton and the *btgene* was introduced into the cotton plant with the help of terminator gene technology. Monsanto also introduced Bt insect protected apple. These apples will contain built-in-insect protection against codling moth. DNA Plant Technology Corporation developed disease resistant bananas. These banans are resistant to the fungus *Cercospora* which causes a disease called sigatoka.

Dow Agro Sciences and HilBred International Inc. developed a rootworm resistant corn. The new plants produce a toxic protein against corn rootworm. Thus eliminating the need for soil applied insecticides. Corn plant with this new gene will have built in protection against Western and Northern corn rootworms. The same company produced a cotton variety better than Bt cotton. This plant gets a broader spectrum insect protection. It is resistant against damaging lepidopteron pests, pink bollworm, tobacco budworm, armyworms and loopers.

Monsanto is planning to release a next generation roundup ready cotton in 2006, which is expected to provide growers with an expanded windows of application. Agritope, Inc. has released several genetically modified fruits and vegetables with features of longer shelf life and delayed ripening.

The above account is only a tip of the iceberg as far as the achievements of plant biotechnology is concerned. In the recent times there is deluge of new transgenic crops.

BIOTECHNOLOGY IN INDIA

Biotechnology is the emerging era for India. It is shortly establishing itself as an area called **knowledge—Intensive Branch with high venture capital**. The focus of the industry and the government is on basic and application-oriented new biology and biotechnology covering agriculture and plants, genomics, molecular medicine, bioinstrumentation, bioinformatics, biofuels, biofertilizers, biopesticides, human resource development, environment and biodiversity. The department of biotechnology has been working to bring a synergy between researchers, academics, industry, policy maker and the public. More than 53 technologies have been transferred to the industry so far by various academic organizations.

The applications of biotechnology are mainly seen in three areas—

1. Agricutlure, 2. Medicine, and 3. Industrial products.

In the areas of agriculture successful results have been produced in banana, cabbage, mustard, mungbean, cotton rice and wheat. Crops have improved with respect to features like, enhanced nutritional value, disease resistance, longer shelf life, etc. In India a transgenic variety of rice called **Indica** has been produced carrying three genes. Similarly a transgenic wheat with a gene to confer stress resistance has been produced. A nutritionally important gene AmA1 isolated from *Amaranthus* has been successfully transferred to potato. International Rice Genome sequencing project (IRGSP) has been involved in sequencing the genome of rice plant. India's responsibility in this project was to sequence a part of chromosome 11; 45 mb. In December 2002, IRGSP announced the completion of a high quality draft sequence of rice genome. Scientists at IARI, Delhi University sequenced a region spanning from 57.3 cm to 116.2 cm on the long arm of chromosome 11 and contributed about 15 Mb nucleotide sequences. This draft sequence will promote ground breaking research in development of high yielding varieties to feed the ever increasing population.

Confederation of Indian Industries (CII) has projected the growth of biotech industry to the tune of US $4.5 billion by the year 2010. During last five years, genetic engineering, immunological techniques, cell culture methods, and hybridoma technology are increasingly being used to manufacture new products and in local research in these areas.

Health care products are dominating the biotech scene in India also. It is estimated that about 40% of the estimated growth will be contributed by the products meant for health care.

The department of biotechnology, Government of India is working closely with state Government to achieve the goals. The first state to have specific policies on biotechnology is Karnataka and it became a role model in this respect not only to other states but also to the central government. Entrepreneurship has been encouraged by some of the leading states like Karnataka, Andhra Pradesh, Tamil Nadu, Delhi, Maharashtra, and Gujarat to set up biotech industries. The states have announced specific policies and set up biotech parks and task forces.

CII has identified enormous partnership and outsourcing opportunities in contract research and development services.

Acute funding constraints faced by the US biotech sector has created a real need for outsourcing lower cost R & D from countries like India and China. Companies in India are placed at the advantageous position as compared to China thanks to better English speaking ability but they have not exploited this strength to competitively attract more contract research services. China on the contrary has been extremely aggressive in obtaining funding and grants from government agencies such as the National Institute of Health (NIH). India needs to emulate this strategy. It is estimated that contract R and D services have the potential of generating revenues in excess of US $1 billion per annum.

Drug development is another large segment where Indian companies can carve out a significant niche. There is tremendous pressure being imposed on Pharma companies to decrease the cost incurred on research and development of drugs. Indian companies have immediate opportunity at their hands in the areas of preclinical and proof of concept 'Clinical Investigation. This is helped by human resources and the wealth of biodiversity that India enjoys. The current global spending on clinical trials is more than US $8 billion and India has a good opportunity to get a major chunk from this amount. Bioinformatics is the backbone of biotech and pharmaceutical research. The dependence on bioinformatics for biological research is increasing day by day. India's software capabilities are becoming stronger and stronger. This is enriching the credibility for overseas companies to outsource bioinformatics skills and capabilities from Indian companies. Companies like Strand Genomics, Kshema Technologies, Bigtec, CDC Linux, Clinigene and others have already forayed into the North American market and have started developing both partnering and contractual arrangements with American companies. It is envisaged that this endeavour will generate revenues for Indian players to the tune of US $500 million.

FDI (Foreign Direct Investment) can be significant in the biotech sector as a large number of companies are keen on tapping the Indian market for their products and technologies. Affymetrix, Sequenom, Incyte, MDS and Agilest have already set up subsidiaries in India.

Stem cell research is going to be a potential area for development of biotechnology if proper support is given by the government and other regulatory agencies. India need to take the early mover advantage through prolific research, proactive funding and fiscal grants.

Several companies in the private sector have thrown the gauntlet by pumping money, for developing infrastructure suitable for the R & D activities Shantha Biotechnics, Biocon, Bharath Biotech, Avastagen, Piramal, Reliance Biotech, Khodays Biotech, Jagdale Biotech, Wockhardt, Ranbaxy, Torrent, Dr. Reddy's Labs, United Breweries, Cadila, Monsanto, RPG and many more have already forayed into biotech field. Even the public sector companies like Central Research Immunity, Haffkine, and BIBCL (Bharath Immunological and Biological Corporation Limited) have ventured in a bigway.

Vaccine research holds tremendous export potential especially to the other developing nations of Asia, Africa and Middle East. The research support for the development of vaccines can always come from All India Institute of Medical Sciences (New Delhi), Indian Institute of Science (Bangalore), Institute of Biotechnology and Bioinformatics (Bangalore), Centre for Cellular and Molecular Biology (Hyderabad) and Institute of Genomics and Integrative Biology (Pune), and National Centre against Biological Sciences (Bangalore). Vaccines against Viral diarrhoea, Malaria, HIV/AIDS, Tuberculosis, Rabies, Japanese Encephalitis, and Leprosy have been developed by using recombinant DNA technology. Some of them are already in the market and some are under clinical trials.

Scientists at Panacea Biotech in association with Biotechnology Consortium of India (BCI) comprising of IISC, NII, AIIMS and Bharath Biotech are involved in developing vaccine for anthrax. Vaccine against cholera is almost ready.

Contribution of biotechnology is highest in health care sector and agriculture is ranked second as far as the contribution is concerned. Biotech research has been focused on developing precision tools for improving and sustaining crop productivity and also for improving nutritional quality. Scientists have been successful in transferring specific genes associated with desirable traits from one organism to other. This gives a broader access to desirable characters and possibility of transferring these with much faster rate and greater precision.

Successful transgenic research has been carried out on banana, cabbage, mustard, mungbean and wheat for crop improvement, increased nutritional value, disease resistance, and longer shelf life.

The Micropropagation Technology Parks at the Tata Energy Research Institute (TERI), New Delhi, and National Chemical Laboratory (NCL), Pune, have produced about 11 million plantlets by 2002 by tissue culture method and transferred the plantlets to the field over an area of 9000 hectare at various locations in 17 states.

In India a network of four gene banks has been set up for *exsitu* conservation of rare, endangered, and commercially important medicinal and aromatic plants. Agrotechnology package has been developed for a new variety of *Artemisia annua* containing 0.5-1.3 percent artemisnin and yielding up to 80 kg of artemision per hectare.

An isoquinoline alkaloid, berberin, which acts as an immunological agent has been identified and isolated from *Berbris aristata* (Berberry). A herbal product which heals burns has been developed from Berberry. *In Vitro* bioscreens have been developed for assaying the medicinal plants used for the treatment of diabetes.

Indian Council of Agricultural Research is involved in the development of transgenic crops mainly using *Bt* genes for insect resistance. Even rice, tomato and brinjal are also introduced with genes coding for resistance against insect pests. The National Research Centre at Indian Agricultural Research Institute has developed *Bt* rice and *Bt* brinjal for protection from lepidopteran insects. Osmotin potato and Annexin *Brassica* have been developed for drought resistance.

After transformation and development of transgenic crops the biosafety and toxicological studies are conducted under the aegis of the Department of Biotechnology, Ministry of Science and Technology, which is the nodal agency for biotechnology and transgenic crops in India. The final release of transgenic plants is approved by the Ministry of Environment and Forests after their effective evaluation.

Directorate of Rice Research, Hyderabd, an Unit of Indian Council of Agricultural Research has been working on Vitamin A enriched Golden rice using Indica varieties. The project is using the transgenic Taipei 309 line as donor to transfer the β-carotene biosynthetic genes into the Indica rice varieties like, Swarna, Tellahamsa, Vijetha, and Sarjoo 52 using molecular markers.

5

GENETICALLY ENGINEERED AND MODIFIED PLANTS

OVERVIEW

A Hungarian engineer, **Karl Ereky**, introduced the term '**biotechnology**' in 1919, as the process that permits production of finished goods from raw materials using living organisms. Biotechnology can be defined scientifically as a technology that has permitted the development of new characteristics to living organisms or enhance a product that has been derived from an indigenous plant, animal or microbial cell. The term can also be defined as "the branch dealing with harvesting the beneficial factors of living organisms and their systems with the help of technology for the welfare of mankind". Likewise there are several definitions. When a broader definition is connotated to biotechnology, however, it cross-cuts areas such as waste recycling, environmental quality, pharmacology, cell fermentation, biomedicine, food and fuel production, farming and forensics. In 1943 the first direct evidence that DNA carried genetic information appeared. "Modern biotechnology" began with the discovery of the structure of DNA (deoxyribonucleic acid) in 1953 by Watson and Crick. A new discipline called **Molecular Biology** has been drawing the interest of thousands of scientists all over the world. It has now become a part and parcel of biotechnology. Now, the genetic code has been deciphered and it is an established fact that the genetic code is universal. Specific genes have been identified and located in living cells; with the result that scientists have developed techniques to manipulate them and thereby transferring them from one organism to another. With the discovery by the Swiss microbiologist Werner Arber in the 1960's of special enzymes, called restriction enzymes (RI) popularly called **molecular scissors**, in bacteria, it became possible to use this category of enzymes to cut the DNA double helical structure strands present within the protoplasm of an organism at precise points. This marked the beginning of **recombinant DNA (rDNA)**

technology, also known as genetic engineering. Techniques have been developed with the help of which genes from other organisms were transferred to bacteria. This discovery led to the first transfer of a human gene to *Escherichia coli*. By the end of the 19th century and thus biotechnology started flourishing. Biotechnology therefore is largely regarded as the manipulation and transfer of genes and can be looked upon as a new science. When the modified plant, animal or microbial cell is produced on a large scale in the external environment for the benefit of mankind through bioindustrial process, it suddenly acquires patented or intellectual property rights for commercial applications.

Molecular biology together with microbiological applications is leading to the development of interferons, monoclonal antibodies, interleukins vaccines, drugs, medical diagnostics, improved fermentation, eco-friendly products, and solid waste treatment. Plant protoplast, cell, tissue, and organ culture techniques are leading to rapid propagation of improved clonal varieties, better yielding, better drought or disease resistant crops, and in the conservation of biological diversity. Consequently in developed and newly industrialized countries, a large number of commercial applications have been developed particularly in the pharmaceutical industry, forensics, in improving agricultural productivity and veterinary services, in cell fermentation technology, fuel production, and environment friendly technologies.

Moving from research in molecular biology to biotechnology leads to bioindustrial development. Biotechnology has application also in the mining industry. The bacterium *Thiobacillus ferrooxidans* can use the molecules of copper present in chalcopyrite to form copper sulfate, which in turn, can be converted chemically to obtain pure copper. This microbial mining process accounts for about ten percent of copper production in the United States. Scientists are now trying to extract gold through plants. Pioneer trials became inspiration for many, as this concept holds bright prospects in the future. It is now possible to extract gold through mining, chemical reactions, by using plants, and utilizing bacteria. Plants are grown on the soil samples from deserted gold mines and gold is extracted by burning these plants into ash. Scientists at Cornell University, USA, have obtained preliminary success in this regard.

Everybody has recognized that sustainable development is possible only through biotechnology. The main application areas with reference to India are three; 1. Agriculture, 2. Healthcare, and 3. Industrial products. Indian biotech industry has come a long way and holds promising future in all these three segments. Achievements in the area of agriculture are laudable. Biotechnology in India is discussed in the earlier chapter.

All these and many more such discoveries and inventions led to a sort of revolution in the fields of agriculture, horticulture, and healthcare. One immediate outcome of this is the production of organisms generally called **genetically modified organisms (GMOs**). It is the general tendency of the human beings that they do not accept anything new so easily unless it is proven to be totally safe. Same is the treatment given to GMOs. Now, there is big list of them, which has been growing all the time. GMOs are entering this world whether one likes them or not. Orchids is no exception to this list. There are several genetically engineered orchids now. Production of genetically engineered orchids is discussed in Chapter 9. In this section the general aspects and ethical and legal controversies of GMOs are only discussed.

Genetically modified organisms are organisms that have been genetically engineered. In simple terms, genetic engineering (GE) is the process of taking genes from one strain of a plant, animal, or virus and inserting them into another, with the goal of reproducing characteristics of the original species in the receiving species. Insect resistant cotton plants, which can kill their pests like bollworm *(Heliothis zea)* produced by Monsanto Co, St Louis, USA caused a sort of revolution (Perlak *et al.,*1990).

There are other controversial issues related to GMOs such as biopiracy and biosafety. According to Vandana Shiva, the founder of "**Navadanya**", a Non-Government Organization (NGO) in India, biotech companies of the western countries are looting the genetic wealth of India and she calls it **Gene Robbery**.

The US Federal Government approved the first pharmaceutical (insulin) made by splicing human DNA into that of bacteria. It is available in the market in the name, **humilin** and is being commercially produced by Elly Lilly Co, USA. Prior to this, the bulk of insulin was derived from the pancreas of pigs, for which the animals had to be slaughtered and more over the insulin thus obtained was far inferior to humilin. Since 1982, several pharmaceuticals have been made with recombinant DNA technology, including human growth hormone (HGH) and many antibiotics. Genetically engineered plants were approved for patenting in 1985, and genetically engineered mice in 1988. However, it was not until the next decade that the first GE/GM foods found their way to US supermarket shelves. In 1994, Chicago consumers became the first in the world to have genetically modified, *Flavr Savr* tomatoes with in their produce isles. These tomatoes carried a gene for longer shelf life. GE/GM crops are not the same as traditionally bred crops. Typically, the process of creating them involves moving fewer genes, however the transferred genes may be from other species. Three federal agencies are involved in the approval process in the USA. They are the United States Department of Agriculture, the Environmental Protection Agency, and the Food and Drug Administration (FDA). Similar agencies work in all nations to monitor the release of such products. The bodies that grant permission to biotech industries in India are directly coming under the Government of India. Each and every state in India has a separate ministry to cater to biotech industries. In India Central Drug Research Institute (CDRI) conducts specified tests and clears the biotech product just like any other new drug to be released into the market. CRDI conducts tests on behalf of the Government of India.

In the USA, USDA grants permission for companies to field test promising crop lines providing facilities to set up buffer zones to prevent the escape of GE/GM pollen and to destroy the crop remains after the test. Prior to commercialization, the company must prove to the USDA that the crop will not have any significant effect on the environment, such as creating herbicide-resistant weeds. In essence, it regulates whether GE/GM plants can be grown and under what conditions. Similar agencies are available in other nations too but due to lack of funding and resources in some developing and underdeveloped countries the systems fail or at least is dysfunctional. In India also there are agencies to cater to this need but more importantly the public awareness is good. India being a agro-based nation needs to be extremely careful in introducing any new crop produced by new technology. *Bt* cotton though was introduced in India but received tremendous opposition from the Indian farmers. In the state of Gujarat alone the crop was grown in 10,000 acres of land. Monsanto the patent holder of the *bt*-cotton seed along with Mahyco, its partner in India, introduced the crop in the name of "**Navabharat-151**".

In India, to regulate the introduction of genetically engineered and modified crops, in 1989 the Government has formulated laws under the Environment Protection Act (1986). The biosafety laws thus formulated are titled " Rules for the manufacture, use, import, export and storage of hazardous microorganisms, GMOs, or cells. The union ministry for environment has also framed a body called, Genetic Engineering Approval Committee (GEAC), which will oversee all the aspects related to it. There is a second committee called Review Committee on Genetic Manipulations (RCGM) under the control of Department of Biotechnology (DBT). RCGM gives the clearance for genetic engineering experiments in laboratories or greenhouses.

The world body, EPA may or not play a role, depending upon whether the crop produces insect-killing proteins, such as *Bacillus thuringiensis* (Bt) toxins. Sometimes the tests conducted by the patent holders and governing agencies are insufficient. Losey *et. al.*, reported in 1999 that monarch butterfly, *Danaus plexipus* was killed by ingesting milkweed leaves dusted with the pollen grains of *Bt*-cotton. However, the effects of potentially marketable GE/GM crops are tested on the same animal groups as chemical pesticides. At present, FDA does not mandate pre-market safety tests. However, it does review companies' voluntary GE/GM toxicology, nutrition, and allergenicity testing results.

REVIEW OF GMOs

1) GMOs,
   sometimes carry terminator genes which may switch off other useful genes.
2) have significantly higher levels of toxicants than present in other edible varieties of the same species.
3) contain marker genes that could produce antibiotic resistance in people who consumed the food.
4) are plants that will be used for animal feeds or animals directly eat the plants especially in countries like India.
5) contain proteins that cause an allergic response. This would, however require additional testing and labeling.
6) have significantly lower or higher levels of important nutrients.
7) are plants developed to make substances like pharmaceuticals or polymers that will also be used for food.
8) contain new and sometimes unknown substances that differ significantly in composition from such substances currently found in food.
9) produce unexpected genetic effects.

Several legislations have been introduced that would require FDA to oversee mandatory labeling of GMO foods. In 2000, the White House in the USA, called upon FDA to develop guidelines for voluntary labeling by the companies. Now in India all herbal food items carry label "a green dot in the centre of a white square". The trend is extended to other products also. Ultimately the controversy surrounding GMOs comes down to whether they are adequately tested for safety or not. As with pesticides and drugs, safety testing for GMOs is done by the companies that produce them. Given that EPA and FDA had to call for the removal of previously approved pesticides and medicines from the market, as some consumers are understandably concerned about GMO safety. In fact, some scientists, including some at the FDA, are uncomfortable with the key assumption underlying federal safety assessments of GMOs. GMO approval in the US is based upon the concept of "substantial equivalence." This concept is the result of several international panels and is used by most so-called developed nations. Applied to GMOs, substantial equivalence maintains that if the resulting crops or foods are chemically equivalent, then safety can be presumed. Some scientists claim that there is no science underlying this assumption of safety. To be proven safe, a GMO would have to undergo long-term animal testing which presents its own inherent problems.

National Research Council (NRC) in the US states that the federal agencies must bolster the mechanisms they use to protect human health and the environment due to the increasing public concern. This is despite of finding that there is no existing evidence to suggest that current GMO-containing foods are unsafe to eat. The NRC recommended USDA, EPA, and FDA to quickly agree on their respective roles for regulating GM plants and advised that any new rules be flexible to reflect improved scientific understanding. It also advocated that the focus of GMO risk assessment be on the properties of the crops themselves, and not the methods by which they are derived.

The specific recommendations of the committee formed by NRC are:

1) closer examination of ecological concerns such as the possibility that transgenic plants could affect organisms which are not the target of a GM plant's pesticidal trait, the potential transfer of novel genes from one type of plant to another, and the evolution of new strains of immune pests.

2) the development of improved methods for identifying potential allergens, "specifically focusing on new tests relevant to the human immune system and on more reliable animal models."

Though it was intended as a balanced scientific review to allay consumer concern, the report created additional controversy. Several factors determine whether a crop gets modified. Chief among these are how far the incorporated trait is functional, as well as how quickly the modification can be made and brought to market. For companies, the issue of benefits comes down to whether enough farmer demand exists to allow the company to recoup its costs and earn a profit. In public sector research, benefits can be less market-driven such as improved nutrition.

Companies as well as, public institutions, should have GMO research programs. In the case of companies, research direction typically stems from market research and corporate objectives. But in some of the national institutes like those in India the research programs are not driven by the trends in the market. However, public research is typically directed by the objectives of funding groups, which can include commodity organizations. Unlike traditional plant breeding, which has at times been financed by farm cooperatives. GMO development requires large amounts of funding. Working via their commodity organizations, farmers have been able to collectively provide significant funding for biotechnology research by the National Science Foundation and the USDA. Additionally, farmers determine which GMOs remain on the market via their seed selections. However, these decisions may be influenced by global consumer preferences. Research in India regarding the development of GMOs is considerably less due to the lack of funding by the Government agencies.

CONSUMER ACCEPTANCE OF GMOs

There has always been a stigma attached to a crop that is modified. The acceptance of this product by the consumer is guided by several factors. Basically public awareness of the product is important. Sometimes, like in India, such novel commodities are readily accepted because of ignorance. Though US consumers have accepted genetically engineered medicines, but genetically engineered food has been a different story. This is undoubtedly due to the high rate of public awareness in the Americans. Since the early 1990s, the biotechnology industry has been trying to convince consumers that genetically engineered crops are just as good, if not better, as traditionally-bred crops. In 1992, two-years before the *Flavr Savr* tomato was to hit supermarket shelves, the industry-sponsored International Food Information Council issued a preferred vocabulary of terms and persuasive messages. Despite the fact that the Council's preferred term "genetically modified" caught on, the *Flavr Savr* tomato did not. Consumers didn't like the taste of the tomato, and many were upset by the fact that genes from other species could be included in their food without labeling. Though consumer advocacy groups addressed the GMO issue for nearly a decade, it was not until a research study showed that some GE/GM corn may harm monarch butterflies that a significant number of consumers demanded to be heard on the GMO issue and informed about the contents of their food. It appears that a majority of US consumers desire mandatory food labels which show GMO content. Given that mandatory labeling would require testing expenses and the fact that GE/GM pollen can drift to fields that are not planted with GE/GM crops, industry has been against this measure. In 2000, FDA was authorized to develop voluntary labeling guidelines for GE/GM ingredients. Voluntary labeling pleases industry despite the fact that its own research shows that consumers oppose it. Meanwhile, the public relations machines will continue to roll on both sides bombarding consumers with conflicting messages. The Grocery Manufacturer's Association will likely conduct a campaign to enlist the medical community to help promote the benefits of GMO foods to its toughest audience: women. At the same time, the Organic Consumers Association/ Bio-Democracy will be carrying posters of piranha-like strawberries at its many demonstrations calling for a moratorium on GMOs.

Globally, the GMO issue has met with a different fate. Due to the combination of public protest and the desire to take over US export markets, many countries including Australia, Japan, Thailand, and much of Europe, and companies have instituted restrictions on GMOs and adopted mandatory labeling laws. Though the majority of these rulings have come within the last two years; as early as 1988, the German government blocked pharmaceutical manufacturer Hoechst from making genetically engineered human insulin.

Scientific evidence continues to emerge regarding the safety of genetically modified crops and products. However, it is not unusual for organizations to interpret scientific data to suit their needs or selectively promote studies that further their political position. Organizations like the Biotechnology Industry Organization and the Grocery Manufacturers of America have weighed in supporting GMOs, while organizations like Greenpeace have been largely opposed. Information originating from individual industry and activist groups will be biased according to group's agenda. Of the non-governmental organizations issuing GMO safety information.

Federal agency and university sources also have tended to issue balanced information; however, both have received criticism for their relationships with the biotech industry in recent months. For example, the National Research Council recently came under fire for an April 2000 research paper on GMOs when it was learnt that an individual associated with the study left the council to become an executive director for the Biotechnology Industry Organization.

Additionally, some new organizations have been better than others in terms of providing balanced information regarding GMOs and safety. Unintended environmental impacts include harming non-target and beneficial species in the case of crops with engineered insecticidal properties, as well as the development of new strains of resistant pests. Additionally, there is a concern that pollen from genetically engineered herbicide-resistant crops could reach wild, weedy relatives of the crop and create so-called **superweeds**.

At present, there is no evidence to suggest that GM foods are unsafe. However, there are no absolute guarantees, either. Unintended health impacts from GMOs concern allergens, antibiotic resistance, decreased nutrients, and toxins. To declare that the GMOs are totally safe a thorough study spanning for several years conducted using wide variety of test animals is necessary. There are several criteria to be considered at the time of evaluating a novel plant or an animal produced by biotechnology.

IMPORTANT FEATURES OF GMOs

Introduced toxins: Residual toxins resulting from introduced genes of the bacteria *Bacillus thuringiensis* in so-called *Bt* crops are unlikely to harm humans. But nevertheless they may surely be dangerous to other fauna as shown by Losey et al., the killing of monarch butterfly. Toxin produced by the bacteria is highly specific to certain types of insects. Prior to its inclusion in GE/GM crops, *Bt* has been used as a biological insecticide, causing no adverse effects in humans consuming treated crops. But this does not mean that the technology is totally safe. The danger is actually with regard to the **terminator gene** used in the process.

Allergens: Protein sequences are changed with the addition of new genetic material, there is concern that the engineered or modified organism could produce known or unknown allergens. A recent National Research Council committee report on GMOs recommended the development of improved methods for identifying potential allergens "specifically focusing on new tests relevant to the human immune system and on more reliable animal models."

Antibiotic resistance: Plant genetic engineers have frequently attached genes they are trying to insert to antibiotic resistance genes. This allows them to readily select the plants that acquire the new genes by treating them with the antibiotic. Sometimes these genes remain in the transgenic crop which has led critics to charge that the antibiotic resistance genes could spread to pathogens in the body and render antibiotics less effective. However, several panels of antibiotic resistance experts have concluded that the risk is miniscule.

Naturally occurring toxins: There is concern that genetic engineering could inadvertently increase naturally occurring plant toxins. However, traditional plant breeding can also result in higher levels of plant toxins.

Decreased nutrients: Because the DNA of genetically engineered plants is altered, there is concern that some GMOs could have decreased levels of important nutrients, as DNA is the code for the production of nutrients. However, it must be noted that nutritional differences have also been documented with traditionally bred crops. This is not a serious matter as nutrient levels can be enhanced by other supplements.

Markets: Unintended market impacts include lower prices and higher costs for farmers, as well as lost premiums and markets. For example, the golden rice produced by the introduction of β-carotene gene into the rice plant, developed to be used to treat vitamin A deficiency (VAD). But it is economically viable or not is the question. Ban on GM imports or moratoriums on approving new GM varieties/hybrids reduces the number of export destinations for co-mingled GM and non-GM crops. This results in depressed crop prices due a greater percentage of crops needing to be used domestically. Additionally, moratoriums on new GM varieties/hybrids and mandatory labeling practices in some countries, including some of the US, larger export partners, may necessitate separate handling of grain. The cost for this is ultimately borne by US farmers. Pollen from GM crops can contaminate non-GM crops, especiality those certified organic, which are subject to a zero-GMO tolerance. Stray pollen could render a crop ineligible for organic or specialty premiums on contracts requiring non-GM varieties. Compounding the problem is the fact that many genetic tests for GMOs result in false positives.

In June 2000, Novartis became the first multinational to announce that it would not be purchasing GM crops/ingredients for any of its products worldwide, including those for the US market.

INTERNATIONAL ORGANIZATIONS AND AGENCIES

There are several organizations to address the needs of the consumers and the manufacturers of the GMOs. But most of them are located in the

US and some European nations. The number of such agencies is very less in India. The following are some of the important ones, which provide information through internet. It can be accessed by all from anywhere in the world.

Biotechnology Industry Organization (BIO)

BIO represents more than 900 companies, academic institutions and biotech centers in 47 states and 26 nations. BIO members are involved in the research and development of healthcare, agricultural, industrial and environmental products. As its name suggests, the BIO site represents the views of industry. BIO is a biased site, but useful for gaining an industry perspective.

Food First

The Institute for Food and Development Policy, better known as Food First, is a nonprofit think tank and activist organization. Its work highlights root causes and value-based solutions to hunger and poverty around the world, with a commitment to establishing food as a fundamental human right.

Institute for Agriculture and Trade Policy

IATP is "a nonprofit research and education organization committed to creating environmentally and economically sustainable communities and regions through sound agriculture and trade policy." IATP researches, produces publications, and consults on community economic development, environment, and trade issues. Though IATP has taken a position against plant patenting, its Resource Center via Sustain contains a library of policy-related documents, organization lists, URLs, and news. IATP offers a different perspective on genetic engineering and intellectual property rights.

International Food Information Council

IFIC's stated purpose is to bridge the gap between science and communications by collecting and disseminating scientific information on food safety, nutrition and health and by working with an extensive roster of scientific experts to help translate research into understandable and useful information for opinion leaders and ultimately consumers.

Organic Consumers Association (Bio Democracy)

OCA (BioDemocracy) is a nonprofit, public interest organization dedicated to building a healthy, safe, and sustainable system of food production and consumption. OCA/BioDemocracy is one of the major grassroot organizers of American consumers against GMOs. Though the organizaton is biased, it offers a good chronology of consumer opposition to GMOs.

Physicians and Scientists for Responsible Application of Science and Technology (PRAST)

PRAST members are seriously concerned about conditions that are hampering impartial, comprehensive and interdisciplinary evaluations of the safety of new applications of science and technology. PRAST is calling for a moratorium on GMOs based upon the charge of inadequate long-term testing.

Rural Advancement Foundation International (RAFI)

RAFI is dedicated to the conservation and sustainable improvement of agricultural biodiversity, and to the socially responsible development of technologies useful to rural societies. RAFI is concerned about the loss of genetic diversity, especially in agriculture, and about the impact of intellectual property on agriculture and world food security. RAFI's website presents the plight of developing countries with regard to GMOs.

Union of Concerned Scientists (UCS)

Founded in 1969, UCS is an independent nonprofit alliance of 50,000 concerned citizens and scientists across the country. UCS provised rigorous scientific analysis with innovative thinking and committed citizen advocacy to build a cleaner, healthier environment and a safer world." UCS is a science advocacy organization which objects to GMOs on the grounds that they do not further agricultural sustainability and may prevent research funds from going to technologies that could do more to address the problem of world hunger. UCS' page addressing world food and biotechnology is particularly insightful.

Navadanya

Navadanya is an NGO based in India and is involved in the protection of biodiversity of India. As per record, India is the second richest nation in terms of biodiversity and it is second only to Brazil. The orchid wealth of India is amazing. Navadanya is led by its founder Vandana Shiva a physicist from the Indian Institute of Science. She has written several articles on environmental issues. All the articles are available in detail on 'Znet Daily Commentary.' The following are some of the articles related to biotech industry:

1) Biotech Wars; Food freedom vs food slavery.
2) Biotech companies are bio-terrorists.
3) Monsanto is expanding monopolies from seeds to water.
4) Bioterror and biosafety.
5) Stopping biopiracy.

All these agencies are involved in basically creating awareness among the people about GMOs, which is necessary for accepting or rejecting any new crop introduced into the markets. Because of the easy availability of the technology to make GMOs, the number of modified crops is increasing in all categories.

The above information is very much relevant to orchids as most of the orchids are used in the floricultural, medicinal, and food sectors. *Vanilla* is one orchid now being cultivated in India and few other Asian nations for extracting a food flavoring substance used extensively for making ice-cream.

6
TRANSGENESIS

OVERVIEW

Transgenic plants contain genetic material (their own, foreign or artificial) stably integrated into their genome that has been introduced by methods other than classical breeding and that is passed on to successive generations. Several strategies are used to generate transgenic plants. One of them is derived from the plant pathogen *Agrobacterium tumefaciens*, which transfers DNA as a nucleoprotein complex. All other methods use naked DNA that enters plant cells during transient permeabilization or by shooting DNA-coated metal particles into the cells. Alternative methods are currently being developed. Transforming DNA integrates randomly in the plant genome by an illegitimate recombination process. Expression of the transgene depends greatly on the site of insertion, the structure of the surrounding DNA and the complexity of the transgenic locus. Plant transgenesis, in addition to its usefulness for crop improvement, has been a key feature in the development of modern plant biology.

Since the first report of transgenic plants in 1983, transformation of plants with foreign genes has been a key feature in the development of modern plant biology. Transgenic plants, apart from their value for crop improvement, provide an indispensable tool for the analysis of function and regulation of plant genes. In addition, studies on the transformation process have played a key role in the development of plant molecular biology. Different methods have been developed to introduce foreign genes into plants. A common feature is that the transforming DNA has to bypass different membrane barriers; it has first to enter the plant cell by penetrating the plant cell wall and the plasma membrane; then it must pass the cytoplasmic barrier. Finally, the newly introduced DNA segment has to establish itself in the plant genome. Once this is achieved, transformation events can be selected for and, if regeneration competence overlaps with transformation competence, transgenic plants can be regenerated. The following sections will focus on the different strategies available for plant transformation per se. While *Agrobacterium tumefaciens*-mediated DNA transfer and direct gene transfer are the most frequently used techniques, alternative strategies, some of which also rely on direct gene transfer, are being developed. Finally, expression of transgenes will also be discussed.

Agrobacterium-mediated Gene Transfer

Agrobacterium tumefaciens is a soilborne bacterium that initiates the development of a crown gall tumor on susceptible plants. This disease is the result of a unique natural inter-kingdom gene transfer event. Within the last two decades, it has attracted much attention because of the mechanism of transfer itself, and its applicability to the generation of transgenic plants. *Ti Plasmid tumors* induced by *Agrobacterium tumefaciens* result from the transfer and stable integration of a piece of bacterial DNA, the T-DNA (transferred DNA) into the plant genome. This first component consists of a Ti (tumor inducing) plasmid of 200 kilobases (kb) or even larger, which encodes most of the enzymes and structural components required for tumor development. T-DNA genes are expressed in plant cells because of the eukaryotic expression signals they carry. Parts of them code for the synthesis of, or the sensitivity to, plant growth factors (called auxins and cytokinins) leading to the uncontrolled growth of the transformed plant cells. T-DNA also carries genes responsible for the synthesis of opines, which are low molecular mass compounds that agrobacteria use as carbon, nitrogen and energy sources. Transformation by the T-DNA results, therefore, in proliferation of plant cells that produce metabolites used preferentially by the inciting bacteria; it thus provides an ecological niche for *Agrobacterium*. The transfer of the T-DNA to plant cells requires the action of chromosomally encoded genes for the early steps of the transformation process. Nevertheless, most of the functions required for the transfer process itself are encoded by a second region of the Ti plasmid: the virulence (vir) region. This consists of at least eight vir operons that encode 22–27 genes, depending on the strain. The virulence region provides in trans all functions required for the excision of the T-DNA from its Ti plasmid context and its transfer to plant cells. Indeed, the T-DNA itself does not carry genes responsible for its transfer. The only elements essential for the mobility of the T-DNA are two 25-base pair (bp) direct repeats, called border sequences, that delimit the T-DNA. The fact that no cis-acting genes are required for T-DNA transfer renders *Agrobacterium* a very attractive tool for plant transformation. Any DNA cloned in the T-DNA between the two borders is efficiently transferred, in a sequence nonspecific manner, to susceptible plants. Originally, susceptible plants were defined as plants on which a defined *Agrobacterium* strain could induce tumor development. With the use of *Agrobacterium* as a transformation vector the host range can be redefined as the spectrum of organisms for which transformation is possible. Successful transformation has been achieved not only for dicotyledonous plants, the natural hosts, but also for some monocotyledonous plants, including agriculturally important cereals such as wheat, maize, rice; gymnosperms; like yeast — and fungi *Saccharomyces cerevisiae*. The host range does not only depend on the ability of *Agrobacterium* to transfer T-DNA. It also depends on the capacity of the transformed cells to accept the T-DNA and to integrate it into its genome (the transformation competence of the plant material). Finally, the host range depends on the capacity of the transformed cells to regenerate into fertile transgenic plants. This regeneration competence of the plant material is in many cases the limiting step.

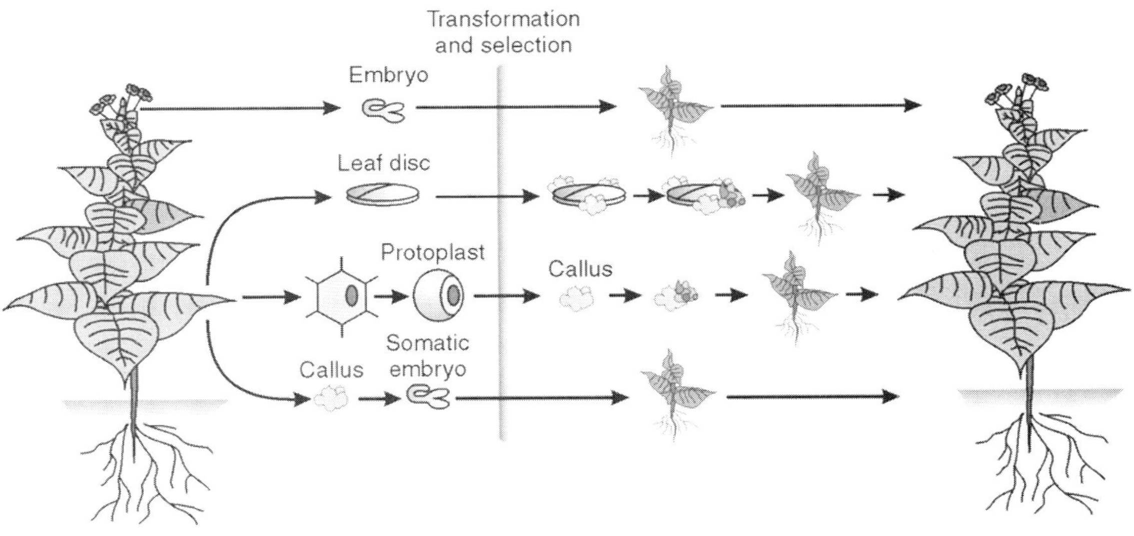

Fig. 6.1 Schematic diagram of biotransformation in plants.

Transfer of T-DNA to plant nuclei

The initiation of *Agrobacterium*-mediated transformation requires wounded plant cells, targets for agrobacteria. In the process of wound healing plant cells release precursors for cell wall regeneration. These are sensed by *Agrobacterium*, leading to transcription of the genes of the virulence region.The products of these genes initiate T-DNA transfer. Two proteins,

VirD2 and VirE2, play a key role. The VirD2 protein recognizes and cleaves the T-DNA at its border sequences. Concomitantly VirD2 becomes covalently attached to the 5′ end of the T-DNA, leading to the release of a single-stranded T-DNA molecule. The VirE2 protein is the most abundant protein synthesized by induced Agrobacteria. It can bind to single-stranded DNA in a sequence-unspecific and cooperative manner, coating the entire T-DNA. The VirD2–single-stranded T-DNA complex and VirE2 proteins are exported to the plant cells where they probably meet to form the T-complex. The export is mediated by a transfer apparatus that delivers different components to the plant cells. Once inside the plant cell, the T-complex has to be targeted to the plant nucleus and has to enter it. Nuclear import of the T-complex through the nuclear pore complex is mediated by its protein components VirD2 and VirE2. Thus, in analogy to virus particles, *Agrobacterium* equips its transfer complex with specificities to gain access to the nucleus. This step represents an important advantage of *Agrobacterium*-mediated gene transfer over other transgenesis methods.

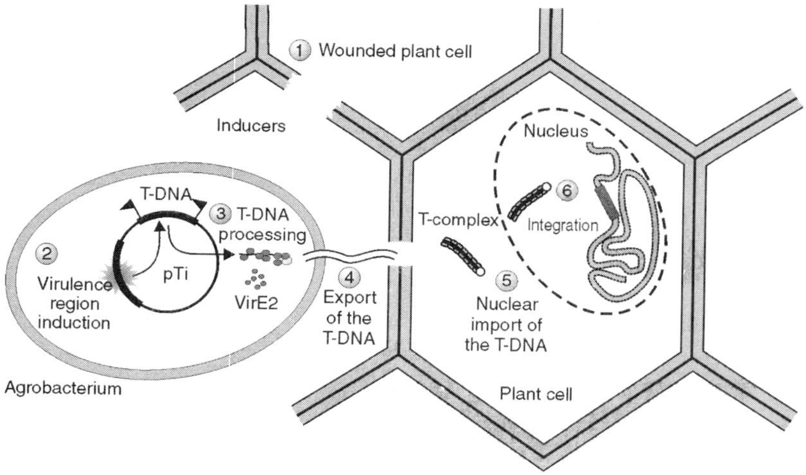

Fig. 6.2 Schematic diagram of *Agrobacterium*-mediated gene transformation.

INTEGRATION OF THE T-DNA

The T-DNA enters the plant cell nucleus, where it integrates into the plant genome. This integration step occurs randomly with respect to the chromosomal insertion site, predominantly into actively transcribed regions, by a process involving plant enzymes. One of the characteristics of *Agrobacterium*-mediated gene transfer is that up to 50% of the transformed cells contain a single copy of the transgenes. This could be due to the fact that only a few T-DNA complexes are transferred to plant cells. In addition, the transfer of the T-DNA as a nucleoprotein complex protects the T-DNA until its integration into the plant genome. Indeed, in about half of the cases, the 5′ end of the T-DNA is preserved to the nucleotide to which the VirD2 protein is attached. The VirE2 protein accords complete protection of the T-DNA over its length and only a few nucleotides of the left border are lost. Thus almost complete T-DNA molecules are integrated into the plant genome. This is in clear contrast to the integration of transgenes following the use of other transgenesis methods.

DIRECT GENE TRANSFER

In addition to *Agrobacterium*-mediated gene transfer, several artificial methods have been developed for plant transgenesis. These rely on the direct transfer of large amounts of naked DNA during transient permeabilization of plant cells. Several agents and strategies are available to achieve this transient permeability of the cells: chemicals, electric fields and bombardment of plant tissues with high velocity particles. These methods can be used to deliver transgenes not only to the nuclear genome but also to the genome of chloroplasts.

PEG-mediated transformation

Chemicals can be used to disrupt transiently the plasma membrane of plant cells. This method requires direct access of the transforming DNA through the plasma membrane to cells deprived of their cell wall, i.e. protoplasts. The protoplasts are subjected to treatment with polyethylene glycol (PEG) and bivalent cations (calcium or magnesium) in the presence of naked transforming DNA. PEG and the cations compact DNA, destabilize the plasma membrane, and render it transiently permeable to DNA. Once in the cell, DNA enters the nucleus, possibly while the nuclear envelope is disrupted during cell division, and integrates randomly into the plant genome. In the PEG-mediated transformation procedure, the cells that are competent for transformation receive large amounts of DNA. Prior to or concomitant to their integration into the plant genome, different DNA molecules can recombine. In addition, the transforming DNA is easily accessible to DNA-degrading enzymes, a process that frequently leads to truncated versions of the transgene. These complex integrated units may be unstable and part of the transgenes may be lost upon recombination between the long stretches of identical sequences. In addition, gene expression from complex loci may be silenced. The fact that protoplasts are used for PEG-mediated transformation has an advantage as well as disadvantage.

Protoplasts can be easily obtained from a wide range of plant material. Millions of them can be handled and transformed simultaneously, leading to the production of hundreds, or even thousands, of transformants. However, the generation of fertile transgenic plants by this technique depends on the availability of a protoplast-to-plant regeneration procedure, the efficiency of which is greatly dependent on the plant used.

Electroporation

One alternative method to the chemical treatment of protoplasts is the application of an electric shock in the presence of a large amount of naked transforming DNA. High voltage (thousands of volts) applied for very short periods of time (milliseconds) also transiently disrupts the plasma membrane and renders it permeable to DNA. As in the case of PEG-mediated transformation, the bulk of transforming DNA enters the nucleus most probably upon its disruption during cell division. This leads to the integration, at one or few genomic loci, of multiple copies of the transgene (complete or truncated) which may be genetically unstable and/or lead to loss of its expression. Originally, electroporation could only be applied to protoplasts. Therefore, the possibility of obtaining a transgenic plant was greatly limited by the availability of protoplast-to-plant regeneration protocols. More recently electroporation has been successfully applied to more complex tissues for which the regeneration procedure can be easier.

Particle bombardment

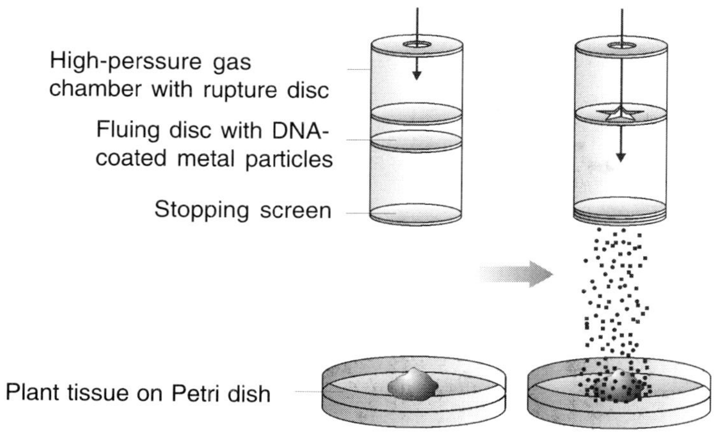

Fig. 6.3 Schematic diagram of particle bombardment.

In the particle bombardment approach metal particles coated with transforming DNA are projected at high speed on to plant cells. This variation of the direct gene transfer method can be used on entire plant organs, thus bypassing protoplast-to-plant regeneration problems. Particles of tenths of micrometres in size, made of gold or tungsten, are coated with DNA, either by coupling it to the particles with spermidine and calcium or by directly precipitating it on to the particles. Coated particles are shot on to the plant tissue with a velocity that allows them to penetrate the cell wall. Originally this was achieved by real shotguns, but now high-pressure gas (helium or nitrogen) is preferred. Shot particles can reach the nuclei and release their DNA directly to them. Alternatively, the particles can land in the cytoplasm, releasing DNA that can enter the nucleus while the nuclear membrane is disrupted during cell division. Generally, DNA delivery by particle bombardment leads to a very low survival rate of the treated cells because of the damage caused upon the entry of the particle. Several factors can reduce this damage and improve the chances of the success of particle bombardment-mediated transformation. For example, a screen placed between the particle bombardment device and the plant tissue reduces the shock wave occurring during shooting and prevents additional damage of the plant tissue. As with other direct gene transfer methods, particle bombardment-mediated transformation often leads to complex transgene integration patterns, with multiple complete or truncated copies of the transgene integrated randomly at one or a few genomic loci. This complexity may result from the fact that the particles shot into the plant are coated with many copies of the transgene. Recently, particle bombardment has also been used successfully to transform the chloroplast genome. Higher plants possess 500–10,000 copies per cell of the circular chloroplast genome, ranging in size from 120 to 160 kb. Transformation by this route probably occurs when a DNA-coated particle directly enters a chloroplast and releases its content. In contrast to transgene integration into the nuclear genome, DNA entering the organelle integrates in the chloroplast DNA by recombination with homologous sequences. The efficiency of plastid transformation is generally 2–3 orders of magnitude lower than nuclear transformation. In addition, the transgene integrates in only few copies of the plastid genome. A major problem in the regeneration procedure is how to obtain cells and finally plants in which all copies of the chloroplast genome contain the transgene. Unfortunately, the success of this method was concerned to the unicellular alga *Chlamydomonas* and to *Nicotiana tabacum* (tobacco), but nowadays it has been used extensively in case of many plant systems including orchids (see for details in Chapter 9).

Despite its technical limitations, chloroplast transformation seems a very attractive strategy. Due to the high copy number of chloroplast DNA, extremely high expression levels of the transgene can be achieved. In addition, the chloroplast genome is of bacterial origin, allowing the direct use of bacterial transgenes without prior modification. Finally, chloroplasts are maternally inherited in most plant species. Therefore plastid transformation allows better transgene containment because it avoids its dispersion by pollen.

Other transformation methods

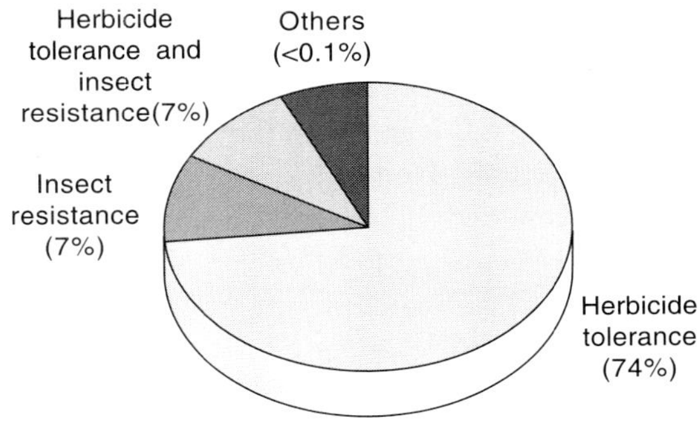

Fig. 6.4 Types of transgenic plants produced.

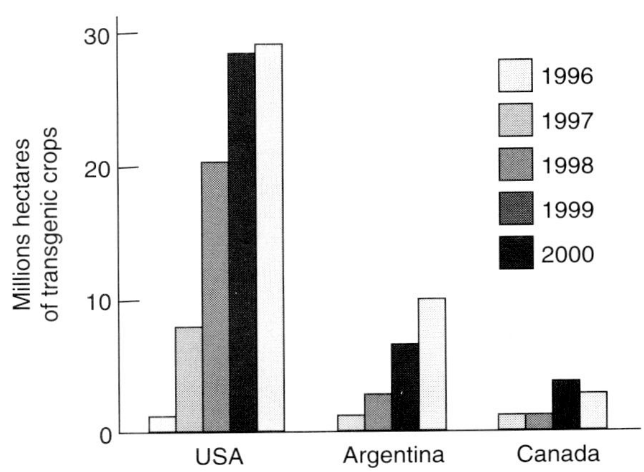

Fig. 6.5 Cultivation of transgenic plants in nations like USA, Argentina and Canada.

The limitations inherent in the previously described methods have led to the development of alternative strategies for plant transformation. These methods have been shown to function for some model plants but require further development to be of general use. In the bacterial and animal systems viruses are key players for gene transfer. In comparison, plant viruses are rarely used despite a high level of interest in the early 1980s. This situation is mainly due to inherent problems. The first attempts to use viruses for plant transgenesis were based on a caulimovirus, Cauliflower mosaic virus, a circular double-stranded DNA virus. Attempts to replace a dispensable gene with a transgene have been only partially successful. This is due to transgene instability in the viral vector, high mutation rate within the virus and transgene size limitation imposed by the packaging constraint within the virus particle. One of the most successful approaches so far is the use of an RNA-containing potex virus, Potato virus X. Because of the nature of the virus, particle size limitations of the viral genome are less strict, allowing addition of genes. Indeed, inoculation of plants with modified Potato virus X led to the spread of the virus and to expression of the new gene to high levels. In addition to its inherent limitations, virus-mediated plant transgenesis raises special biosafety problems, and serious concerns exist about the possible release of viruses with new pathogenic characteristics into the environment. Nevertheless, this technique allows the effect of over- or under-expression of genes to be tested very rapidly. It may also be of interest to engineer plants as bioreactors. Direct injection of naked DNA into plant cells is used much less than it is in animal cells. This is due to several plant peculiarities. Plant cells are surrounded by a cell wall that imposes a very strong mechanical resistance, thus restricting this method to protoplasts or to very few cell types. In addition, plant cells are extremely rich in polysaccharides, which limit the use of microinjection by blocking the injection capillary. Due to these difficulties, as well as the technical limitations inherent to the technology itself, microinjection is used mainly for specialized purposes and not for routine transformation.

Expression of Transgenes

The strategy used to generate a transgenic plant depends on the availability of previously described protocols and on empirical factors. In all methods, the regeneration of a transgenic plant depends on the presence of the transforming DNA of a selectable marker gene that is expressed in the plant cells. This gene, generally an antibiotic or herbicide resistance gene, allows the very few plant cells in which the transgene has been stably established to become resistant to the corresponding selective agent.

Subsequent propagation of the transformed plant material and regeneration of complete plants on media containing the selective agent lead to the death of all non-transformed plant cells. Only the few transformed ones survive the selective treatment and will eventually regenerate into a transgenic plant. This regeneration process can last from a few weeks up to a year, depending on the plant material used. Furthermore, transgenic plants directly emerging from the transformation and regeneration processes can rarely be used. An additional period of time, of variable length depending on the life cycle of the plant, is required to propagate these plants and to obtain offspring that can be studied. The creation of transgenic plants is thus a time-consuming process. This is especially true in the case of transgenic plants generated for agricultural purposes. Up to 10 years can pass between first transformation experiments and the release of plants for cultivation. Regardless of the method used, the transforming DNA integrates randomly into the plant nuclear genome via an illegitimate recombination process, i.e. no extensive homologies exist between the transgene and the site of the plant genome at which it integrates. Nevertheless, the pattern of transgene integration differs greatly depending on the transformation procedure used. In up to 50% of cases, *Agrobacterium*-mediated gene transfer leads to insertion of a single complete transgene copy at a single locus of the plant genome. Direct gene transfer methods can also lead to insertion of DNA at a single locus within the plant genome. However, in almost all cases multiple transgene copies, some of which are truncated, are integrated. The copy number of transgenes has an enormous influence on the genetic stability of the transgenic locus but also on its expression level. Multiple copies of a transgene can be expressed at high level. Nevertheless, multiple copies are frequently unstable due to the long stretches of identical sequences that can be rearranged, leading to the loss of part of the transgenic locus. In addition, multiple transgene copies can be subject to gene silencing, a process by which the expression of transgenic locus is switched off. Finally, whatever transformation method is used and whatever number of copies is integrated, the expression of a transgene depends greatly on the context of the surrounding DNA. Because transgene integration is random in the plant genome, the level of expression of a transgene is impossible to predict. Most of these problems can be circumvented by selecting the most promising transformants from the pool of transformed plants that have been regenerated. The ability to introduce and to modify the expression of specific genes in plants has been a key feature in the development of basic modern plant biology. It allows the functions of a gene to be determined *in vivo*. This is usually done by placing the gene of interest under the control of a strong constitutive promoter. Conversely, transformation of plants with an antisense copy of a gene (reverse complement copy of a gene; the RNA of which can pair with the normal sense RNA, rendering it inaccessible to translation) can lead to the reduction or complete loss of its expression, also providing insight into its possible function *in vivo*. Plant transgenesis is also used to study the regulation of genes. For this purpose plants are transformed with DNA-containing regulatory sequences of the gene to be analysed, controlling the expression of a reporter gene. The reporter gene, easily monitored for its expression, allows rapid determination of the sequences important for the temporal, spatial and environmentally influenced regulation of the expression of a gene. Finally, plant transformation is now widely used for insertional mutagenesis. Because transgenes integrate randomly into the plant genome it is possible to create a collection of plants each of which contains a transgene integrated at a different locus. By disruption of a gene, the insertion abolishes its expression, often leading to a mutated phenotype. Such collections of plants, in which theoretically every single gene could be disrupted in the sum of the individuals, are currently being assembled for the model plant *Arabidopsis thaliana*, which has a small genome size. These collections provide plant scientists with which it is possible to study the gene function. Better known to the general public is the use of plant transformation for applied purposes and specifically for agriculture. So far, most transgenic plants released have been engineered for increased resistance to herbicides, insect predators or diseases. This has been generally achieved by adding one single gene to the original plant genome. In the future more complex transgenic plants, requiring the introduction of many genes, can be envisioned. This will lead to the creation of plants with improved nutritional values and new characteristics. The first example of such a plant is rice with an increased content in pro-Vitamin A. Its consumption in Asia may result in the reduction of blindness (vitamin A deficiency). Finally, plants can also be envisioned as bioreactors producing industrially useful organic compounds. Transgenic plants developed for agricultural purposes are meant to be released into the environment. They are subject to particular concerns with respect to the nature of transforming DNA and the expression of the genes it encodes. One concern is the presence of selectable marker genes on the transforming DNA. These selectable marker genes theoretically may spread to microorganisms living in the soil, although this seems improbable and has never been demonstrated so far. However, transgene dissemination has been shown to occur through pollen traveling to wild relatives in the neighborhood. Their offspring will express the transgene, acquiring the new characteristic it encodes. Several strategies are being developed to circumvent this problem. One possibility is the co-transformation of plants with two different DNA molecules, containing either the selectable marker gene or the transgene of interest. In some cases the DNA molecules integrate at different loci of the plant genome. Crossings of such transgenic plants with their non-transgenic parents allow the two genomic loci to be segregated, making it possible to select plants containing only the transgene of interest. Alternate strategies are being developed in which the selectable marker gene is specifically excised after regeneration of transgenic plants. Despite these limitations, transgenic plants have several advantages. Results from the first 7 years of large-scale culture show that transgenic plants engineered for increased resistance to herbicides, insect predators or diseases have in most instances met their goals. They allow a substantial reduction of chemical treatments, thus limiting pollution of the environment. These plants also have an increased yield not only due to better protection provided by the expression of the transgene but also due to reduced soil disturbance; thanks to reduced field treatment. Therefore, plant transgenesis also provides farmers with direct financial returns. These advantages explain why the use of transgenic plants is increasing. In 1998 transgenic plants were cultivated on 26 million hectares world-wide. Almost 42% of the soybean and 34% of the cotton cultivated in the USA were of transgenic origin. The USA has been one of the first countries to use these genetically modified plants on a large scale. Nevertheless, the status of transgenic plants varies depending on the country. In Europe a more careful and restrictive approach has been taken and it was only in 1997 that the first transgenic plants received approval for agricultural use. Plant transgenesis is still in its infancy, but very few technologies have experienced a rapid and promising development. It seems that plant engineering is only limited by technology, biologists' imagination and their ethical stance.

7

APPLICATION OF BIOTECHNOLOGY IN HORTICULTURE

OVERVIEW

Recent progress made in the field of biotechnology has been boosting the improvement in plants of agricultural and horticultural importance. This has made horticulture a potential area of research and making it one of the most dynamic and promising field of experimental biology. *In vitro* culture techniques are now made important tools for the study of various fundamental problems of physiology, morphgenesis, cell biology, genetics and genetic engineering. The improvement of techniques has allowed the study of different aspects previously speculative and inaccessible and have turned the dreams of earlier scientists. Studies made in the area of cell biology have enabled researchers to understand the processes like totipotency, differentiation, cell division, nutrition, metabolism, cryopreservation of cells, mutagenesis, etc.

Micropropagation of plants has helped horticulturalists to grow plants directly from tissues from almost all regions of the plants. Now it is possible for us to produce large number of plants within limited spares either by direct organgenesis or by somatic embryogenesis. It helps in reproducing even those plants which fail to reproduce sexually. It is also possible to produce at various ploidy levels. The interspecific or intergenic hybrids, which otherwise sometimes remain sterile can also be multiplied, provided they carry significant characters of economic importance. *In Vitro* methods, like *in vitro* pollination fertilization have been successfully employed to overcome sterility of pollen, anther or ovule culture methods can be used to produce haploid plants. These are also the potential procedures in the studies of genetics. Techniques are now being developed that may have been potential in manipulating plant systems at the cellular level for use in asexual plant improvement. Some species respond readily to one or more of these approaches. *In vitro* propagation of horticultural plants has stimulated the commercial success in the area. For example, tissue (meristem) culture of orchids, which was successfully achieved by Morel (1965) has made a tremendous impact. Followed by this many ornamental plants like zinnia, marigold, sweet pea, stock, petunia, aster, snapdragon, gladiolus, lily, easter, iris, daffodil, caladium, peony, dahlia, tulip, chrysanthemum, rose, carnation, geranium, poinsettia, azalea, orchids and many other are now being multiplied mainly by using tissue culture techniques only. The reason is that the technique allows clonal propagation and almost disallows any modification in the features of the plants or its organs, which is almost crucial for it to be sold in the market.

The monetary benefits in horticulture is largely based on use and maintenance of clean stocks derived from good quality explants by using sophisticated procedures for achieving the growers' needs to be acquainted with the required technology. This is why perhaps, in the area of horticulture only the advanced nations like the USA, the Netherlands, Denmark etc. are far ahead of others. Products from most Asian nations including India are treated as second category. But there are some exceptions like Taiwan, which has monopolized orchid trade almost completely.

Grower who wishes to make a mark in the horticulture by supplying quality products needs to maintain a substantial stock of well documented genotypes, as the technique cannot be considered a short cut or an inexpensive source of "instant" new varieties. Creation of new varieties from existing lines by the treatments of colchicines must be carefully controlled because there is a danger that even slight chromosomal variations may lead to aberrant genotypes and phenotypes.

Thus micropropagation is not merely a multiplication of plants but it also covers improvement of plants and production new varieties. There are various techniques available to induce variations *in vitro*. It may be the use of colchicine and 8-HQ to produce polyploids or use of radiations, mutagenic gases and chemicals to induce matations there have been constant efforts by scientists to introduce new and improved varieties. Every consignment of orchids dispatched from Taiwan to Holland which occurs more than once in a week, there are atleast a few new varieties of orchids always introduced. This constant endeavour has made it possible for that small nation to take over the orchids trade completely.

Isolation of protoplasts and their subsequent culturing and fusion are some of the most significant developments in the improvement of economical plants. This techniques of far reaching implications in the improvement of cell modification and somatic hybridization was developed by Cocking (1972). Isolated protoplasts also offer a method of tackling various fundamental and pragmatic research problems (See chapter no. 8 for details).

Modification of cells by genetic engineering renders genes manipulation in ways which were not possible before. This opens up new avenues and a host of scientifically interesting prospects for plant improvement. Plant protoplasts have remarkable property of being able to take up macrom lecules and particles such as ferritin, polystyrene, proteins, DNA and organisms like viruses, bacteria or even whole nucleus and isolated chloroplasts. This property of protoplasts can be utilized for the incorporation of desirable characters into plants of horticultural importance.

The most significant application of protoplast culture has been the somatic hybridization. This could solve the problem of producing a hybrid by cross between two sexually incompatible organisms. This idea gained importance followed by the initial successes related to somatic cell hybrids in animals achieved by Barski *et al*. (1960) and later the production of a "man-mouse" hybrid cell. Genetic manipulation measures combination of genetic material from different sources. There have been innumerable reports of such research activities. New techniques have been developed to combine the genetic material of taxonomically unrelated species. The techniques involve second two steps. First, the uptake of exogenous of genetic material and in the step expression of the introduced foreign genetic material is achieved.

Plant tissue culture techniques are highly valuable for studying the biosynthesis of secondary metabolites and many also eventually provide an efficient increase of producing commercially important plant products. Brown and Tennis Wood (1975), reported the isolation of innamic, caffeine and ferulic acids from tobacco callus. Similarly, chlorogenic acid was also isolated from the cultures of tobacco by Sargent (1960) and *Haplopappus gracilis* by Stickland and Sunderland (1972a). The advances in understanding of useful product production from plant tissue culture have been tremendous and encouraging.

Biotechnology has seen successes in the *in vitro* production of colouring agents, condiments, perfumes, food emulsifiers, sweeteners, insecticides, insulin like compounds, hormones, aplirodisiacs, drugs, vitamins, etc.

Genetic factors present in a plant species may be expressed differently in tissue culture and thus affect both the concentration and identity of products produced. Either this property is exploited *in vivo* or the tissues produced *in vitro* are induced to produce the products in excess quantities. These two are the main ways in which production of secondary metabolites can be utilized efficiently.

A plant tissue that is isolated from a plant of horticultural importance and cultured *in vitro* by the application of relatively simple techniques offers great advantages:

1. to a horticulturist, who can multiply in large scale in a relatively quick time and achieve clonal propagation.
2. to a cytologist who can make a cytogenetic analysis of the cells and study the structure and behaviour of chromosomes.
3. to a morphologist who can explore the developmental potential of single somatic cells and understand the morphogentic pathways.
4. to a physiologist who can study various metabolic pathways, to assertain the nutritional requirements, and understand the role of growth hormones.
5. to a geneticist, who can treat millions of isolated cells as individual mutable units and select for rare variants in a few Petri dishes.
6. to a radiation biologist, who can use his skills and treat the isolated cells with radiations and induce variations and look for the creation of a break through variant.
7. to a biologist, who can study the basic physiology of a population of essentially identical cells.
8. to a biotechnologist, who can do a lot of things like isolating protoplast, fusion of protoplasts, production of artificial seeds, gene cloning, introduction of organelles and organisms, etc.

Thus biotechnology can be of immense importance to the improvement of all organisms including horticultural plants.

8

CELL AND PROTOPLAST CULTURE AND SOMATIC HYBRIDIZATION

Contribution of Cocking in 1970 is significant and has become a milestone in the field of biotechnology. It is an impetus to attempt plant improvement by cell modification and somatic hybridization. Subsequent to the development of techniques related to general protoplast culture and somatic hybridization, tremendous amount of work has been done to investigate various problems in plant physiology, radiobiology, cell modification and uptake of macromolecules, organelles and organisms.

This fascinating field has already played an important role in opening of new vistas and has awakened the interest of plant physiologists, pathologists, cytologists, biotechnologists, and molecular biologists. At present protoplasts are used frequently as research tools for studies in molecular biology and genetic engineering for plant improvement (Fig. 8.1)

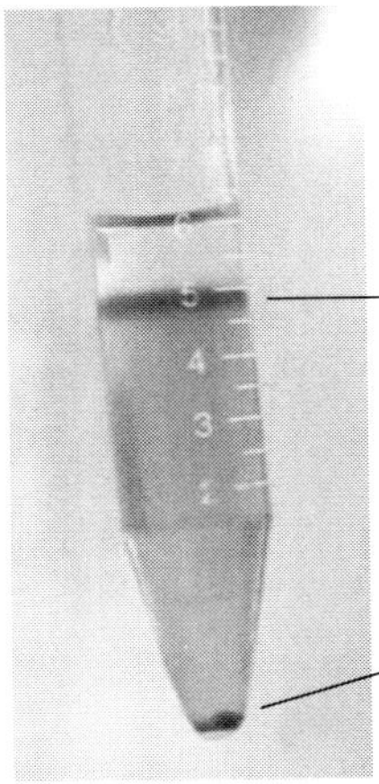

Fig. 8.1 Isolation of protoplasts from leaf.

ISOLATION OF PROTOPLASTS

Protoplasts are isolated conventionally mainly by two methods: 1. Mechanical and 2. Enzymatic.

The mechanical isolation of protoplast is usually avoided. It is the first method developed for isolation of protoplasts. The advantages offered by second method (enzymatic) are: 1. Protoplasts can be obtained in large quantities 2. The chance of protoplast damage is less 3. Somatic shrinkage is much less. The historically important mechanical method of isolation of protoplasts is occasionally used as it isolates proto-plasts without the unknown effects of enzymes on protoplasts. Sometimes both the methods are used in combination to retain the merits of both. Cells are first mechanically separated and then enzymatically transformed into protoplasts. Isolation of protoplasts has been reported from almost all important genera using all possible parts of the plant. Some notable examples are listed in the Table: 8.1.

Some of the somatic hybrids is listed in the Table 8.1.

List of hybrids produced by protoplast fusion	
Parents	**References**
Nicotiana glanca × *N. langsdorfii*	Carlson (1972)
Solanum tuberosum × *S. chacoense*	Kuchko (1985)
Solanum tuberosum × *Lycopersicon esculentum*	Melchers (1978)
Glycine × *Nicotiana*	Chenetal (1982)
Arabidropsis Haldiana × *Brassica compestris*	Gleba and Hoffmann (1978)

Table 8.1 list of hybrid produced by protoplast fusion.

ENZYMATIC ISOLATION OF PROTOPLASTS FROM LEAVES

It involves four steps:

1. Sterilization of leaves
2. Peeling of the epidermis
3. Enzymatic treatment and
4. Isolation and clearing of protoplasts

Completely developed healthy leaves from young plants in vegetative state are selected and are subjected to surface sterilization by dipping them in

70% ethyl alcohol for 5 minutes. Then the leaves are treated with 0.1% mercuric chloride by dipping in the solution for 30 seconds. Finally, the leaves are rinsed 5 times with autoclaved distilled water. Further steps are followed under aseptic conditions. Watts et al., (1974) have discussed the problems arising during the isolation of protoplasts.

The leaf is cut into small pieces after removing the lower epidermis. The leaves are relatively easier to peel off if the water supply to the plants is limited before excision of the leaves, or if the leaves are allowed to become flaccid after sterilization. Protoplasts can be obtained from the cells of both mesophyll and the epidermis. Isolation is carried out either by direct method (one step) or sequential method.

Direct method: In the direct method the stripped and sterilized leaf segments are placed in a Petri dish with their lower epidermis facing downwards. The petridish should contain sterile mixture of enzymes (0.5% macerozyme and 2% cellulase in 13% sorbitol or mannitol. The pH is maintained at 5.4. The leaf segments with Petri dish containing enzyme mixture are incubated at 25°C for about 15-18 hours. The protoplasts are then gently teased out after the period of incubation. These are then filtered through fine wire guaze to eliminate the debris. The filtrate is kept in a dessicator. Later, it is transferred to a water bath and subjected to slow shaking. The temperature is maintained at 25°C. This process is continued for 15 minutes and later the enzyme mixture is carefully replaced by a fresh mixture and incubated the leaf segments with enzymes for 1 hours. After the completion of incubation period the mixture is filtered through a nylon mesh. The filtrate is centrifuged at 100g for 1 min and washed for three times using 13% mannitol to get a pure sample of isolated cells.

The isolated cells are incubated with a mixture of enzymes containing 2% cellulase in 13% mannitol at pH 5.4 for about 90 min at 30°C. The cells are then centrifuged at 100 g for 1 min. The debris floats as supernatant and protoplasts form pellets at the base of the tube.

Proplasts are carefully transferred to tubes with caps. The tubes are centrifuged at 100g for 1 min. The debris forms a supernatant while the protoplasts are obtained as a pellet. The supernatant containing debris is completely removed. This process is repeated at least three times for obtaining protoplasts without debris and the pellets are washed with 13% sorbitol. Finally, centrifugation is carried out using 20% sucrose instead of sorbitol solution and debris settles at the bottom. The protoplasts are then pipetted out and stored for further use.

Sequential method: In the sequential method, first the cells are isolated and then the protoplasts are transferred into enzyme mixture for the separation of cells. Leaf segments without lower epidermis are placed in a petridish with a mixture of 0.5% macerozyme, 0.3% potassium dextran sulphate in 13% mannitol at pH 5.8 and vacuum infiltered for 5 minutes.

The pellets containing protoplasts are removed and cleaned with mannitol. The protoplasts are further purified by floating them on 20-25% sucrose solution. Polyethylene glycol and dextran, i.e. a two-phase aqueous system can also be used for the separation of protoplasts.

The yield of protoplasts though higher in one step method, the protoplasts isolated by the two-step method are definitely better for culture studies. The enzyme treatment for 15-18 hours duration in the one step method usually affects the plasma membrane of protoplasts, adversely, and hence the subsequent growth and development is also affected. In addition to this, there is one more disadvantage of the one-step method that is the protoplasts tend to fuse with each other. It is noticed by the formation of multinucleated cells. Perhaps the only disadvantage of two-step method is that it involves two separate experimental steps and is elaborated.

With minor variations protoplasts can be isolated from other sources also. The procedure is almost similar if other plant organs are used as sources of protoplasts. For obtaining protoplasts from cell suspensions, the cell suspensions are filtered and treated with 3% cellulase in 6.6 M mannitol for about 5 hours and incubated at 30°C in a water bath which is subjected to slow shaking. While isolating the protoplasts from pollen grains one will have to encounter with the complex pollen wall. This wall can be dissolved. Further the pollen is subjected routine treatments and protoplasts are isolated. (Fig. 8.2)

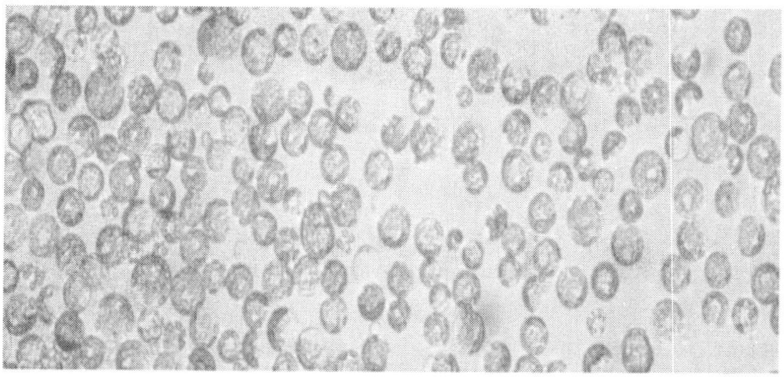

Fig. 8.2 Purified plant protoplasts.

PROTOPLAST CULTURE

Plant cells are totipotent and have the capacity to unfold their morphogenetic potential to develop into whole plants (Braun, 1959). Similarly, Isolated protoplasts regenerate a cell wall around them and form reconstituted cells in culture, which later undergo repeated divisions to form callus. Later the callus could be induced by nutritional and physiological manipulations to regenerate into plants.

Interestingly the protoplasts exhibit a property of uptaking large bimolecules, viruses, ongenelles, bacteria, and DNA. This is the property exploited by genetic engineers to achieve biotransformation and to produce transgenic plants.

Protoplasts can be cultured by suspending them in a hanging drops of liquid medium in flasks without shaking. This method is called hanging drop technique and was developed by Bawa and Torrey (1971). Protoplasts can also be similarly cultured by keeping them in microchambers as described by Vasil (1973). The microchambers are prepared by placing two cover glasses about 18mm apart on a drop of heavy mineral oil placed on the slide. A drop of medium containing protoplasts is placed in between the two cover slips. A third cover slip is then gently placed on the medium and finally sealed with mineral oil. Once the protoplast undergoes a few divisions they are then transferred to an agar medium for further growth to proceed.

Bergmann (1960) developed a method, in which protoplasts are directly placed on solid agar medium. The technique was later modified by others and was successfully used. It has been used for culturing large number of protoplasts conveniently as plating ability can be determined easily by observing under microscope. 2-3 ml aliquots of protoplasts are mixed in liquid medium and are poured into Petri dishes and gently mixed. The temperature of the medium should be maintained below 45°C. The Petri dishes are sealed using parafilm to prevent desiccation. The Petri dishes are then incubated by keeping them in inverted positions at 25°C. Protoplast density is crucial for plating ability. Aliquots of protoplasts should ideally have protoplast at 10^5/ml. At higher densities there is a chance of the origin of chimeras often caused by coalescence of adjacent cells. It makes it difficult to isolate the colony of single cell origin. Kareh *et al*. (1973) used a method in which non-dividing X-irradiated protoplasts were used as feeder layer to support the division of viable protoplasts plated at a density as low as 10^2/ml. This method is convenient for the selection of colonies formed from single mutant cells originated experimentally.

Cell Culture

Tissue culture is based on totipotency, which is the capacity of each and every cell of an organism to divide and develop into a complete organism. In principle every isolated cell should behave as a zygote. This feature of a cell is exploited fully in suspension or cell culture (Fig. 8.3).

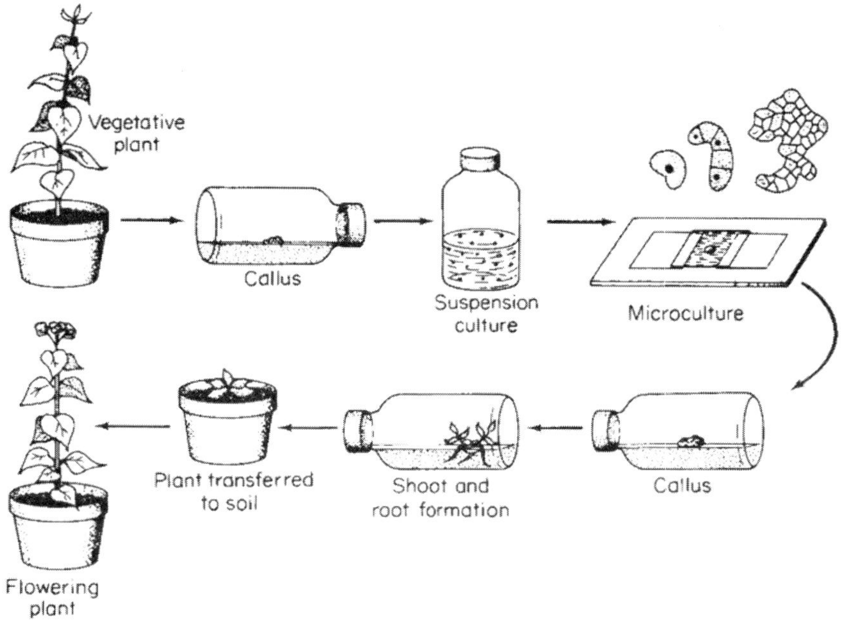

Fig. 8.3 Schematic outline of single cell culture.

The procedure for the isolation of cell remains the same as described in the previous section. The isolated cell is then transferred into the nutrient medium preferably to a liquid medium. Cells can be separated by fragmenting callus. About 500mg callus is fragmented in 500ml liquid medium and agitated aseptically to make the cells free. Usually all cells may not be separated instead in the suspension there are isolated cells, cell aggregates and dead cells (Dodds and Roberts, 1985).

A good suspension is one with high frequency of single cells produced with optimum environmental factors. There are several methods of cell separation described in the literature. Perhaps a better method for a particular plant has to be determined by the researcher.

The isolated cells in the suspension containing the suitable nutrient medium start dividing. The division of cells follows the routine sequence of events.

Initially, the cells will be in the lag phase in which there is slow cell division. The rate of cell division increases linearly but the rate decelerates when it comes to stationary phase. At this stage to keep the cells viable, it is essential to subculture the cells. By using plate technique, cell lines can be raised and it is crucial for the rate of success. Growth of culture depends on a critical cell density below which culture will not grow.

Inherent dualistic nature of cultured cells is a well known phenomenon. On one hand they possess necessary genetic information for existence on cell level to remain at any one of the stages like, production, growth maturity apoptosis, etc. while on the other hand, the cultured cells retain supplementary informations that determine synthesis of substances necessary for integrative functions and biocenotic interactions of the plants.

Cultured cells can be used for the induction of somatic embryos and shoots, raising mutagenic plants, mutants, production of secondary metabolites and transformation.

SOMATIC HYBRIDIZATION BY PROTOPLAST FUSION

One of the important applications of protoplast isolation is undoubtedly protoplast fusion. It is advantageous for those who wish to produce hybrids by using almost unrelated parents. They can produce interspecific and intergeneric hybrids more freely now.

This technique is a boon for horticulturists interested in the fusion of protoplasts of genetically different plants. In case the plants show physical or chemical incompatibility in normal sexual crosses this technique is quite useful. Some plants however still pose new challenges to plant breeders interested in crop improvement.

Coalescence of two cytoplasms of different origin occurs due to the fusion of protoplasts. But the two nuclei may or may not fuse together resulting in the formation of heterokaryons and heterocytes. If the two nuclei fuse, it results into the formation of a synkaryocyte. Cybrids are cytoplasmic hybrids. They are formed when two cytoplasms are fused, but the nuclei are not fused, one of the nuclei remains in the cell and the other is lost or inactivated.

The process of production of cybrids is called **cybridization** in contrast to hybridization, which is the fusion of two nucleated cells. Male sterility, resistance to phytotoxins, drug resistance and a few other characters are often controlled by cytoplasmic genes. Therefore cybridization process can be used to transfer genes coding for these characters.

PROTOPLAST FUSION

Protoplast fusion is always followed by the isolation of protoplasts. They are either cultured as such or they may be induced to undergo fusion by using fusogenic chemicals, i.e., chemicals promoting protoplast fusion. (Fig. 8.4, 8.5 and 8.6).

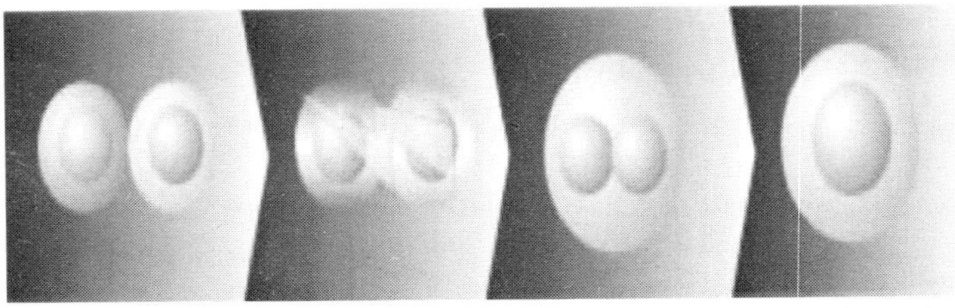

a. Allignment of protoplsts | b.Electric pulse for 15 micro seconds | c. Heterokaryon phase | d. Synkaryon; protoplasts fused

Fig. 8.4 Diagrammatic representation of protoplast fusion.

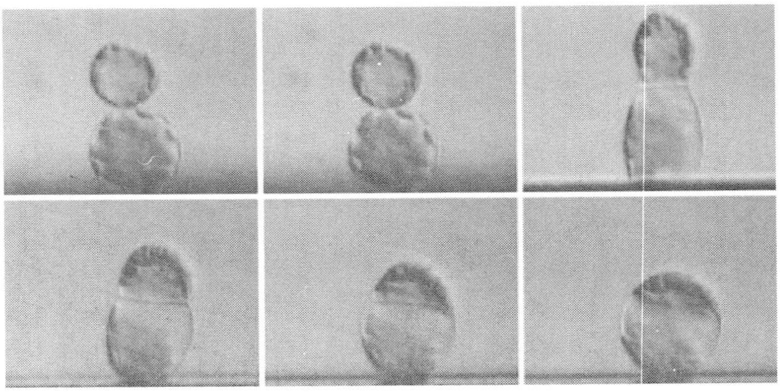

Fig. 8.5 Various stages of protoplast fusion.

Isolated protoplasts are taken in a tube and mixed with 20% (w/v) polyethylene glycol (PEG) and sodium nitrate. The pH is adjusted to 10.5 and centrifuged at 37°C. PEG induces fusion of protoplasts to form heterokaryon. The fused protoplasts further regenerate the cell walls and thus hybrid cells are produced. The hybrid cells later can be selected, subcultured and organogenesis is induced to obtain hybrid plants.

Geneborg and Nobors (1987) have described a number of variations in cell fusion and other related events. Protoplasts of *Nicotiana langsdorfii* and *N. glauca* were isolated and were fused to produce somatic hybrids by Carlson (1972). It is the first interspecific hybrid produced. Similarly,

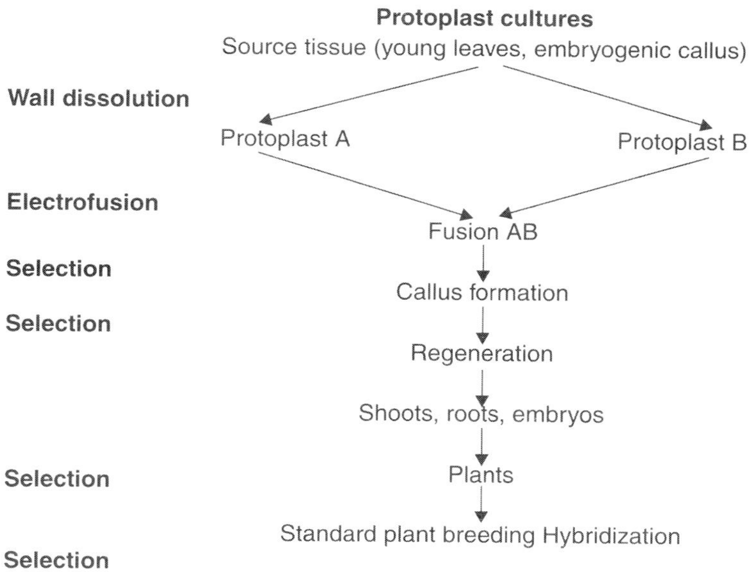

Fig. 8.6 Schematic representation of protoplast fusion.

Kuchko (1985) succeeded in fusing the protoplasts of cultivated (*Solanum tuberosum*) and a wild (*Solanum chacoense*) varieties of potato.

Pomato produced by the fusion of protoplasts of potato (*Solanum tuberosum*) and tomato (*Lycopersicon esculentum*) is the first intergeneric hybrid. It was achieved by Carlson *et al*. (1972).

9

BIOTRANSFORMATION OF ORCHIDS

OVERVIEW

Biotransformation is being widely carried in various plants and animals. It is basically a method involving several techniques of molecular biology for bringing about changes in phenotypic characters in organisms of commercial importance. A gene coding for a particular character found in a plant can be isolated and is transferred into another organism of interest. The whole process can be conveniently called genetic engineering.

The endeavour of transforming organisms by transferring genes from one organism to another organism began in microorganisms. One of the first successes in this area was achieved by a group of scientists involving Dr. Saran Narang from Canada who synthesized and transferred insulin gene from man into a bacterium *Escherichia coli* (Guo G H *et al.* 1984). In case of plants there have been several success stories. Transfer of ***Bt*** (*Bacillus thuringenesis*) gene into cotton plant (*Gossypium herbaceum*) to produce most infamous ***Bt***-cotton. Similarly, golden rice was also produced by the transfer of β-carotene genes into rice plant (*Oryza sativa*). The plants and animals are popularly known as genetically modified organisms (GMOs). The GMOs and transgenic plants are discussed in chapter no. 5.

There are several approaches to bring about biotransformation. These techniques are necessarily employed in the improvement of orchids in various ways as there are innumerable number of opportunities to bring about desirable changes in the phenotypic characters of this wonderful group of plants, which have always been admired by all for their exquisite beauty, colours, fragrance and the medicinal values. Orchids are undoubtedly the most advanced group and this advancement is obviously due to their genetic nature, which can be identified and probably transferred to the genes responsible into other plants of interest. Some of the procedures especially employed in the biotransformation of orchids are discussed in this chapter.

AGROBACTERIUM-MEDIATED BIOTRANSFORMATION IN ORCHIDS

The general outline of this technique has already been given in chapter 6. Biotransformation of higher plants using Ti (Tumor inducing) plasmids from *Agrobacterium* has been proven to be an efficient method for dicotyledonous plants, and has been gradually established in monocots also. Chan *et al.* (1992) have reported the transformation of *Oryza sativa* mediated by *Agrobacterium*. Based on the same method Yu *et al.* (2001) succeeded in transferring a gene into a *Dendrobium* hybrid. Disease and pest resistance and other desirable traits could be conveniently transferred from one organism to the other by molecular engineering. This becomes necessary in case of orchids, especially for those which have good demand in the floriculture industry.

A gene for coniferyl alcohol has been identified as a virulence gene in some orchids like *Dendrobium* similar to the function of several phenolic compounds in dicots. Prior to the success achieved by Yu *et al.* there have been several efforts made to develop efficient *Agrobacterium* mediated transformation systems for a few orchid hybrids (Nan *et al.*, 1998). In 1998 Nan *et al.*, reported the incorporation of only partial T-DNA into *Dendrobium* orchid genome.

Orchid plant material required for the transgenic studies are cultured and maintained *in vitro*. The entire cDNA with blunt ends are obtained by PCR amplification using DNA polymerase like VentR and primers like 5´CTGTGTCCACCACCGAGTC3´ (forward primer) and 5´ AATTAACCCT CACTAAAGGG3´ (reverse primer).

Antisense Vector (pBI´121-DOH-1) can be constructed by cutting the PCR product with *Xba 1* to produce the 5´ cohesive end and then cloning into the *Xba 1* and *Sac 1* sites of the pBI 121 binary vector downstream of the cauliflower mosaic virus (CaMV) 355 promoter in the place of the b-glucuronidase (GUS) gene. Before the ligation, the *Sac 1* cohesive end in the pBI 121 vector is fitted in with the Klenow fragment. The construct sequence can be analysed to eliminate any PCR introduced mutations. The pBI 121 vector containing a promoterless GUS cassette and the pBI 121 vector containing the CaMV355 promoter-GUS fusion are used as negative and positive controls, respectively.

These constructs are introduced into the disarmed octopine type *Agrobacterium tumefaciens* strain LBA 4404 by tri-parental mating.

The next step in the transformation process is the regeneration of transformants. *Agrobacterium tumefaciens* is grown on specific nutrient media like, Luria-Bertani(LB) medium (Sambrook *et al.*, 1989) and incubated at 28°C. The overnight culture is diluted with modified Knudson C(KC) medium. The dilution can be made to 3×10^8 cells/ml.

LB medium should be added with 50 mgl^{-1} Kenamycin and 25 mgl^{-1} streptomycin. Similarly, KC medium is supplemented with 2% (W/V) sucrose, 15% (V/V) coconut water and 5 mgl^{-1} BA.

Before transformation, 1mm thick sections of protocorm-like bodies (PLBs) are cultured on KC (liquid) medium for three days. The cultures of PLBs sections are then immersed in the culture of *A. tumefaciens* for 30 min. The explants are blotted on to sterile filter paper, and then transferred onto modified solid KC medium without antibiotics. The co-cultivation is maintained for 3 days in 16h photoperiod. The explants are then transferred onto another modified KC medium containing 5 mgl^{-1} carbenicillin, which inhibits the fast growth. *A. tumefaciens* and this co-cultivation is continued for 3-4 weeks. Further, the calli obtained from this culture are thoroughly rinsed using liquid KC medium containing 200 mgl^{-1} Kenamycin and Carbenicillin. The calli are then sliced into small pieces and sub-cultured for 3 weeks on modified KC medium with 200 mgl^{-1} Kenamycin and Carbenicillin. The calli will proliferate in this medium, which are further sub-cultured for next three weeks on the same medium.

This process is performed for further screening. After repeating the above process for three times, the calli are sub-cultured once in every 2 weeks. The PLBs obtained from the calli are isolated and are screened by PCR and Southern blot analysis and maintained on modified solid KC medium with 200 mgl^{-1} Kenamycin and Carbenicillin.

DNA from both the wild type orchid and PLBs of transformants can be isolated by any one of the standard procedures. But the methods used by Yu and Goh (2000b) and Junghans and Metzlaff (1990) are considered to be better. There are, however, several methods available for the isolation of plant DNA.

In the routine procedure for DNA isolation the leaf and apical meristems are taken ideally. The tissues are ground in liquid nitrogen with pestle and mortar to isolate the cells. The cell membranes are disrupted to release the DNA, the buffer by using a detergent like Sodium dodecyl sulphate (SDS) or Cetyl trimethyl ammonium bromide (CTAB).

Ethylene diamine tetra acetic acid (EDTA) is used to protect the isolated DNA from endogenous nucleases. EDTA is a chelating agent that binds magnesium ions which are considered necessary co-factors for most of the nucleases. To denature and to separate the proteins from the DNA, the tissue/buffer mixture is emulsified with chloroform-phenol. The DNA is precipitated using either ethanol or isopropanol. DNA preparation can be made RNA free by digestion with ribonucleases.

DNA, thus obtained from the two sources is initially analysed by PCR arrays. Specific oligonucleotide primer that amplifies the 700 bp fragments spanning the 3´ end of the CaMV 355 promoter and 5´ end of the inserted DOH´ antisense gene usually is 5´CCTCCTCGGATTCC A TTGCCCAGCTAT 3´. PCR amplification is performed with 30 cycles at 94°C for 1 min, at 65°C for 1 min and at 72°C for 1 min. The resulting products are separated on 1% (W/V) agarose gel and visualized by staining with Ethidium bromide.

Southern blotting of the DNA from transformants is done for further analysis. DNA extracted from putative transformants is digested with *Sma* I and fractioned onto 0.7% (W/V) agarose gel. Southern hybridization and detection is carried using digoxigenin-labeled DNA probe, which is prepared from the full-length *DOH1* cDNA.

The presence of the *DOH1* antisense gene can be confirmed by initially amplifying by PCR. Southern blotting analysis further confirms the integrity of the *DOH1* antisense gene in transgenic orchid plants. All the transgenic lines will have one or two copies of the *DOH1* antisense gene, which are represented by the extra bands in addition to that of the endogenous *DOH1* gene. Only a single band is detected in wild type plants.

Orchid transformation by partial bombardment results in the integrity of the multiple copies of the antisense gene into the orchid genome. The expression of the *DOH1* antisense gene can be enhanced by CaMV355 promoter. The amounts of *DOH1* antisense on RNA in vegetative shoots from transgenic plants can also be analysed and the analysis can be compared with that of wild type orchid.

DOH1 gene is found to play a role in the maintenance of basic plant architecture in orchids. Abnormal development of orchid shoots was observed through the introduction of the *DOH1* antisense gene into the genome of *Dendrobium* Madame Thong-In (Yu *et al.,* 2002).

Similarly, a transformation procedure for *Phalaenopsis* orchid was established Mishiba K *et al.,* 2005 by using immature protocorms for *Agrobacterium* infection aimed at the introduction of target genes into individuals with divergent genetic backgrounds. Protocorms obtained after 21 days of culture on liquid New Dogashima medium were inoculated with *Agrobacterium* strain EHA101(pIG121Hm) harboring both beta-glucuronidase (GUS) and hygromycin resistance genes. Subculture of the protocorms on acetosyringone-containing medium 2 days before *Agrobacterium* inoculation can use highest transformation efficiencies (1.3-1.9%) based on the frequency of hygromycin-resistant plants produced. Surviving protocorms obtained 2 months after *Agrobacterium* infection on selection medium containing 20mg l^{-1}) hygromycin are cut transversely into two pieces before transferring to recovery medium without hygromycin. Protocorm-like bodies (PLBs) proliferated from pieces of protocorms during a 1-month culture on recovery medium followed by transfer to selection medium. According to Mishiba *et al.,* hygromycin-resistant *Phalaenopsis* plants that regenerated after the re-selection culture of PLBs showed histochemical blue staining due to GUS. Transgene integration of the hygromycin-resistant plants was confirmed by Southern blot analysis. A total of 88 transgenic plants, each derived from an independent protocorm, was obtained from seeds 6 months after infection with *Agrobacterium*. Due to the convenient protocol for *Agrobacterium* infection and rapid production of transgenic plants, this procedure can be utilized to assess expression of transgenes under different genetic backgrounds, and for the molecular breeding of any orchid taxa.

Liau, CH *et al.,* (2003) developed a similar protocol for establishing a routine transformation procedure via *A. tumefaciens* for an important *Oncidium* orchid cultivar. An expression vector containing hptII and gusA genes driven by the cauliflower mosaic virus (CaMV) 35S promoter was successfully introduced into the *Oncidium* orchid genome by *A. tumefaciens*. Protocorm-like bodies (PLBs) derived from protocorms, were the target explants for transformation. The transformation was performed through two stages of co-cultivation, the first stage occurring on antibiotic-free medium for 3 days, and the subsequent stage on medium containing 100 mg/l timentin for 1 month. Among 1,000 inoculated PLBs, 108 putatively transformed PLBs were proliferated on 5 mg/l hygromycin selection medium. A total of 28 independent transgenic orchid plants were obtained, from which six transgenic lines that were positive for β-glucuronidase were randomly selected and confirmed by Southern, Northern and Western blot analyses. These results indicated that the foreign DNA was successfully integrated into the orchid genome and expressed transcriptionally and translationally in these orchid plants. The present transformation method reported is suitable for improving the *Oncidium* orchid through genetic engineering.

Cymbidium Mosaic Virus (CymMV) and *Erwinia carotovora* have been reported to cause severe damage to orchid plants. To enhance the resistance of orchids to both viral and bacterial phytopathogens, gene stacking was applied on *Phalaenopsis* orchid by double transformation by Chan, YL *et al.*, 2005, PLBs originally transformed with CymMV coat protein cDNA (*CP*) were then re-transformed with sweet pepper ferredoxin-like protein cDNA (*Pflp*) by *Agrobacterium tumefaciens*, to enable expression of dual (viral and bacterial) disease resistant traits. A non-antibiotic selection procedure in the second transformation minimized the potential rate of 'stacking' antibiotic genes in the orchid gene pool. Transgene integration in transgenic *Phalaenopsis* lines was confirmed by Southern blot analysis for both *CP* and *pflp* genes. Expression of transgenes was detected by Northern blot analysis, and disease resistant assays revealed that transgenic lines exhibited enhanced resistance to CymMV and *E. carotovora*. According to Chan *et al.*, this is the first report describing a transgenic *Phalaenopsis* orchid with dual resistance to phytopathogens.

Similarly, any desirable gene can be transferred into the orchid plant genome in order to introduce the novel characters. It is a very well known fact that there are wide varieties of orchids having some wonderful characters pertaining to fragrance, shapes and colours. In fact, a gene from any other source can also be transferred into orchid genome by way of which designer orchids can be developed. This approach gives way for making the wonderful world of orchids ever more fascinating.

GENETIC TRANSFORMATION OF ORCHIDS BY PARTICLE BOMBARDMENT

Genetic improvement of orchids has been going on increasing substantially in the recent past vowing to the growing importance of orchids in the multibillion dollar floriculture industry. Genetic transformation of orchids for disease and stress resistance, precocious flowering, improvement in flower colour, changing of flower colour, changing of floral architecture, introduction of genes coding for fragrance, is actually of important commercial value.

All said and done plant improvement by conventional breeding techniques has been limited mainly due to the prohibitively long reproduction cycles, slow seed maturation and probably the highest rate of promiscuity in orchid. Therefore the application of genetic engineering techniques appears to be an attractive alternative. Particle bombardment using biolistic process as a delivery systems to introduce foreign DNA into plants has been used successfully in several cases (See Table 9.2).

In one of the first examples of the use of particle bombardment techniques of genetic transformation, Kuchnle and Sugii (1992) reported transgenic orchids, *Dendrobium*. Transformation was confirmed by Kenamycin selection and PCR analysis of the putative transgenic orchids. Similarly, Chia *et al.,* (1994) transformed another *Dendrobium* orchid by using the same technique. But they had used an alternative selection method as the sensitivity of Kenamycin is very low even at 500 mgl^{-1}. Chia *et al.,* introduced luciferase gene from firefly. Light emitting cell clumps carrying luciferase gene were screened and isolated using photon - counting video camera-photomultiplier system and a high power dissection microscope. Pure transgenic cell lines could thus be easily isolated. The disadvantage with this technique is the requirement of expensive equipment and time consuming microscopic screening of transgenic cell.

However, Yang *et al.,* (1999) developed an improved protocol for the genetic transformation of orchids by particle bombardment. They employed a liquid culture system to stimulate active proliferation of the meristematic tissues in pre and post-bombardment periods and developed a method that minimizes the toxic influence of dying cells on neighbouring transformed cells.

Procedure for biotransformation:

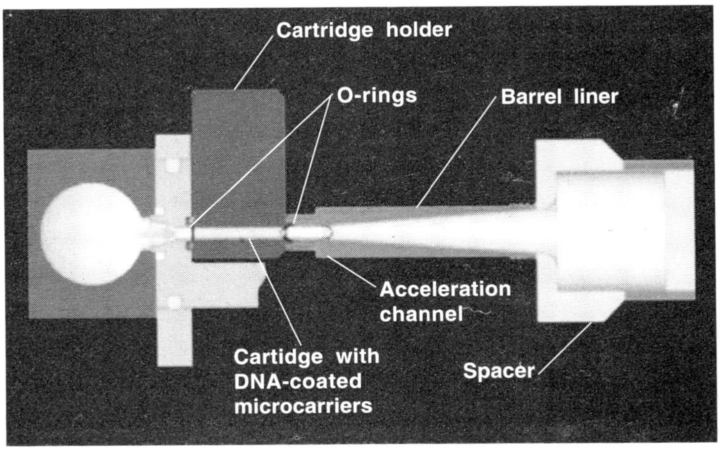

Fig. 9.1 Schematic diagram for Gene gun Delivery.

Fig. 9.2 Particle bombardment chamber.

PLBs are raised by culturing apical meristam or leaf explants (the technique for this has been described in the earlier section). PLBs thus produced are sub-cultured and maintained of Knudson C medium with 5% sucrose, 10 mgl^{-1} NAA, 0.1% charcoal, and 0.7% agar.

PLBs are transferred into modified liquid MS medium without hormones and kept on a gyratory shaker with 50-100 rpm. PLBs are maintained in this condition for a week and are transferred onto a Petri dish containing solid MS medium which served as a target surface. Plasmid DNA is precipitated in the presence of gold particles. The procedure for the precipitation is elaborated by Klein *et al.*, (1987). Then the pre-cultured PLBs are bombarded using a Model Biolistic PDS-1000 / He or any other method (see Table 9.3 for details). The instructions given by the manufacturer have to be followed carefully.

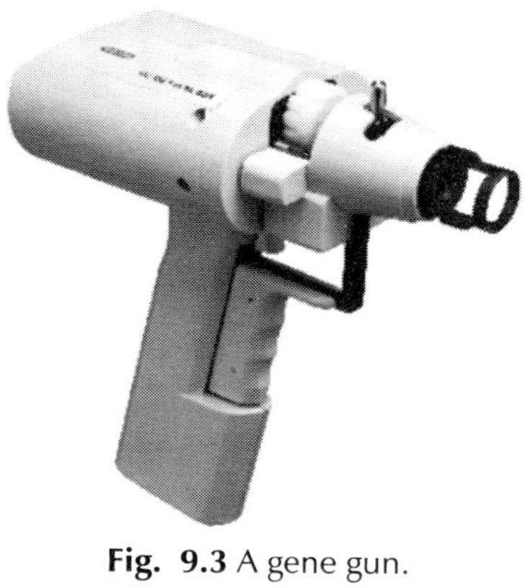

Fig. 9.3 A gene gun.

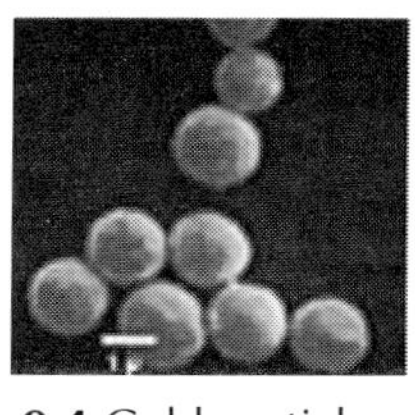

Fig. 9.4 Gold particles used for gene gun delivery.

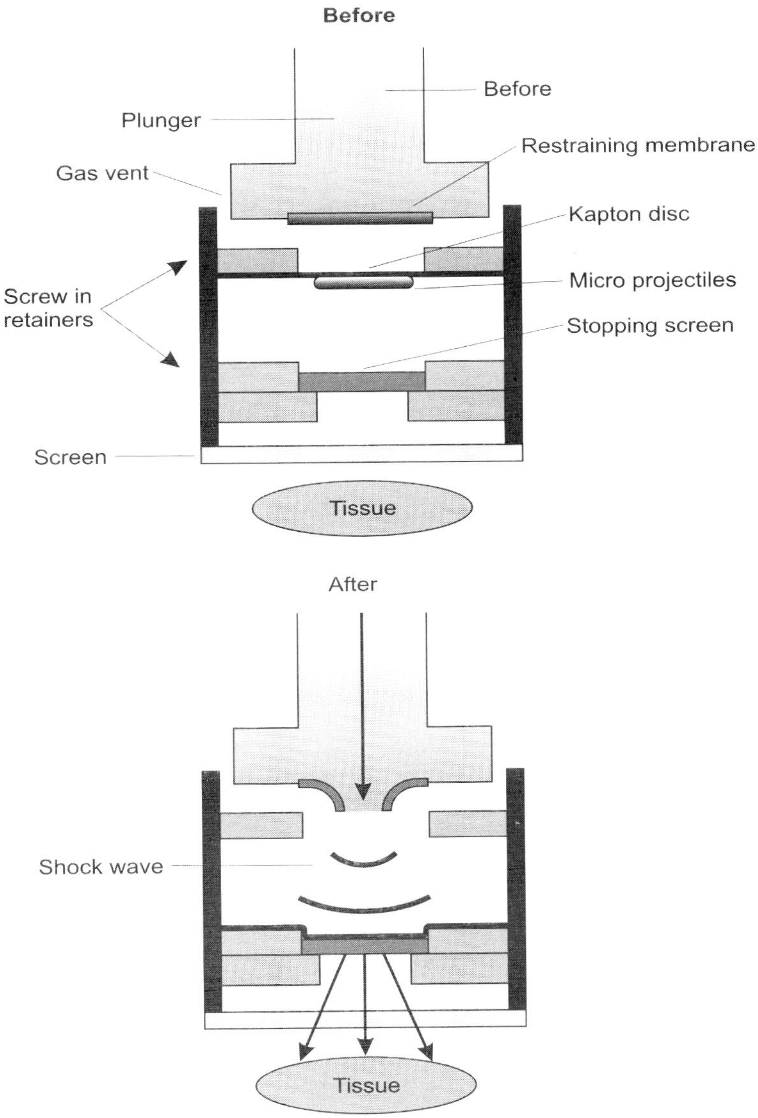

Fig. 9.5 Schematic Diagram of particle bombardment.

The bombardment chamber has to be evacuated at a pressure of 28 inches of mercury. Each bombardment delivers approximately 500 mg of particles associated with 0.42 mg. Bombardments are carried out with pressurized helium in the range of 1100 - 1550 psi at a range of approximately 60mm from the stopping plate to the surface of the treated tissue.

After the process of bombardment using biostic procedures, the bombarded PLBs are cultured for a period of 10-45 days on liquid MS medium kept on a gyratory shaker with speed of 50-100 rpm. This induces proliferation of PLBs. The proliferated PLBs are to be transferred to Petridishes containing solid MS medium with 0.7% agar and growth regulators like BA and NAA in concentrations recommended for a particular orchid taxa and Kenamycin (100-200 mgl^{-1}). Sub-culturing can be made once in week. For the induction of rooting the PLBs are transferred into MS medium without growth regulators.

Gold particles used for coating with plasmid DNAs can be of different sizes, ranging from 0.6 to 1.6 mm. The bombardment of particles into the PLBs at different helium gas pressures (1100, 1350 and 1550 psi). Even Kenamycin concentrations can also be varied from 0-500 mg l^{-1}. Kenamycin is used for effectively killing or suppressing the growth of non-transformed PLBs.

After the bombardment, to test the expression of the inserted genes like GUS, the transgenic plants are assayed at various times ranging from 1 day to 9 months after the bombardment. GUS expression is assayed by histochemical staining. Defferson *et al.,* (1987) has elaborately described this technique.

Genomic DNA for PCR work can be isolated by various methods (the technique is described in the previous section). Amplification of the gene (e.g., npt II) is generally performed with *Taq* polymerase at three temperatures, 94°C, 55°C and 72°C (1min each). Using specific primers like 5´ATTGCACGCAGGTTCTCCGG3´ and 5´ AGAACTCGTCAAGAAGCGGA 3´ (Franche *et al.,* 1997). The PCR product can be confirmed both by Southern blot analysis using the gene, npt II as a probe and sequencing by an ALF automatic sequencer.

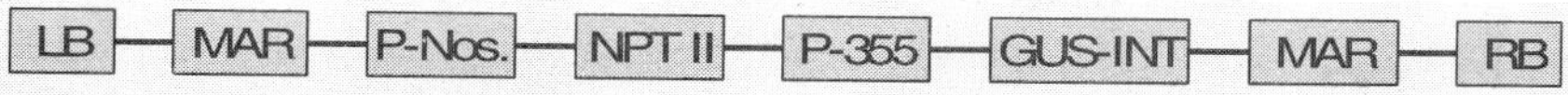

Fig. 9.6 Diagrammatic representation of the T-DNA region of the plasmid pKH200. The relative position of GUS-INT gene and the npt II gene are flanked by the tobacco matrix regions (MARs) and shown with respect to the left border (LB) and right border (RB) of the T-DNA MAR [matrix attachment region]. (Yang *et al.,* 1999).

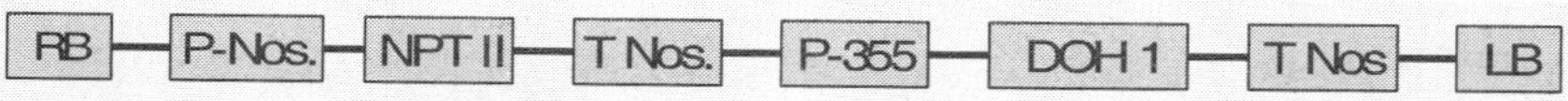

Fig. 9.7 Diagrammatic representation of the T-DNA region in the Pbt121 – DOH1 as construct. The original GUS cassette in the pBI121 plasmid is replaced by DOH1 gene [P355-CaMV promoter, LB-Left border, RB-right border, P Nos. nopaline synthase promoter, T Nos. -polyadenylation signal of the nopaline synthase gene]. (Yu, *et al.,* 2001)

GUS expression can be observed in transient-gene expression state following particle bombardment-mediated transformation and in virtually all the plant tissues after selective regeneration. The expression can be observed even after 2-3 days after the introduction of pKH200 by bombardment. Almost all bombarded PLBs will show GUS activity. Usually high transient GUS activity of more than 100 GUS foci per PLBs is seen for 3 days, but at generally declines gradually. This decline in GUS foci is a major obstacle in the selection of tranformants in an orchid system that is characterized by slow growth in tissue culture. Thus, the culture systems that stimulate rapid cell division and transform cells to PLBs has to be developed. Long term and stable expression of the introduced GUS gene can be achieved by differentiating transformed PLBs and tissues rapidly into Kenamycin resistant plantlets that have acquired the desired gene.

An additional care can be taken while selecting the right kind of gene for transformation experiments. Use of an intron-containing GUS gene, generally preferred as the gene with intron, is expressed in plant cells but not in any contaminating microorganisms.

Use of matrix attachment regions (MAR) significantly increases the expression of flanked genes, compared to the genes not flanked by the MAR (Han *et al.*, 1997b). Generally transgenic PLBs survive the Kenamycin selection whereas the untransformed PLBs are effectively killed with as low as 50 gl^{-1} Kenamycin.

Men *et al.*, (2003) reported the transformation of two species of orchid, *Dendrobium phalaenopsis* and *D. nobile*, by biolistic bombardment. Calli or PLBs were used as target explants. Gold particles (1.0 micron) coated with plasmid DNA (pCAMBIA1301) encoding an intron-containing beta-glucuronidase gene (gus-int) and a hygromycin phosphotransferase (hpt) gene were introduced into the PLBs or calli using the Bio-Rad PDS-1000/He Biolistic Particle Delivery System. Calli and PLBs were then chopped up and pre-cultured in 1/2-strength MS medium supplemented with 0.4 M mannitol for a 1-h osmoticum treatment before bombardment. Immediately after bombardment, the calli and PLBs were transferred to half-strength MS medium without mannitol for recovery. Putatively transformed plantlets were obtained by selection and regeneration on medium supplemented with 30 mg/l hygromycin. The highest efficiency of transformation was obtained when selection was conducted at 2 days post-bombardment. For *D. phalaenopsis* and *D. nobile*, respectively, about 12% and 2% of the bombarded calli or PLBs produced independent transgenic plants. Integration and expression of the transgenes were confirmed by Southern hybridization and Northern hybridization. No nontransformed plants were regenerated, indicating a tight selection scheme. However, separate incorporation of the GUS gene and the hpt gene was observed, and in one transgenic line the GUS gene was integrated into the genome of the transgenic plant, but not expressed.

Orchids are substantially different from other plants in their requirements for a transformation system. Because of the following reasons, the orchid cells have a lower rate of proliferation as compared to other cells. Generally orchids do not regenerate from differentiated tissues. It is a well known fact that the orchid cells in tissue culture exude a large quantity of phenolics that become toxic to the cells when oxidized. Hence, specific techniques for inducing transformation in orchid are required to be developed and employed. By direct delivery of DNA into the cells of organized tissues (Nehra *et al.*, 1994), bombardment of meristematic tissues and selfing the treated generation and selecting for the transformed traits are considered to be the appropriate approaches for obtaining homogeneous and non-chimeric transformations in orchids.

OTHER BIOTRANSFORMATION STUDIES IN ORCHIDS

The orchid floral homeotic gene, DOMADS1, is a marker gene specifically expressed in the transitional shoot apical meristem during floral transition in *Dendrobium* **Madame Thong-In**. DOMADS1 is not detectable in vegetative tissues except a weak expression in the stem. Its transcript is uniformly localized in both of the inflorescence meristem and floral primordia, and later expressed in almost all of the floral organs. Yu *et al.*, (2002) isolated and sequenced a 3.5 kb DOMADS1 promoter fragment upstream of the transcription start site, demonstrating the location of several putative DNA-binding sites, through which MADS-box and class I *knox* genes may modulate the DOMADS1 expression. To gain insight into the molecular basis of the regulation of DOMADS1, deletion analysis of the DOMADS1: β-glucuronidase (GUS) gene fusions have been performed by means of the stable orchid transformation systems. The studies show that the full-length upstream promoter sequence confers the same spatial and temporal GUS staining pattern as that of the distribution of DOMADS1 RNA during orchid development. It is also identified the distinct cis-acting regulatory regions required for the control of DOMADS1 expression in vegetative and reproductive tissues, as well as the shoot apical meristem during floral transition.

Some angiosperms are limited to a range of possible flower colors. This limitation can be due to the lack of an anthocyanin biosynthetic gene or to the substrate specificity of a key anthocyanin biosynthetic enzyme, dihydroflavonol 4-reductase (DFR). *Cymbidium hybrida* orchid flowers primarily produce cyanidin-type (pink to red) anthocyanins and lack the pelargonidin-type (orange to brick-red) anthocyanins. To investigate the underlying molecular mechanism of this flower's color range, Johnson *et al.,* (1999) cloned a *Cymbidium* DFR gene and transformed it into a DFR-*Petunia* line. They found that the *Cymbidium* DFR did not efficiently reduce dihydrokaempferol (DHK), which is an essential step for pelargonidin production. Phylogenetic analysis of a number of DFR sequences indicate that the inability to catalyze DHK reduction has occurred at least twice during angiosperm evolution. These results indicate that developing a pelargonidin-type orange flower color in Cymbidium may require the transformation of a DFR gene that can efficiently catalyze DHK reduction.

Hieber *et al.,* (2005) investigated the genetic basis of colour in the orchid *Oncidium* **Gower Ramsey**. Orchids are one of the most unique and highly evolved flowering plants, with many being valuable floricultural crops. Spatial localization of pigments within the flower of the commercially important bi-color *Oncidium* **Gower Ramsey** demonstrated a mixture of carotenoids and anthocyanins concentrated in the adaxial epidermis. Chromatography identified the predominant yellow pigment to be an equal mixture of all-trans and 9-cis isomers of violaxanthin, with esterification specific to the 9-cis isomer. Red ornamentation comprises the anthocyanins cyanidin and its methylated derivate, peonidin. Five key pigment biosynthesis genes encoding dihydroflavonol 4-reductase (DFR), phytoene synthase (PSY), phytoene desaturase, carotenoid isomerase, and the downstream 9-cis epoxycarotenoid dioxygenase have been isolated and their expression profiles determined. Northern analyses showed both phytoene desaturase and carotenoid isomerase expression to be up-regulated in floral tissue relative to leaves whereas PSY is not. Three closely related DFR genes have been isolated, including one with an insertion in the 3´ coding region. DFR expression occurred throughout flower development in Oncidium, unlike in *Dendrobium* and *Bromheadia* orchids. A number of the isolated anthocyanin and carotenoid genes showed variations due to insertion events. These findings raise questions about the genetic stability in interspecific crosses in orchids, such as the tri-specific **Oncidium Gower Ramsey**.

Table 9.1 Isolation and fusion of Orchid protoplast

Sl. No.	Orchids	Nature of Study	Reference
1.	Renantanda Rosalind Cheok	Isolation and Culture	Teo CKH & Neumann KH (1978a)
2.	Renantanda	Isolation and Fusion	Teo CKH & Neumann KH (1978b)
3.	Dendrobium	Isolation and Fusion	Price GR and Earle ED (1984)
4.	Aranda Noorah	Isolation and Culture	Loh CS & Rao AN (1985)
5.	Aranda Hybrids	Isolation and Culture	Koh MC, Goh CJ and Loh CS (1988)
6.	Aranda Chark Kaun	Isolation and Culture	Kanchanapoom IC and Tongseedam P (1994)
7.	Phalaenopsis from protoplasts	Regeneration of plansts	Sajise JU and Sagawa Y (1991)
8.	*Dendrobium*	Isolation and fusion	Kanchanapoom *et al.*, (2001)

Table 9.2 Transformation followed by Particle bombardment

Plant Species	Tissue Bombarded	Regeneration of plants	Reference
Arabidopsis	Roots	Yes, progeny	Seki M *et al.*, (1991a)
Bean	Meristems	Yes, progeny	Russell DR *et al.*, (1993)
Cotton	Meristem	Yes, progeny	Mc Cabe and Martinell BJ (1993)
Cranberry	Stems	Yes	Serres R *et al.*, (1992)
Liriodendron	Embryogenic culture	Yes	Wide HD *et al.*, (1992)
Maize	Suspension culture	Non-regenerable*	Klein TM *et al.*, (1989a)
Maize	Embryogenic culture	Yes	Goff SA *et al.*, (1990) Gordan-Kamm WJ *et al.*, (1990)
Maize	Embryos	Yes, field test	Koziel MG *et al.*, (1993)
Oat	Embryogenic callus	Yes, progeny	Somers DA *et al.*, (1993)
Orchid	Protocorm	Yes	Kuehnle AR and Sugii N (1992)
Papaya	Somatic embryos	Yes	Fitch MM *et al.*, (1990)
Picea	Embryogenic callus	Yes	Ellis DD *et al.*, 1993
Picea	Somatic embryos	No	Robertson D *et al.*, (1992)
Populus	Cultured nodules	Yes	Mc Cown BH *et al.*, (1991)
Rice	Embryos	Yes	Christon P *et al.*, (1992)
Rice	Embryogenic callus	Yes, progeny	Li L (1993)
Sorghum	Suspension culture	No	Hagio H *et al.*, (1991)

Plant Species	Tissue Bombarded	Regeneration of plants	Reference
Soybean	Embryos	Not attempted	Christon P *et al.*, (1988)
Soybean	Apical meristem	Yes	Christon P *et al.*, (1989), Morikowa H *et al.*, (1989)
Sugarcane	Embryogenic callus	Yes	Bower B and Birch RG, (1992)
Sweet Potato	Leaves	No	Prakesh CS and Varadarajan V (1992)
Tobacco	Suspension culture	Non-regenerable*	Klein TM *et al.*, (1988c), Russell DR *et al.*, (1992a) and (1992b)
Tobacco	Leaves	Yes	Klein TM *et al.*, (1988c), Tomes DT *et al.*, (1990)
Tobacco	Leaves, chloroplasts	Yes, progeny	Svab Z *et al.*, (1990) Svab Z and Maliga P, (1993)
Wheat	Embryogenic callus	Yes, progeny	Vasil V *et al.*, (1992)

Tissue type	Species regions	Promoter/Coding	Reference
Cells	Tobacco	35S / GUS	Klein TM *et al.*, (1989a), (1989b) and Morikowa H *et al.*, (1989)
	Tobacco	35S / LUC	Morikowa H *et al.*, (1989)
	Tobacco	35S / CAT	Daniell H *et al.*, (1990)
	Tobacco	Chloroplast promoter / CAT#	Daniell H *et al.*, (1990)
	Soybean, Wheat, Rice	Adh / GUS or 35S / CAT	McElroy D *et al.*, (1990), Oard JH *et al.*, Wang YC *et al.*, (1988)

*Plants can not be regenerated from these lines of suspension culture cells even prior to bombardment.

Table 9.2 Continued

Tissue type	Species	Promoter/Coding regions	Reference
	Corn	Adh / GUS	Klein TM *et al.*, (1988b), Wang YC *et al.*, (1988)
	Corn	35S / CAT	Klein TM *et al.*, (1988a)
	Corn	PAT / GUS	Bruce WC *et al.*, (1989)
	Barley	35S / CAT	Kartha KK *et al.*, (1989)
	Barley	GUS, NPTII	Mendel RR *et al.*, (1989)
	Larix	GUS	Duchesne LC *et al.*, (1993)
	Picea	35S / GUS	Charest PJ *et al.*, (1993)
	Picea	EM / GUS	Charest PJ *et al.*, (1993)
Leaves	Tobacco	35S / GUS	Klein TM *et al.*, (1988c) Twell D *et al.*, (1989)
	Corn	35S / R^{++}	Ludwig SR (1990)
	Wheat	Adh / GUS	Oard JH *et al.*, (1990)
Seedling	Corn	35S / R	Ludwig SR (1990)
	Egg plant	35S / luciferace	Morikawa H *et al.*, (1989)
	Rice, Barley, Oats, Cucumber, Tobacco, *Arabidopsis*	35S / CAT, maize ubiquitin / LUC, or rice phytochrome / CAT	Bruce WB *et al.*, (1989)
Seed tissues	Corn aleurone or embryo	Adh / GUS	Oard JH *et al.*, (1990) Klein TM *et al.*, (1988b)
	Corn aleurone or embryo	35S / GUS or *R*	Ludwig SR (1990)
	Corn aleurone or embryo	Adh / LUC or CAT	Klein TM *et al.*, (1989c)

Table 9.2 Continued

Tissue type	Species	Promoter/Coding regions	Reference
	Corn aleurone or embryo	Bzl or Al§ / LUC	Goff SA *et al.*, (1990) Klein TM *et al.*, (1989c)
	Corn aleurone or embryo	Al or Bzl genomic clones**	Klein TM *et al.*, (1989c)
	Barley embryos	35S / CAT	Kartha KK *et al.*, (1989)
	Corn, Rice embryos	Adh / GUS	Cao J *et al.*, (1990)
	Wheat, apical	Adh / GUS	Oard JH *et al.*, (1990)
Epidermal Cells	Onion	35S / CAT	Klein TM *et al.*, (1987)
	Corn	35S / R	Ludwig SR, (1990)
Floral Tissue	Floral Tissues Tobacco anthers or petals	35S / GU	Twell D *et al.*, (1989)
	Wheat meristem		Ledue N *et al.*, (1993)
Pollen grains	Tobacco	Pollen promoter or 35S / GUS	Twell D *et al.*, (1989)
	Barley	Viral mediated, wheat dwarf virus	Oard JH *et al.*, (1990) Wang YC *et al.*, (1988)
Orgenelles	Tobacco chloroplast	Chloroplast Promotor or 35S / GUS	Daniell H *et al.*, (1990)

Abbreviations: 35S, 353 promoter from cauliflower mosaic virus ; Adh promoter from the Alcohol dehydrogenase 1 gene of maize; GUS, β-glucuronidase; CAT, chloramphenicol acetyltransferase; LUC, luciferase, NPTII, neomycin phosphotransferase II.

++ The R gene from maize is one of the regulatory genes of the anthocyanin biosynthetic pathway. Production of the product of the R gene in cells of mutants lacking a functional R gene results in expression of the pathway and therefore pigment production in the affected cell.

§ Al and Bzl are structurai genes of the maize anthocyanin biosynthetic pathway.

** Introduction of the Al or Bzl genomic clones into tissue from the appropriate mutant plant results in pigment production in bombarded cells.

Construct specifically expressed in the chloroplast.

10

PROTOPLAST ISOLATION, FUSION AND CULTURE IN ORCHIDS

OVERVIEW

The potential of plant regeneration using isolated cells and protoplasts provides the basis for possible cell selection, somatic cell hybridization and genetic manipulation. To pursue this, efficient techniques for isolation and fusion as well as plant regeneration are necessary. In case of orchids, protoplast isolation and culture has been considered to be difficult. However, Teo and Neumann (1978a) have reported the successful isolation and culture of protoplasts from the orchid hybrid, *Renantanda* Rosalind. They subsequently developed methods for fusing the protoplasts of orchids. Similarly, Price and Earle (1984) also succeeded in fusing the isolated protoplasts isolated from *Dendrobium*. After these initial results several others also reported the isolation and fusion of orchid protoplasts.

PROCEDURE FOR THE ISOLATION OF PROTOPLASTS FROM ORCHID LEAVES

Plantlets grown aseptically under *in vitro* conditions are preferred to those plants grown in the greenhouses to obtain the explants. The greenhouse grown plants can also be used but the explants have to be thoroughly cleaned before taken for the study. If *in vitro* grown plants are used as explants however this step can be avoided.

Leaves taken from the *in vitro* grown plants with the size of about 2.5 cm can be used as source of explants. All leaves are pre-plasmalyzed for 1 hour in salts and 0.7 M mannitol in the dark and then are used after chopping transversely into strips of about 1-2mm in width. Approximately 0.3g of leaf strips are used for 0.5 ml of filter sterilized enzyme solution in 50mm Petridishes. The enzymes used are 1% cellulase, 0.5% driselase and 1% macerozyme. All enzymes are dissolved in 0.4 M Mannitol at 5.7 pH. The leaf enzyme mixture is then placed on a rotary shaker at 80 rpm at 30°C in the dark for 3 hrs. After 3 hrs of incubation, protoplasts are sieved through 61 and 43mm stainless steel mesh to remove any clumps of undigested tissues and debris. The filtered protoplasts are centrifuged at 40×g for 5 min. The filtrate is removed using a Pasteur pipette. The pellet is re-suspended and washed twice on 0.4 M mannitol. Finally, the pellet material is transferred to the top of the 15ml screw capped centrifuge tubes containing 6 ml of 0.5 M sucrose solution. Centrifugation is performed at 80×g for 10 min and protoplasts are observed on the surface of the sucrose solution while the remaining cells and debris is sedimented to the bottom of the tube. Protoplasts are gently collected and washed 2-3 times with 0.4 M mannitol. Protoplasts thus obtained are used further according to the need.

Assessment of protoplasts yield can be made as follows. Protoplasts can be counted by haemocytometry. The viability of isolated protoplasts can be monitored using flourescein diacetate (FDA). Formation of wall material can also be determined. Fluorescence of the labeled protoplasts can be viewed using Epifluorescence Microscope (Leitz or Nikon Optinot).

To obtain optimum results in the isolation of protoplasts the governing parameters are light and dark condition, size of the donor plants, type and concentration of enzymes, incubation period and osmolarity. The effect of light and dark condition is not so significant. But size of leaves is an important factor affecting the yield and quality of protoplasts both in *Vitis vinifera* (Theodoropoulas and Ranbelakis - Angelakis, 1990) and *Aranda* (Koh *et al.*, 1988). Older and longer leaves contained numerous crystals (calcium oxalate) or raphides which deflate and kill the protoplasts. Orchids store carbohydrates synthesized in the form of oxalic acid, which eventually forms calcium oxalate (Oshiro and Steinhart, 1991). To clean up protoplast suspension from raphides, the sucrose floatation and sedimentation procedure can be employed. For releasing protoplasts cellulase onozuka R-10 and Macerozyme R-10 are found to be very effective by (Kanchanapoom, 2001). The yield of protoplasts can be further enhanced if the enzyme Driselase is added to the mixture. It actually enhances cell wall digestion (Loh and Rao, 1985; Koh *et al.*, 1988). Optimum yield of protoplast can be obtained by incubating the explants in the enzyme mixture for 3 hrs. 0.4 Mannitol can be used either in the enzyme mixture as an osmoticum or in the washing solution.

PROTOPLAST FUSION

If the protoplasts are allowed as such for a while in the culture medium they randomly and indiscriminately fuse with other resulting in chimeras. But controlled fusion of protoplasts can be induced by specific procedures. In one such method used by Kanchanapoom *et al.*, (2001) one drop of dense protoplast suspension was pipetted onto the middle of the microscope slide and allowed the protoplasts to settle to the bottom for 5 min. Fusion was induced by adding a drop of 40% polyethylene glycol (PEG) solution to the settled protoplasts. Treated protoplasts are incubated on the slide for at least 20 min at room temperature. For observing the process of protoplast fusion, protoplast suspension is placed on a slide and converted with a cover slip. The protoplasts are washed by perfusing the protoplasts with washing solution across the slide. Fusion of protoplasts can be examined by keeping the slide on an inverted microscope (Carl Zeiss or Olympus) equipped with fluorescence attachment. The microscope can even be connected to a video camera so that the process of protoplast fusion can be viewed on the TV screen and recorded at the same time. It is thus possible to follow and catch up the fusion events.

On the addition of PEG solution the protoplasts are instantly shocked and lose their round shape. The two protoplasts come close in contact with each other (Fig. 8.5a). The protoplasts membranes of the two entities fuse and a connection is formed between the two cytoplasm (Fig. 8.5c and d). Finally, the two protoplasts formed a spherical fusion product (Fig. 8.5f). A high aggregation of protoplasts occurs if the protoplasts are used immediately after isolation. Diluted levels of PEG always favoured the process of fusion. This fact has been shown by many researchers (Kao and Michayluk, 1974; Willin *et al.*, 1974 and Kanchanpoom *et al.*, 2001). This indicates that osmotic changes upon dilution of PEG causes fusion to occur.

11

CYTOLOGICAL TECHNIQUES

OVERVIEW

Cytology of orchids is an another fascinating facet because of the high rate of promiscuity that this group of plants exhibits. It has been discussed earlier that hybridization is one of the easiest methods that can be done in orchids. Interspecific and intergeneric hybridization could be easily achieved and as a result of which there are innumerable number of hybrids available. Interestingly, there are also quite a good number of natural hybrids among orchids. This indicates the fact that orchids exhibit normal behavior without undergoing much of phenotypical changes even when there is genetic variability. Because of the demand in flouriculture market and the need for clonal propagation of plants of choice with desired combination of features, the high quality *in vitro* propagation has become an inevitable choice for multiplication of orchids. There is no doubt that *in vitro* micropropagation yields clones without much genetic alteration or phenotypic changes. But genetic material of the tissues, cells, protoplasts grown *in vitro* can undergo many rearrangements. These include changes in chromosomal number, and also structural aberrations like deletions, duplications, inversions and translocations. Such cytological changes to a great extent can be evaluated by conventional cytogenetic procedures that involve staining the chromosomes and evaluating their morphology at metaphase of mitosis and meiosis. Even the chromosomal behavior both at mitosis and meiosis can also be studied.

Materials for cytological studies can be obtained from various methods and sources. For the mitotic studies generally, orchid seeds are inoculated on a nutrient medium suitable for the taxa and allowed to germinate. Once the roots are formed, the seedlings are isolated and tips are excised for the use. Root tips can also be obtained from the plants grown in green houses. Protocols discussed here are suitable for orchids. Additional information if needed could be obtained from the book, *The handling of*

Chromosomes by Darlington and La Cour (1976). Leonardo Pessoa Félix and Marcelo Guerra have studied the cytology of Cymbidoid orchids from Brazil by using various cytological techniques. Similarly, Oncidiinae group (Odontoglossum, Miltonia and Oncidium) orchids, have been widely studied cytologically (Sinotô, 1962; Charanasri *et al.*,1973).

Callus is raised by inoculating the explants on the suitable medium. Meristem is considered to be the best explant. Once the callus is produced it is isolated and used for the study. The callus which has been derived freshly from the explant is ideal for cytological analysis. However, if one is desirous of studying the effect of *in vitro* treatments on callus culture, callus from different batches of subculture are obtained serially and analysed.

For the meiotic studies, however, the source available is the flower buds. Plants are grown in greenhouses and immature buds are picked up and anthers are isolated for making use in cytological studies. The sporogenous tissue found inside the anthers undergoing meiosis is used for studying the meiotic behavior of chromosomes.

Isolated cells and protoplasts grown in suspension culture can be used directly for cytological studies. In fact the evidence for protoplast fusion and **cybridization** can be obtained by cytological analysis of cells and protoplasts in culture. Subsequently, cytology of resulting plants from cells and protoplasts, fused or unfused will reveal the chromosomal details.

PROCEDURE FOR THE PREPARATION OF STAINS

Stain	Reagents	Procedure	Comments
Modified Carbon Fuchsin	Basic Fuchsin Ethyl alcohol 70%, 5% phenol, glacial acetic acid, formaldehyde sorbitol	**Stock solution A**: 3g basic fuchsin in 100ml of 70% ethylalcohol **Stock solution B**: 10 ml of solution A added to 90ml of 5% phenol in distilled water **Carbol fuchsin**; 40ml of solution B is added to 6ml of glacial acetic acid and 6ml of 37% formaldehyde.	Ideal for species with small chromosomes used generally for mitotic studies For staining cell and protoplasts Stain can be kept for upto 2 years

Stain	Reagents	Procedure	Uses
		Modified Carbol fuchsin solution: 2 - 10ml of carbol fuchsin solution is added to 90 - 98 ml of 45% acetic acid and 1.8g sorbitol	Stain better after aging for a few months
Aceto - Carmine	Glacial acetic acid, distilled water, carmine powder	2g of carmine is added to the mixture of 45 ml of glacial acetic acid and 55ml of distilled water and gently boiled. Cool, shake, filter, label and store	Used generally for meiotic studies. Stains only the proteins of chromosomes. DNA is not stained
Alcoholic Carmine	Carmine, distilled water, HCl(conc), 85% ethyl alcohol	4g of carmine is mixed in 15ml of water and boiled gently to which 1 ml of conc. HCl is added. After cooling 95ml of ethyl alcohol is added. Shake, filter, label and store	Used for mitotic studies. Stains only the proteins of chromosomes. DNA is not stained
Aceto Orcein	Orcein, 45% acetic acid	2g orcein is mixed in 100ml of distilled water and gently boiled. Filter the solution and dilute by making a 1: 1 of mixture of the above solution and 45% acetic acid	1% of the stain is used for meiotic studies. Only proteins of chromosomes take this stain.
Lacto - Propionic Orcein	Orcein, Lactic acid, propionic acid, distilled water	45ml of an equal mixture of lactic acid and propionic acid and 55ml of water is taken and to it 2g of orcein is added	Used for meiotic studies. Used instead of orcein if it leaves heavy stains on cytoplasm
Feulgen	Fuchsin in HCl, $K_2S_2O_5$, Distilled water	Mix 3g of Fuchsin in 600ml water and boil the solution gently, shake, cool, filter and mix in 90ml of 1N HCl and 9g of $K_2S_2O_5$. Store in the dark room at room temperature. To this add 3g of carbon is added. Shake, filter and store in dark coloured bottle at 2°C.	For mitotic studies and when chromosomes are larger. It stains only DNA. **Note**: Direct inhaling should be avoided and the stain should be colourless.

PROCEDURE FOR THE PRE-TREATMENT OF PLANT MATERIAL

Root tips

1-2 cm long roots are collected by cutting them with fine tweezers. The root tips can be directly used for squashing or they can also be stored in the pre-treatment solution for arresting the cell divisions. Duration of treatment for arresting cell division varies from species to species. The following are the commonly used fixatives:

a. 0.1M Colchicine for 4 hrs at room temperature and store at 2°C over night.

b. 2mM aqueous solution of 8-hydroxyquidine (8-HQ) for 4 hrs at room temperature and store at 2°C over night.

c. A mixture of ethanol and glacial acetic acid in 3:1 ratio. Treatment for 1 hour before processing.

Root tips stored in any of the above fixatives at 2°C can be used for a long period of time.

Colchicine and 8-HQ when used for treating the plant materials, they induce endoduplication resulting in the increase of chromosome number. It also helps in obtaining large chromosomes.

Flower Buds

Fixing of flower buds to be used for meiotic studies is done on the similar lines as that of root tips. Immature flower buds are collected and stored in a fixative (6:3:1 mixture of ethylalcohol: chloroform: acetic acid). After keeping the flowers for 1 day, the anthers can be dissected from the flowers. It is necessary to distinguish between immature and mature anthers. Meiotic stages will be visible only in the sporogenous tissue which eventually forms pollen grains.

Callus

Callus is collected from the source within 2-4 days after the initiation of callusing from the explants on the fresh medium. Callus collected from the old culture or after several sub-cultures may show genetic abnormalities. Hence for correct genetic assessments fresh callus is preferred, however for the purpose of studying chromosomal aberrations callus from different stages is obtained. The callus collected can also be treated just like the root tips, but the callus grows very fast thus it is difficult to find a high frequency of dividing cells in callus. At the same time the cell population in the callus is highly variable in stage of the cell and often it rapidly gets differentiated into meristematic areas and non-dividing areas. These areas are composed of different cell populations and may differ in chromosome number and structural abnormalities, hence if possible they should be analyzed separately.

Isolated Cells

Cell suspension cultures are added with Colchicine and 8-HQ solution directly and incubated at 25°C for 2 hrs on a rotatory shaker. Transfer the cell suspension to centrifuge tubes and centrifuge at 1100 rpm for 10 min. Remove the supernatant, add 20ml of fixative to the pellet, shake gently and keep overnight at 2°C.

Centrifuge the fixed cells, remove the fixative with a pipette, resuspend the pellet in 0.1M sodium acetate with pH 4.5 and again centrifuge at 1100 rpm for 10 min, discard the buffer and transfer the pellet to a flask.

Then later, add 20ml of enzyme mixture containing 0.25 g onozuka cellulase R10, 0.25g macerozyme R10, and 49.5 ml of 0.1 M sodium acetate with pH 4.5. Incubate the mixture at 25°C for 2 hrs. Wash the cells in 0.1M sodium acetate buffer, centrifuge and resuspend pellet in 45% acetic acid. Pipette out 20ml of suspension and place it on a slide cleaned with 1 N HCl. Place a siliconized cover slip and apply pressure gently to flatten the cells.

Finally, remove the cover slip after freezing on dry ice. Later, add the solution of 1% macerozyme R10 to cover the cells and to soften the cell walls. Drain the macerozyme solution, add a few drops of modified carbol fuchsin and allow to stain for several minutes. Keep a clean cover slip to cover the area carrying cells and gently press to flatten the cells. Now the slide is ready for observation. It can be scanned under microscope to evaluate Karyotypic changes.

Protoplasts

Protoplasts are isolated and cultured as described in the earlier chapter. To the suspension containing protoplasts 5ml of 0.2% Colchicine is added and incubated at 25°C for 2 hrs on a gyratory shaker.

The suspension is centrifuged at 1100 rpm for 10 min. Remove Colchicine and resuspend pellet in 1 ml of fixative, and store overnight at 4°C. Centrifuge the solution again at 1100rpm for 20 min, remove the fixative i.e. supernatant, and resuspend pellet in 2 drops of 45% acetic acid. Add a few drops of carbol fuchsin stain for 2 min. Place a cover slip, flatten by pressing gently. Now the slide is ready for microscopic study.

Schedules Of Staining

Carbol Fuchsin

(a) Cells and protoplasts: The stain is used directly for staining the cells and protoplasts as described earlier.

(b) Root tips: Take out the root tips from the fixatives and wash them 2-3 times. Keep 5-6 root tips on a watch glass containing 1 N HCl and gently warm it at 60°C for 5-10 min. Period of hydrolysis may be shorter or longer depending on the fixation duration. The longer is the fixation period the shorter should be the hydrolysis time. Hydrolysis softens the tissues. Wash root tips again in water for 5 min and stain as described.

Procedure For Squashing

Place a hydrolysed root tip on a slide with 45% acetic acid cut and select only the extreme meristematic region of the root tip and discard the remaining part. Macerate the tissue by using a needle or a sharp teflon instrument to gently crush the tissue. Do not allow the acetic acid to dry up. Place a cover slip and press gently, stain the tissue by placing one drop of carbol fuchsin solution at the edge of the cover slip and allow it to run under cover slip. Add more drops until stain has reached all areas under the cover slip.

Acetocarmine

Root tips, callus segments and anthers can be directly placed on a watch glass containing few drops of acetocarmine and incubated at 25°C for 15-20 min. Then, the tissues are squashed by placing them on a slide in acetic acid as explained earlier. In case of anthers the anther wall and other debris must be removed. A clean cover glass is placed and pressed gently. The slide is ready for observation.

Note: Alcoholic carmine is considered to be suitable for the staining of meiocytes in anthers.

Aceto-orcein

It is used very much the same way as that of acetocarmine. It is known to be more suitable for anthers than root tips and callus segments. Hydrolyze the tissue, macerate with few drops of aceto-orcein. Place a cover slip and gently warm the slide. Heating causes the cells to swell and the chromosomes to spread so that the overlapping of cells and chrommosomes is minimized.

Lacto-Propionic Orcein

It is used in conditions where aceto-orecin leaves heavy stains on the cytoplasm. It is for better meiotic studies.

Feulgen

Hydrolyze tissues for 10 min and wash them with water for 5 min. Place the tissues for 1 hour in a watch glass containing several drops of feulgen. The stain is taken up by DNA. Acid hydrolysis removes the purine bases producing free aldehydes and thus provides specific sites to which the dye can bind. Hydrolysis also softens the tissue by breaking down the middle lamella and cell wall, allowing the cells to be separated and squashed more easily.

Feulgen staining is suitable for the root tips. This stain is not ideal in cases where the size of the chromosome is very small.

CHROMOSOMAL STUDIES

Each species will have a specific number of chromosomes. As mentioned earlier, ploidy level highly varies in orchids due to the ease with which hybridization occurs both naturally as well as artificially. The gross morphology of the chromosomes in all species exhibits a standard format. The chromosome consists of two arms with a centromere.

The location of the centromere may vary giving different morphology for different chromosomes. One of the several pairs of chromosomes will carry a satellite. This short segment of chromosome located distally. Sometimes the satellites are separated from the main chromosomes and are counted as separate chromosomes by mistake. However, if the pretreatment has marked well such mistakes do not occur.

For the chromosomal studies, the slide prepared is placed on the stage of any suitable microscope and scanned carefully. Initially all those cells showing metaphase and having properly spread out chromosomes are marked. The approximate location, size and shape of the chromosome is sketched on a drawing roughly or neatly. This drawing is used to count the number of chromosome. For this purpose Camera Lucida technique can also be used. On the basis of these counts the correct number of chromosome is determined. By comparing the sketches made by observing different cells of the same species, numerical and structural aberrations can be determined. This also gives an idea of the extent of deviation from the normal number and structure of chromosomes. Even if an abnormality is discovered the results need to be confirmed by thorough analysis as this could also be due to pool slide preparation.

Meiotic analysis requires a knowledge of meiotic stages and an ability to recognize different meiotic stages. Though the majority of the studies are focused on metaphase I, the other stages also provide valuable information. It is also necessary to study all stages of meiosis as this focuses light on the meiotic behavior of chromosomes in the tissues derived from tissue culture procedures.

Meiotic behaviour in normal diploids greatly varies from that of polyploids. In diploid plants the chromosomes normally form bivalents at meiosis. Bivalents are formed by pairing of two homologous chromosomes and can appear as a ring or rod-type configuration. This kind of normal bivalents do not appear in case of polyploids and the configuration varies depending on the level and type of polyploidy. We may come across trivalents quadrivalents etc. Pairing is usually disrupted if the chromosomes have undergone structural abberations like deletion, duplication, inversion and translocations. This result in unpaired, incompletely paired chromosomes and heteromorphic bivalents. All these and many more anomalies are of common occurrence in case of *in vitro* derived tissues.

These cytological results, in comparison with phenotypical observations, can lead to the location of genes, but this is only preliminary discovery which has to be further confirmed by using the techniques of molecular biology.

MICROSCOPY

Since Anton van Leeuwenhoek (1632-1723) has observed the microorganisms for the first time there has been a revolution in the microscopic world. He gets the credit of constructing simple microscope capable of magnifying up to 300 times. Now with so much of advancement in all areas of technology microscopes have come of age. The following is a list of some of the most important microscopes used.

TABLE 11.1 DIFFERENT TYPES OF MICROSCOPES

Microscopes	Remarks
1) Simple light microscope	It is based on light and it consists of one lens.
2) Compound light microscope	It uses light and two or more number of lenses.
3) Dark field microscope	Light is concentrated and transmitted directly through the specimen.
4) Phase contrast microscope (PCM)	It is used to observe unstained living organisms.
5) Differential interference contrast microscope	It is like PCM but gives much higher resolution.
6) Fluorescence microscope	Ultraviolet light is used as the source. Materials will give a sharp image.
7) Electron microscope	It uses electron beams. Objects smaller than 0.2 micron can be seen.
8) Transmission electron microscope	It uses electron and gives better image of the internal structure.
9) Scanning electron microscope	It creates images of the surfaces of objects and can resolve up to 20nm.
10) Scanning tunneling microscope	A wire probe made up of platinum or viridium is used to trace the surface ofobject. Electron clouds from the surface of objects are used for scanning.
11) Atomic force microscope	Gives 3D images of the object. Used for studying DNA and other biomolecules.
12) Acoustic microscope	It uses the sound to generate images.

12

MICROPROPAGATION OF ORCHIDS

OVERVIEW

Orchids are propagated vegetatively as well as by seeds. The progeny from a cultivated orchid will be extremely heterogenous, seldom identical to the parent plant. With vegetative propagation, the progeny is identical to the parent plants. Monopodial orchids are propagated by stem cuttings with an intact growing apex, a few leaves and young roots. In sympodial orchids, propagation is by separation of pseudobulbs (Henderson and Addison, 1956). Orchid cloning *in vivo* is a very slow process requiring about 10 years before a clone of suitable size is obtained (Pierik, 1987). Thus, in principle, cultivated orchids are useful if propagated vegetatively *in vitro*.

The micropropagation of orchids, by means of tissue culture, has a complex history that is not free of controversy and includes unusual episodes (Arditti and Ernst, 1993). It is evident through the survey of literature that most parts of orchids have been used as explants. Ovaries were the first orchid flower segments to be cultured. This was carried out by Ito at the Kyoto Prefectural University, Japan (Ito, 1960). In an earlier paper he reported the culture of immature *Dendrobium* seeds (Ito, 1955). The first orchids in which young flower buds or inflorescence was cultured were *Ascofinetia, Neostylis* and *Vascostylis* (Intuwong and Sagawa, 1973).

The idea of orchid root culturing probably originated independently several times. The first printed suggestion appeared in the theoretical article of Beechey in 1970 (cf. Arditti and Ernst, 1993). *Phalaenopsis* roots were cultured successfully by Tanaka *et al.*, (1976). Similarly, rhizome tips were also cultured by Ueda and Torikata (1972).

The tendency of juvenile leaves on protocorms to produce PLBs led to the development of micropropagation methods through the culture of leaf bases (Champagnat *et al.*, 1970). The first unambiguous and well-documented report that leaves can produce PLBs was made in cultures derived from *Cymbidium* shoot tips (Wimber, 1965). The culture of *Arundina* stem sections was first mentioned in 1966, at the Fifth World Orchid Conference in California (Bertsch, 1966). Knudson, who formulated the asymbiotic germination of orchid seeds, was indirectly responsible for the first attempt to asexually propagate an orchid *in vitro*. Thomale (1957) cultured bud containing explants from tubers of *Orchis maculata,* and obtained shoots and roots. He clearly described the potential of tissue culture as a means of mass rapid clonal propagation, but his work was overlooked. Though his work was published three years earlier than the report of *Cymbidium* shoot tip cultures by Morel (1960), perhaps, because his book deals only with orchids for German amateurs. The credit of developing orchid meristem culture goes to Georges Morel. His first paper in 1960, on orchids, described the production of virus-free *Cymbidium* through the culture of apical meristems. Subsequently, a revolution took place in the vegetative propagation of orchids. Cloning of orchids by meristem culture became the first commercial application of vegetative propagation *in vitro*. This method, originally developed for *Cymbidium,* was later modified and used for *Cattleya* and many other orchids. 'Mericloning' of orchids (vegetative propagation by meristem culture) is now carried out on a large scale world wide.

India is a major repository of orchid germplasm. More than 1200 species grow in the wild (Vij, 1997). The climate here is better for the growth of orchids, as there are no severe winter seasons like in European countries. There is no dearth for space and manpower in this country. If suitable micropropagation techniques are developed and made available for all those interested in establishing tissue culture laboratories for producing orchid plants and cut flowers of international standards, the orchid industry can develop very well and produce the desired results. The study of *in vitro* morphogenesis greatly helps in understanding the nutritional requirements of orchids at different stages of their development , which is necessary for developing the most suitable medium for proper growth. Hardening of ex-agar plants is a crucial stage, since the success of tissue culture lies in producing plants which can be easily propagated *in vivo* for commercial purposes. The term plant tissue culture is generally used for the aseptic culture of cells, tissues and organs under *in vitro* conditions (Thorpe, 2000). Plant tissue culture has opened a number of avenues in biotechnology. As a result, enormous literature has been accumulated and several authors have reviewed the literature, time and again (see, Withner, 1959; Marston, 1966; Arditti, 1967; Murashige, 1974; Rao, 1977; Arditti and Ernst, 1993; Bhojwani and Razdan, 1996).

PLANT TISSUE CULTURE: A BRIEF RETROSPECT

To some degree, the history of tissue culture involves the entire history of Botany, though, its origin is lost in the antiquity (Kyte and Kleyn, 1996). The major events in plant tissue culture research are given in the Table 12.1. Gottilieb Friedrich Johann Haberlandt (1902) made the first attempt to culture isolated leaf mesophyll cells of *Lamium purpureum* and other species on modified knop's solution, but failed to induce any cell division (see Krikorian and Berquam, 1969). Winkler (1908) cf. (Gautheret, 1985) attempted to culture string bean segments and reported cell divisions, but no substantive proliferation occurred. Simon (1908) reported the formation of callus, buds and roots from popular explants for the first time. Lewis Knudson (1919) cultured root cap cells which failed to divide and eventually died. Robbins (1922) succeeded in culturing corn roots and maintained them for a long time. The first successful indefinite cultures of roots tips were of tomato by White (1934a).

Initially, the *in vitro* propagation of orchids was only confined to the use of seeds as the source of explants, and plants were raised by culturing seeds only. But gradually, other sources such as leaf, meristem, etc. were used. In respect of orchids, the first innovative horticultural method for orchid seed germination was developed by David Moore (Morel,1849). Fifty years after Moore's discovery, Noel Bernard (Bernard,1899, 1909) made another quantum jump when he developed a method for the symbiotic germination of orchid seeds *in vitro*. This is the first method derived for the *in vitro* propagation of any plant. Knudson (1921, 1922) was the first to propose a medium for the *in vitro* propagation of plants in axenic culture. He developed a method for the asymbiotic germination of orchid seeds i.e., without fungal association. Burgeff (1936) has listed nutritional requirements regarding seed germination in 25 orchid genera. The medium proposed by Vacin and Went (1949) is widely used for the germination of orchid seeds. After the successful studies of Knudson (1951) on *Cymbidum* were published, many workers followed his techniques and started to grow orchid seedlings under laboratory conditions. Various workers have demonstrated that even immature seeds can be germinated (Ito, 1955; Withner, 1955; Nimoto and Sagawa, 1961; Rao and Avadhani, 1964). Morphogenisis of orchids in seed cultures of different orchids has been studied by various authors (Chennaveeraiah and Patil, 1975; Kusumoto and Furukawa,1977; Sharma and Tandon,1990; Singh,1993; Nagaraju and Parthasarathy,1995). Variations seen in the germination of orchid seeds, under *in vitro* and *in vivo* conditions were also documented (Clement, 1973a, b). History of plant tissue culture was discussed by Guatheret (1985). Arditti *et al.* (1982) showed that mature seeds of *Epipactis atrorubens* germinated on Nostog medium, whereas immature seeds of *Epipactis gigantia* germinated on Curtis medium. Seeds of *Cymbidium elegans, Coelogyne prolifera, C. cristata, C. porsecta, Aerides multiforum, Sarcanthus pellidus, Bulbophyllum cosmosus* and *Thunia alba* were incubated in tubes containing KC or VW medium. Best germination for all species was observed on KC medium. Protocorms formed with root initials in 10-12 weeks (Sharma and Tandon, 1987). Mamura and Saitou (1987) reported that seeds of *Aerides japonicum* showed the best germination after storage at 18°C, for 380 days. Seedlings grew well on a culture medium containing 0.8% agar with pH 4.5. Baker *et al.* (1987) achieved seed propagation in *Pontiera* and *Cattleya*. Effect of different media and growth regulators on seed germination and seedling growth of *Cymbidium* was studied by Paek *et al.*, (1989). Effect of media on the germination of seeds and growth of five species of *Dendrobium* was studied by Devi *et al.*, (1990). *Bletilla striata* seeds showed 99 percent germination on MS medium. Seedling growth was best in MS medium, supplemented with Kinetin and BA (Suner, 1995). Seed germination and seedling growth technique in *Encyclia boothiana* was developed. Seed germination was achieved on KC medium, but for seedling growth MS medium was found suitable (Stenberg and Kane, 1998). Morphogenesis of *Cyperidium acaule* was studied *in vitro* (Leroux *et al.*, 1997). Selection of best medium for *in vitro* propagation of *Dendrobium* was studied, and MS medium was found to be the best (Kaur and Sharma, 1997). Seeds of *Oncidium ceboleta* were germinated on MS medium. PLBs of *Darwinia* were successfully preserved without the loss of viability in the dark (Kishi and Takagi, 1997). Addition of BAP resulted in an increased rate of germination and development of seedlings (Alloufa *et al.*, 1998). Embryology of orchids has been studied by Swamy (1949 a, b).

NUTRIENT MEDIA USED WITH PARTICULAR REFERENCE TO ORCHIDS

Media used for orchid tissue culture and seed germination may reflect both the special requirements of each species and the preferences of the researchers. Different media used for the propagation of different orchids is listed in the Table 12.2. Occasionally a particular recipe is proposed for a given genus or species and improvised by modifying the medium, by adding coconut water, tomato juice, banana pulp, different fruit juices, fish emulsion, leaf extract, potato extract, and even beer (Alberts, 1953; Withner, 1959; Arditti, 1967; Chow, 1986; Villalobos *et al.*, 1994).

Table 12.1 Major milestones in plant tissue culture and molecular biology

1838	M J Schleiden hypothesised cell theory, suggesting totipotency of cells
1859	Charles Darwin hypothesised that animal and plant populations adapt over time to best fit the environment
1865	Gregor Mendel investigated how traits are passed from generation to generation-called them factors
1869	Johann Meischer isolated DNA from the nuclei of white blood cells
1882	J Sachs reported that plants synthesising organ-forming substances that are polarly distributed
1902	G Haberlandt made the first (but unsuccessful) attempt at tissue culture; Walter Suton coined the term "gene" and proposed that chromosomes carry genes (factors that Mendel said that could be passed from generation to generation)
1904	B Hanning attempted embryo culture (crucifers)
1909	E Kuster achieved fusion of protoplasts although the products did not survive
1910	Thomas H. Morgan proved that genes are carried on chromosomes; coined the term "biotechnology"
1921	M Molliard cultured fragments of plant embryos
1922	L Knudson germinated orchard seeds *in vitro*; WJ Robbins cultured root tips
1924	Blumenthal and Meyer achieved callus formation on carrot root explants through the use of lactic acid
1925	Laibach used embryo culture to rescue interspecific hybrids of *Linum*
1926	FW Went demonstrated that there were growth substances in coleoptiles from *Avena*
1930-1940	Plant hybridisation was used widely in plant breeding
1934	RJ Guatheret unsuccessfully cultured cambial tissues of trees and shrubs; PR White maintained long-term cultures of tomato roots; Kogl *et al.*, identified the first plant growth regulator, IAA
1936	CR LaRue cultured embryos of conifers

1938	Proteins and DNA studied by x-ray crystallography; Term "molecu lar biology" coined
1939	Successful continuously growing cambial cultures of carrot and tobacco was done by Gautheret
1940	Gautheret cultured cambial tissues of *Ulmus* to study adventitious shoot formation
1941	Van Overbeek used coconut milk to support growth and development of very young *Datura* embryos. Braun cultured crown gall tissues *in vitro*.
1942	Gautheret observed secondary metabolites in callus cultures.
1943-1950	Braun elucidated the tumour-inducing principle of crown galls
1944	Folke Skoog used tobacco cultures to study adventitious root formation
Mid-1940's	Transition from animal power to mechanical power on farms
1946	E Ball produced the first whole plants from shoot tip cultures of *Lupinus* and *Tropaeolum*
1948	Formation of adventitious shoots and roots in tobacco culture (Skoog and Tsui)
1949	Culture of fruits *in vitro* (Nitch)
1950	Ball regenerated organs from sequoia; Morel reported the first successful cultures of a monocot using coconut milk; Edwin Chargaff determined that there is always a ratio of 1:1 adenine to thymine in DNA of many different organisms
1951	Nitch cultured excised ovaries *in vitro*; Skoog demonstrated chemical control of growth and organ formation *in vitro*
1952	First virus-free plants produced from meristem culture; first successful micrografts achieved by Morel and Martin.
1953	James Watson and Frances Crick identified the helix structure of DNA. Haploid callus was produced from pollen grains of *Ginkgo* by Tulecke.
1954	Muir *et al.*, regenerated calli from single cells, using nurse cultures
1955	Synthesis of kinetin was done by Miller *et al.*

1957	Discovery of the regulation of organ formation by changing the ratio of auxin: cytokinin (Skoog and Miller). Culture of excised anthers of *Allium* by Vasil Francis Crick and George Gamov explain how DNA functions make proteins
1958	Regeneration of somatic embryos *in vitro* from the nucellus of *Citrus* ovules by Maheshwari and Rangaswamy. Proembryo formation in callus and cell suspensions of carrot by Reinert and Steward.
1960	First successful test tube fertilisation in *Papaver rhoeas* by Kanta. Use of the microculture method for growing single cells in hanging drops in a conditioned medium, developed by Jones *et al*. Enzymatic degradation of cell walls to obtain large numbers of protoplasts, achieved by Cocking. Filtration of cell suspensions and isolation of single cells by plating, was achieved by Bergmann.
1962	Development of the famous Murashige and Skoog medium.
1964	Production of first haploid plants from pollen grains of *Datura*, by Guha and Maheshwari
1965 1965	Vasil and Hildebrant regenerated tobacco plants from single cells. Production of protocorms from orchids *in vitro* by Morel
1966	Marshall Nirenberg and Severo Ochoa determined that a sequence of 3 nucleotide bases determines each of the 20 amino acids
1967	Haploid plants produced from pollen grains of tobacco by Bourgin and Nitch.
1968	Enzymes involved in cleaving DNA, termed restriction endonucleases, are shown by Meselson and Yuan.
1970	Selection of biochemical mutants *in vitro* by the use of tissue culturederived variation was made by Carlson. First achievement of protoplast fusion was done by Power et al. Hybrid embryo culture and subsequent chromosome elimination for haploid production in barley was made by Kao and Kao. Discovery of first restriction endonuclease from *Haemophillus influenzae* Rd. It was later purified and named *Hind*I (Smith)
1971	Preparation of first restriction map, using *Hind*I to cut circular DNA of SV 40 into 11 specific fragments (Nathans); Regeneration of first plants from protoplasts by Takebe *et al*.

1972	First report of interspecific hybridisation through protoplast fusion in two species of *Nicotiana,* done by Carlson *et al*. Restriction fragments can be joined by DNA ligase, regardless of their origin if they are cut with the same restriction enzyme by Mertz and Davis. Isolation of reverse transcriptase by Temin
1973	Use of the Lobban and Kaiser technique to develop hybrid plasmid - insertion of EcoRl fragment of DNA molecule into circular plasmid DNA of bacteria, using DNA ligase Gene from African clawed toad inserted into plasmid DNA of bacteria, by Herbert Boyer and Stanley Cohen. It is the beginning of genetic engineering and it is first recombinant DNA organism Cytokinin found to be capable of breaking dormancy in Gerbera by Pierik *et al*. Somatic hybridisation of tomato and potato, resulting in pomato was achieved by Melchers *et al*.
1974	Regeneration of haploid Petunia hybrid plants from protoplasts (Binding); Biotransformation in plant tissue cultures (Reinhard); Discovery that the Ti plasmid is the tumour inducing principle of *Agrobacterium* (Zaenen *et al*.; Larebeke *et al*.); Induction of branching by cytokinins in Gerbera shoot tips
1975	Positive *in vitro* selection of maize callus cultures resistant to T toxin of *Helminthosporium maydis* wa done by Gengenbach and Green. Development of the high resolution two dimensional gel electrophoresis procedure, which led to the development of proteomics was achieved by O'Farrel Moratorium, on recombinant DNA techniques, was imposed
1976	Shoot initiation from cryopreserved shoot apices of carnation was done by Seibert Octopine and nopaline synthesis and breakdown found to be genetically controlled by the Ti plasmid of *A. tumefaciens,* was shown by Bomhoff *et al*.
1977 1977	Successful integration of the Ti plasmid DNA from *A. tumefaciens* in plants was achieved by Chilton *et al*., A method of DNA sequencing was developed by Maxam-Gilbert Discovery of split genes was made by Sharp Roberts First practical application of genetic engineering -human growth hormone (HGH) was produced by bacterial cells
1978	Genentech, Inc.USA, used genetic engineering techniques to produce human insulin called humulin in *E. coli,* became first biotech company on New York stock exchange

1979	Co-cultivation procedure developed for the genetic transformation of plant protoplasts with *Agrobacterium* by Marton *et al*.
1980	Use of immobilised whole cells for biotransformation of digitoxin into digoxin was made by Alfermann *et al*. Commercial production of human insulin through genetic engineering in bacterial cells was done by *Eli Lilly and Co*. Studies on the structure of T-DNA cloning the complete EcoRl digest of Ti, tobacco crown gall DNA into a phage vector, thus allowing. A isolation a detailed study of T-DNA border sequence was done by Zambryski *et al*., US Supreme Court allowed patenting of manmade microbes.
1981	Introduction of the term somaclonal variation (Larkin and Scowcroft)
1982	Naked DNA transformation of protoplasts was done by Krens *et al*. Electrofusion of protoplasts was achieved by Zimmerman.
1983	Co-integrate type vectors for *Agrobacterium* transformation were designed by Zambriski *et al*. The idea of polymerised chain reaction (PCR), a chemical DNA amplification process was conceived by Kary Mullis. Transgenic plants were first created in the early 1980s by four groups working independently at Washington University in St. Louis, Missouri, the Rijksuniversiteit in Ghent, Belgium, Monsanto Company in St. Louis, Missouri, and the University of Wisconsin. On the same day in January 1983, the first three groups announced at a conference in Miami, Florida, that they had inserted bacterial genes into plants. The fourth group announced at a conference in Los Angeles, California, in April 1983, that they had inserted a plant gene from one species into another species. The Washington University group, headed by Mary-Dell Chilton, had produced cells of *Nicotiana plumbaginifolia*, a close relative of ordinary tobacco, that were resistant to the antibiotic kanamycin. Jeff Schell and Marc Van Montagu, working in Belgium, had produced tobacco plants that were resistant to kanamycin and to methotrexate, a drug used to treat cancer and rheumatoid arthritis. Robert Fraley, Stephen Rogers, and Robert Horsch at Monsanto had produced Petunia plants that were resistant to kanamycin. The Wisconsin group, headed by John Kemp and Timothy Hall, had inserted a bean gene into a sunflower plant. A chimaeric antibiotic resistant gene, as a selectable marker for plant cell transformation, was isolated by R.B. Flavell, and M.D. Chilton.

1983	Phaseolin gene from bean is expressed after transfer to sunflower via tumour-inducing plasmid vectors in an experiment by A. Stock, R.F. Barker, J.D. Kemp, and T.C. Hall
1985	Infection and transformation of leaf discs with *Agrobacterium tumefaciens* and regeneration of transformed plants was done by Horsch RB *et al*. Development of disarmed Ti plasmid vector system for plant transformation was achieved by Fraley RT *et al*. Gene transfer in protoplasts of dicot and monocot plants by electroporation was achieved by Fromm ME. US declared that plants can be patented
1986	TMV virus-resistant tobacco and tomato (transgenic plants developed using cDNA of coat protein gene of TMV) achieved by Powell-Abel *et al*. Transformation of tobacco protoplasts by direct DNA microinjection was done by Crossway First field trials of transgenic plants were carried out
1987	Development of biolistic gene transfer method for plant transformation was achieved by Sanford *et al*., and Klein *et al*. Isolation of Bt gene from bacterium (*Bacillus thuringiensis*) was done by Barton *et al*. First monocot (*Asparagus*) transformation by *Agrobacterium tumefaciens* was achieved by Bytebier *et al*.
1988	Recovery of stable transformants through particle bombardment was achieved by Klein *et al*. Automated mass propagation with organogenesis and embryogenesis were developed by Levi R *et al*.
1990	Formal launch of the Human Genome Programme was done. Plant transformation by microinjection of intact plant cells was achieved by Neuhaus Electroporation of intact plant tissues for direct DNA delivery was done by Dekeyser *et al*. Silicon carbide fibre-mediated DNA delivery into plant cells was achievevd by Kaeppler *et al*.
1993	Genetically engineered *Flavr Savr* tomatoes sold to the public
1997	Sequencing of *E. coli* genome was completed by Blattner *et al*.
1998	Sequencing of the genome of a multicellular organism *Caenorhabditis elegans* was achieved
2000	Arabidopsis draft sequence completed

2001	Sequencing of the human genome draft completed. John Suston received Nobel prize for his work on HGP.
2003	Of the soybeans grown in the US, 64% are transgenic; EU Union has had a 5 year ban on GMOs.

Subsequent to defining the medium, other aspects were developed such as suitable pH, addition of growth regulators, exploration of the use of alternative carbon sources. Use of activated charcoal and antioxidants for minimising browning effect, etc. (Yam *et al.*, 1989; Paek and Jun, 1994; Kumar and Kumar, 1998) is an important example in this regard.

In the earlier studies solid media were used. Later on, researchers started using a liquid medium either in static or in moving conditions. In the later case, a shaker was used to provide aeration for growing tissues (Heller, 1949, 1953; Steward *et al.*, 1958; White, 1963). A distinction was also made by some authors between the starting, standard maintenance and rooting media, based on their suitability; Meristems of *Cymbidium* were inoculated on a liquid medium, on which PLBs were formed. The PLBs were then transferred on to a solid medium to obtain plantlets (Wimber, 1963, 1965). The choice of liquid or solid medium may depend on the type of explant and the objective of the culture. Some workers have used liquid media, with the inocula supported on a filter bridge above the media. A solid medium was found to be economically viable, and convenient as it produces consistent results (Holdgate, 1977). Extremely complex media are usually unnecessary for general plant propagation. Variations of the media developed by Knudson (1922), Vacin and Went (1949), Murashige and Skoog (1962), or by White (1963), are generally effective at both macro and micronutrient levels.

After 1960, a revolution took place in the clonal propagation of orchids. Morel (1960, 1964, 1965) inoculated meristems obtained from shoot tips on KC agar media to obtain PLBs. Later, he cut these PLBs and subcultured them several times to obtain thousands of plants, from one shoot tip, in one year. Successful results were also obtained in *Dendrobium, Lycaste, Miltonia, Odontoglossum* and *Phaius* (Rao, 1977), by employing Morel's procedure. Wimber (1963) applied the essentials of Morel's procedure to a liquid system for the propagation of *Cymbidium*. The method of free cell culture in liquid media, developed for *Asparagus* (Steward and Mapes, 1971a), was applied to *Cymbidium* successfully to obtain plantlets through PLBs (Steward and Mapes, 1971b).

Thus, the techniques developed by Morel, Wimber, Steward and Mapes provided a foundation for the micropropagation of orchids that is still relevant today. Flower stalk buds of *Phalaenopsis* were cultured on modified knop's solution to obtain plantlets (Arditti *et al.,* 1977). Fernando (1979) used different media to produce plants, using apical and axillary buds of *Dendrobium* **Caesar Red Lip**. They used VW liquid medium for the initial culture and KC solid medium for producing plantlets from PLBs. Huang (1984) proposed alternative media and methods for *in vitro* propagation of *Cattleya* cultivars, as he failed to get success in the commercially available, "Murashige Multiplication Medium". He developed starting solution, shoot multiplication medium and rooting medium, and successfully cultured, *Cattleya guttata, C.* **J.A. Carbone 'orbit 10'**, *C.* **J.A Carbone 'Soleil'**, *C.* **Percia 'Cannizano'**, *Laeliocattleya* and *Potinara* hybrid clones. This was one of the earliest attemps to employ subculture technique, and to use genus specific media to obtain maximum yield.

Singh and Prakash (1985) developed a suspension culture technique for the micropropagation of *Epidendrum radicans*; Tanaka and Sakanishi (1985) used modified MS solid medium for the initiation of PLBs and then cultured the PLBs on modified KC solid medium for producing plantlets in *Phalaenopsis ambalis*. *Phalaenopsis* hybrid **Nos. 1603** and **2723** were cultured on modified VW medium. A medium containing Thomale's macronutrients, Ringe and Nitsch's minor elements and organic supplements was used to culture *Phalaenopsis* (Homma and Asahira, 1985). *Phalenopsis* and *Doritaenopsis* were cultured on VW medium. PLBs were formed within 30 days and they grew into plantlets after 60 days of culture (Lin, 1986); *Aranda* **cv. Christine** was cultured using VW liquid medium, shaken at 80 rpm. (Widiastoety *et al.,* 1986); PLBs were produced from the explants of *Aranthera cv.* **James Storie** used VW liquid medium (Widiastoely, 1986). After four weeks of culture on a liquid medium, meristem tissues of *Cymbidium vallambrosa* and *C.* **cv. Vieux Rose,** *C.* **cv. Dell Park** were cultured on Peptobak - Bacutil medium. In this medium, pearl barley, oak flakes or perlite, and agar were used as substrates. This is one of the first uses of alternative substrates. Protocorms developed in all cases, but shoot and root development was best on medium, containing pearl barley (Kukulczanka *et al.,* 1987).

Soedjono used Bayfolan medium, which is cheaper than VW medium for a successful micropropagation of *Dendrobium* **Walter Odumae**. VW liquid medium was used to culture *Dendrobium* explants (Hew, 1988); *Cymbidium ensifolium* and *C. goeringii* were cultured on MS medium (Kraus and Monteiro, 1989). *Phalaenopsis* **Celebration** was cultured on Kyoto medium and a rapid clonal propagation procedure was developed (Wang, 1989). Clonal propagation of *Aranda* **Deborah** was achieved by Goh and Wang (1990). KC liquid medium was used for initial development. Improved growth and proliferation were seen on VW medium. Plantlet development was faster in VW solid medium. *Aeridis multiflorum, Luisia teretifolia, L. trichorhiza, Neofinetia falcata, Papilionanthe teres, Rynchostylis retusa, Satyrium nepalense, Vanda cristata* and *V. testacea* were grown on Mitra, MS and KC media (Vij and Pathak, 1990). To demonstrate the compatibility of the media for culturing any orchid, cut flower orchid *Dendrobium* **Joannie Ostenhault** was cultured on liquid and solid VW media (Sharon and Vasundhara, 1990). MS medium was used to culture *Bletilla striata, Cleisostoma fordi* and *Pholidota chinesis* (Yam, and Weatherhead, 1991). To produce PLBs and plantlets from explants of *Oncidium* **cv. Gower Ramsey,** *Dendrobium* **cv. Miss Hawaii**, VW and modified MS media were used. (Nurani and Mohammed, 1992). Rate of protocorm proliferation was studied in *Dendrobium candicum*. Optimum proliferation occurred on even half strength MS medium (Zhang *et al.,* 1992). MS medium was used to produce PLBs from the explants of *Dendrobium wardianum*. The encapsulated PLBs were placed on MS agar medium to obtain plantlets (Sharma *et al.,* 1992). New Dogashima medium was used to obtain PLBs from the explants of *Phalaenopsis* and *Doritaenopsis* (Tokuhara and Mii, 1993). *In Vitro* conservation, through clonal propagation of a rare and endangered orchid *Renanthera imshootiana,* was achieved using VW medium. PLBs obtained by using agar medium were encapsulated (Sharma, 1994). Callus induction, PLB formation and plantlet production were achieved through culturing the explants of *Dendrobium* **cv. Madame Pompadour** on a modified MS medium (Mujib and Jana, 1994). PLBs were obtained from the explants of *Cymbidium* **cv. Nativity,** through culturing the explants on MS medium (Begum *et al.,* 1994). Multiplication rate of protocorms of *Dendrobium* **cv. Madame Pompadour,** by culturing PLBs on liquid KC medium, was achieved (Sounderrajan and Lokeswari, 1994). Explants of the bamboo orchid, *Arundina bambosifolia* were cultured on Heller's, MS, Raghavan, Torrey or VW medium. Growth pattern varied from medium to medium (Nagaraju and Parthasarathy, 1995). Induction of shoots was achieved by culturing the explants of *Phalaenopsis* on VW medium (Chen and Piluek, 1995). A rapid *in vitro* regeneration system was developed for mass propagation and plant quality improvement of *Spathoglottis plicata*. PLBs were induced from explants by culturing them on half strength MS medium and subsequently plantlets were also produced

(Wheilan *et al.,* 1997). *Bletilla striata* cultures were initiated in VW, Mitra-Prasad-Roychowdhury, Ichihashi or modified Ichihashi medium. It was found that modified Ichihashi medium favoured a good rate of PLB formation (Vij and Dhiman, 1997). Clonal propagation of *Dendrobium moschatum* and *Cymbidium aloifolium* was tested on five different media, of which Nitsch medium was found to be the best for the formation and proliferation of PLBs (Devi, 1997). Clonal propagation procedure was developed for efficient multiplication of *Vanilla planifolia*. Explants were cultured on semisolid MS medium (George and Ravishankar, 1997). Embryogenic callus was derived from the explants of *Phalaenopsis* **Wedding Promenade** and *Phalaenopsis* **Hanabonshi** × *Phalaenopsis* **Equestris 'Ilocos'**, by culturing them on New Phalaenopsis Medium. Subsequently, PLBs were initiated from callus. (Islam *et al.,* 1998). Callus induction and plantlet regeneration through somatic embyogensis in *Phalaenopsis* **Richard Shaffer 'Santa Cruz'** was achieved by using VW medium (Ishii *et al.,* 1998). Plant regeneration was achieved from callus cultures of *Cymbidium ensifolium,* by using half - strength MS medium (Chang and Chang, 1998). Explants of *Cymbidium* were cultured on MS liquid medium, and the number of PLBs produced per explant was studied (Fujii, 1999). Differentiation of PLBs *in vitro* was studied by culturing explants of *Cymbidium*.

The different media employed by various authors for culturing different orchids has been compilled in the Table 12.2.

MEDIA COMPONENTS

The success of tissue culture in plant propagation is greatly influenced by the nature of the culture medium used (George and Sherington, 1984). A systematic study of nutrient media used in orchid propagation has been studied by Ichihashi (1987). It is known that the components of a plant tissue culture medium include macronutrients, micronutrients, a separate iron supplement, vitamins, a carbon source, plant growth regulators, aminoacids, nitrogenous compounds and complex organic supplements (Dodds and Roberts, 1995).

MACROELEMENTS

The prefix macro or major, when applied to elements (nutrients), refers to the fact that these substances are needed in relatively large amounts. They include calcium, magnesium, nitrogen, phosphorus, potassium and sulphur. Depending on the medium, several salts may be used to supply each mineral (Arditti and Ernst, 1993).

Table 12.2 List of media and explants used in orchid tissue culture

Sl. No.	Orchids Studied	Explants Used	Media Employed	References
01	*Acampe praemosa*	Protoplast	VW	Seeni and Abraham (1986)
02	*Acampe rizida*	Leaf	Heller's and Ichihashi and Yamashita	Yam (1989); Yam and Weatherhead (1991)
03	*Aceridachnis* Bogor 'Apple Blossom'	Apical and axillary buds	VW	Lim-Ho (1981)
04	*Aeridachnis* Alexandra	Apical and axillary buds	VW	Lim-Ho (1981)
05	*Aeridachnis* Elizabeth Howe	Apical and axillary buds	VW	Lim-Ho (1981)
06	*Aeridis multiflorum*	Leaf	Mitra, MS and KC	Vij and Pathak (1990)
07	*Aerides ringens*	Protoplast	VW	Seeni and Abraham (1986)
08	*Anacamptis pyramidalis*	Meristem	MS	Morel (1970)
09	*Anagraecum giryamae*	Protoplast	BS	Price and Earle (1984)
10	*Anoectochilus formosanus*	Lateral buds	MS	Chow *et al.* (1982)
11	*Aroectochilus elatus*	Protoplast (isolation)	Not cultured	Gopalakrishnan and Seeni (1987)
12	*Arachnis hookeriana*	Axilary buds	MS	Vajrabhaya and Vajrabhaya (1976a)
13	*Arachnis* Capama 'Merah'	Apical and axillary buds	VW	Lim-Ho (1981)
14	*Arachnis* Maggie Oei	Cell (isolation)	Not cultured	Hew and Yip (1986)
15	*Arachnostylis* Charcalood	Apical and axillary buds	MS	Lim-Ho(1981)
16	*Aranda* Deborah	Shoot tip	VW	Goh (1973)
17	*Aranda*	Leaf tip	VW	Loh *et al.* (1975)
18	*Aranda* Wendy Scott 'Green field'	Shoot tip	VW	Cheah & Sagawa (1978)
19	*Aranda* Christine No. 5	Shoot tip buds	VW + Heller's microelements	Khaw *et al.* (1978 a, b) and
20	*Aranda* Christine No. 9	Shoot tip and buds	VW + Heller's microelements	Khaw *et al.* (1978 a, b)

Sl. No.	Orchids Studied	Explants Used	Media Employed	References
21	*Aranda* Christine No. 80	Shoot tip and buds	VW + Heller's microelements	Khaw *et al.* (1978 a, b)
22	*Aranda* Tay Swee Eng	Shoot tip and buds	VW + Heller's microelements	Khaw *et al.* (1978 a, b)
23	*Aranda* Majulah Chong Chee Yuen	Shoot tip and buds	VW + Heller's microelements	Khaw *et al.* (1978 a, b)
24	*Aranda* Kook Choo	Shoot tip and buds	VW + Heller's microelements	Khaw *et al.* (1978 a, b)
25	*Aranda* Noorah Alsagaff	Shoot tip and buds	VW + Heller's microelements	Khaw *et al.* (1978 a, b)
26	*Aranda* Lum Chin Lily Chong	Shoot tip and buds	VW + Heller's microelements	Khaw *et al.* (1978 a, b)
27	*Aranda* Christine No. 27	Leaf segments	VW, MS and KC	Fu (1978, 1979 b)
28	*Aranda* Christine No. 130	Leaf segments	VW, MS and KC	Fu (1978, 1979 b)
29	*Aranda* (17 hybrids)	Buds	VW, MS and KC	Lim-Ho (1981)
30	*Aranda* Noorah Alsagoof	Protoplast	VW	Loh and Rao (1985)
31	*Aranda* Christine No. 130	Shoot tip	VW	Widiastoety (1986)
32	*Aranda* Hilda Galistan	Leaf Segments	VW, MS and KC	Manorama *et al.* (1986)
33	*Aranda* Lucy Ladycock	Leaf Segments	VW, MS and KC	Manorama *et al.* (1986)
34	*Aranda* Mei Ling	Leaf Segments	VW, MS and KC	Manorama *et al.* (1986)
35	*Aranda* Deborah	Flower Stalk	KC	Goh and Wong (1990)
36	*Aranda* Tay Swee Eng	Flower Stalk	VW	Lam-Chan and Lee (1996)
37	*Aranthera* James Storei	Shoot with nodal buds	VW and KC	Irawati *et al.* (1977)
38	*Aranthera* James Storei	Shoot with single bud	VW	Cheah and Sagawa (1978)
39	*Aranthera* (6 hybrids)	Buds	VW	Lim-Ho (1981)
40	*Aranthera* Beatrice Ng 'Yellow'	Leaf base	VW	Lim-Ho (1981)

Sl. No.	Orchids Studied	Explants Used	Media Employed	References
41	*Arundina bambusifolia*	Shoot tip	Raghavan and Torrey	Mitra *et al.* (1971)
42	*Arundina bambusifolia*	Shoot tip and stem disc	VW	Banik *et al.* (1986)
43	*Ascocenda* Hilo Rose	Whole leaf	MS and KC	Fu (1978, 1979 b)
44	*Ascocenda* Tan Chai Beng	Shoot meristem	RM and Hyponex	Ichihashi (1979)
45	*Ascocenda* Meda Arnold	Shoot meristem	RM and Hyponex	Ichihashi (1979)
46	*Assocenda* Canival	Shoot meristem	RM and Hyponex	Ichihashi (1979)
47	*Ascocenda* (2 hybrids)	Buds	VW	Lim-Ho (1981)
48	*Ascocenda* Heah Hock Heng	Young leaves	MS and VW	Lim-Ho (1981)
49	*Ascofinetia*	Floral buds	VW	Intuwong and Sagawa (1973)
50	*Barlia robertiana*	Protoplast (Isolation)		Not cultured Pais *et al.* (1983)
51	*Bletilla striata*	Stem nodes	Knop's solution and KC	Yam (1989)
52	*Bletilla striata*	Stem nodes	Knop's solution and KC	Yam (1989)
53	*Bletilla striata*	Root tips	MS (Half Strength)	Yam and Weatherhead (1991 b)
54	*Brassia maculata*	Protoplast (Isolation)	Not cultured	Capesius and Meyer (1977)
55	*Brassocattleya* Princess Patricia	Buds	MS	Green (1983)
56	*Bulbophyllum neilgherrense*	Protoplast (Isolation)		Not cultured Seeni and Abraham (1986)
57	*Burkilla* Ong Thye Chiew	Buds	MS	Lim-Ho (1981)
58	*Calanthe discolor*	Protoplast (Isolation)	Not cultured	Yasugi *et al.* (1986)
59	*Catasetum trulla X* C. Berthrand	Root tip	KC	Kergauy (1984 a)
60	*Catesetum pileatum*	Root tip	MS	Kraus and Monteiro (1989)
61	*Cattleya*	Meristem	Reinert and Mohr	Reinert and Mohr (1987)
62	*Cattleya*	Vegetative shoot	VW	Scully (1967)
63	*Cattleya*	Meristem	MS and KC	Morel (1970)

Sl. No.	Orchids Studied	Explants Used	Media Employed	References
64	*Cattleya*	Axillary shoot	Lindemann *et al.* and KC	Lindemann *et al.* (1970)
65	*Cattleya*	Leaf bases	Heller's	Champagnat *et al.* (1970)
66	*Cattleya*	Leaf tip	Heller's and MS	Churchill *et al.* (1971)
67	*Cattleya*	Shoot tip	Reinert and Mohr and KC	Ichihashi and Kako (1973)
68	*Cattleya*	Protoplast (Isolation)	Not cultured	Capesius and Meyer (1977)
69	*Cattleya*	Dormant floral buds	KC + Nitsch micronutrients	Vajrabhaya (1978)
70	*Cattleya*	Protocorm	Hyponex	Yoneda *et al.* (1979, 1989)
71	*Cattleya*	Protoplast	Not cultured" (Isolation)	Capesius and Meyer (1977)
72	*Cattleya guttata*	Shoot apex	Huang's Starting and shoot multiplication solution	Huang (1984)
73	*Cattleya* J.A. Carbone 'Orbit 10'	Shoot apex	Huang's Starting and shoot multiplication solution	Huang (1984)
74	*Cattleya* J.A. Carbone 'Solcii'	Shoot apex	Huang's Starting and shoot multiplication solution	Huang (1984)
75	*Cattleya* Porcia 'Cannizaro'	Shoot apex	Huang's Starting and shoot multiplication solution	Huang (1984)
76	*Cattlya nobilior* x *C. loddigesii*	Root tip	VW	Kerbauy (1991)
77	*Cattleya aurantica*	Protocorm	VW and KC	Villabobos *et al.* (1994)
78	*Cymbidium*	Shoot tip	KC	Morel (1960, 1963, 1964 b, 1970, 1974)
79	*Cymbidium*	Shoot tip	Tsuchiya	Wimber (1963, 1965)
80	*Cymbidium*	Shoot tip	KC and VW	Sagawa *et al.* (1966)
81	*Cymbidium*	Isolated Cells	White's	Steward and Mapes (1971)
82	*Cymbidium*	Shoot tip	KC and VW	Fonnesbech (1972 a, b)
83	*Cymbidium*	Lateral bud	KC	Dalla Rosa and Laneri (1977)

Sl. No.	Orchids Studied	Explants Used	Media Employed	References
84	*Cymbidium*	Shoot tip	KC	Kusumoto and Furukawa (1977)
85	*Cymbidium*	Shoot tip	KC	Lavrentyeva (1980)
86	*Cymbidium*	Meristem	White's and MS	Wang *et al.* (1981)
87	*Cymbidium*	Shoot tip	MS	Kim and Kako (1982)
88	*Cymbidium*	Shoot tip	Tsuchiya	Cvitanic and Arditti (1984)
89	*Cymbidium*	Buds	KC	Van Renburg and Veelar (1984)
90	*Cymbidium goeringii*	Meristem	MS	Wang (1984)
91	*Cymbdium*	Gynosternia, Ovaries, Flower Stalk, Sepals, Petals, Labellum	MS	Kim and Kako (1984)
92	*Cymbidium faberi*	Rhizome	Kyoto	Hasegawa (1985)
93	*Cymbdium goeringii*	Shoot tip	Linsmaier and Skoog	Hasegawa (1987)
94	*Cymbidium*	Pseudobubs	VW	Gu, *et al.* (1987)
95	*Cymbidium Kanran* "Makino"	Shoot tip	MS and Linsmaier-Skog	Hasegawa and Goi (1987)
96	*Cymbidium* Nepolian	PLBs	White's, KC, VW, RM Tsuchiya, Morel, and Lindemann	Takano *et al.* (1990)
97	*Cymbidium* Molody Fair 'Marillyn Monroe'	PLBs	White's, KC, VW, RM Tsuchiya, Morel, and Lindemann	Takano *et al.* (1990)
98	*Cymbidium* Minidream 'Goldencolor'	PLBs	White's, KC, VW, RM Tsuchiya, Morel, and Lindemann	Takano *et al.* (1990)
99	*Cymbidium goerngii*	Floral buds	MS	Shimasaki and Uemoto (1991)
100	*Cymbidium* Nativity	PLBs (outer tissue)	MS	Begum *et al.* (1994)
101	*Cymbidium aloifolium*	Shoot tip	MS, KC and VW	Devi (1997)
102	*Cymbidium ensifolium*	Rhizomes	MS	Chang and Chang (1998)
103	*Cymbidium* Lucky Rainbow 'Lapine Dancer'	PLBs	MS	Fuji *et al.* (1999)

Sl. No.	Orchids Studied	Explants Used	Media Employed	References
104	*Cymbidium* Thanksgiving "Nativity"	PLBs	MS	Fuji *et al.* (1999)
105	*Cymbidium* Lucky Rainbow 'Lapine Dancer'	PLBs (Outer tissue)	MS	Fuji *et al.* (1999)
106	*Cymbidium* Thanksgiving 'Nativity'	PLBs (Outer tissue)	MS	Fuji *et al.* (1999)
107	*Cymbidium punctatum*	Root tip	VW	Sanchez (1988)
108	*Dactylorchis fuchii*	Dormant shoots	KC	Stokes (1974)
109	*Dendrobium*	Buds	VW	Sagawa and Shoji (1967)
110	*Dendrobium*	Flower stalk	VW	Singh and Sagawa (1972)
111	*Dendrobium* Ng Eng Chew	Shoot tip	VW	Singh (1976)
112	*Dendrobium* Alice Chong	Bud	VW	Khaw *et al.* (1978 a, b)
113	*Dandrobium* Caesar Red Lip	Meristem	VW and KC	Fennando (1979)
114	*Dendrobium* Alice Spalding	Leaves	VW	Fu (1978, 1979 b)
115	*Dendrobium phalaenopsis*	Shoot tip	MS	Gandawijaja (1980)
116	*Dendrobium* (17 hydrids)	Shoot tip	VW	Lim-Ho (1981)
117	*Dendrobium antennathum*	Bud	RM	Kukulczanaka and Wojciechowska (1983)
118	*Dendrobium* Jaquline Thomas "'White"	Shoot tip	VW	Soediono (1983b)
119	*Dendrobium* Beach girl	Protoplast (isolation)	Not cultured	Price and Earle (1984)
120	*Dendrobium* Louis Bleriod	Protoplast (isolation)	Not cultured	Price and Earle (984)
121	*Dendrobium chrysanthum*	Pseudobulb	MS	Vij and Pathak (1989)
122	*Dendrobium loddigesii*	Stam nodes	Knop's solution	Yam (1989)
123	*Dendrobium transparens*	Stem nodes	Knoop's solution	Yam (1989

Sl. No.	Orchids Studied	Explants Used	Media Employed	References
124	*Dendrobium* Joannie Ostenhault	Meristem	VW	Sharon and Vasundhara (1990)
125	*Dendrobium* Formidable	Shoot tip and lateral buds	Takashi and Taikichi	Takashi and Taikichi (1990)
126	*Dendrobium wardiamum*	Shoot tip (synthetic seeds)	MS	Sharma *et al.* (1992)
127	*Dendrobium wardianum*	Shoot tip (synthetic seeds)	MS	Sharma *et al.* (1992)
128	*Dendrobium* Miss Hawaii	Protocom	VW	Nurani and Mohammed (1992)
129	*Dendrobium moniliformae*	Flower stalk	VW	Duan and Yazawa (1994)
130	*Dendrobium* Madame Pompadour	Meristem	MS	Mujib and Jana (1994)
131	*Dendrobium* Madame Pompadour	Protocorn	KC	Sounderrajan and Lokeswari (1994)
132	*Dendrobium* Sonia	Flower stalk	VW	Lam-Chan and Lee (1996)
133	*Dendrobium* Miss Hawaii	Protocorn	VW	Nurani and Mohammed (1992)
134	*Disa uniflora*	Meristem	Thomale GD	Haas and Luckel (1977)
135	*Disa uniflora*	Shoot apex	Thomale GD	Van Waes and De Geest (1983)
136	*Doritaenopsis* Elizabeth Waldhein	Floral buds	MS and VW	Lim-Ho (1981)
137	*Doritaenopsis*	Internodes	VW	Lin (1986)
138	*Doritaenopsis*	Shoot apex	ND and MS	Tokuhara and Mii (1993)
139	*Doritaenopsis*	Callus	NP	Ichihashi and Islam (1999)
140	*Doritis tiny*	Flower stalk	Hyponex	Duan and Yazawa (1994)
141	*Encyclia belizensis*	Protocorn	VW and KC	Villalobos *et al.* (1994)
142	*Epidendrum*	Leaf tip	MS, KC, and Linsmaier and Skoog	Churchill *et al.* (1970)
143	*Epidendrum*	Root tip	Ojima - Fujiwara	Churchill *et al.* (1972)
144	*Epidendrum o'brienianum*	Flower stalk	MS, VW and KC	Stewart and Button (1978)

Sl. No.	Orchids Studied	Explants Used	Media Employed	References
145	*Epidendrum o'brienianum*	Root tips	Ojima-Fujiwara, Schenk and Hilderbrand and KC	Steward and Button (1978)
146	*Epidendrum radicans*	Shoot tip	MS	Kusumoto (1981 b)
147	*Epiphronitis vietchii*	Shoot tip and axillary bud	MS	Kusumoto (1981 b)
148	*Eulophia hormusjii*	Rhizome	MPR	Vij *et al*. (1989)
149	*Grammatophyllum elegans*	Protoplast (isolation)	Not cultured	Seeni and Abraham (1986)
150	*Holttumara*	Buds	MS	Lim-Ho (1981)
151	*Kagawara* Firebird	Buds	MS	Lim-Ho (1981)
152	*Ludisia discolor*	Nodes	KC	Teo (1978)
153	*Luisia trichorhiza*	Leaf	MS	Vij and Pathak (1988 a)
154	*Luisia teretifolia*	Leaf	Mitra, MS and KC	Vij and Pathak (1990)
155	*Mokara* Sally Lim	Buds	MS	Lim-Ho (1981)
156	*Mokara* Wai Liang	Flower buds	MS and VW	Lim-Ho *et al*. (1984)
157	*Mokara* Mak Chin On	Flower buds	MS and VW	Lim-Ho *et al*. (1984)
158	*Mokara* Chark Kuan	Floral stalk tips	VW	Karim *et al*. (1992)
159	*Mokara* Khaw Phaik Suan	Flower stalk	VW	Lam-Chan and Lee (1996)
160	*Mormodes histrio*	Pseudobulb	KC	Holters and Zimmer (1990a)
161	*Mormodes histrio*	Root tips	KC	Holters and Zimmer (1990b)
162	*Neofinetia falcata*	Leaf	Mitra, MS and KC	Vij and Pathak (1990)
163	*Neofinetia falcata*	Callus (seed derived)	MS	Niimi *et al*. (1995)
164	*Neostylis*	Flower stalk	VW	Sagawa and Kunisaki (1982)
165	*Nervilia aragoana*	Tuber	MS	Chow (1986)
166	*Nigritella*	Tubers	MS (1/3) strength)	Haas (1997 a, b)
167	*Oncidium papilio*	Flower stalk	KC and MS	Fast (1973 a)
168	*Oncidium*	Shoot apex	VW	Khaw *et al*. (1978 a, b)
169	*Oncidium varicosum*	Root tips	KC	Kerbauv (1984, b)

Sl. No.	Orchids Studied	Explants Used	Media Employed	References
170	*Oncidium* (7 species and hybrids)	Dormant buds of flower stalk	VW	Lim-Ho and Lee (1987)
171	*Oncidium* Gover Ramsey	Scape nodes	VW	Nurani and Mohammed (1992)
172	*Ophrys apifera*	Tuber	MS	Champagnat and Morel (1972)
173	*Ophrys fuciflora*	Tuber	MS	Champagnat and Morel (1972)
174	*Ophrys apifera*	Shoot apex	P and S	Khaw *et al.* (1978 a, b)
175	*Paphiopedilum* (11 hybrids)	Buds	MS	Mubeck (1973)
176	*Paphiopedilum villosum*	Shoot tip	Heller's	Stewart and Button (1975, 1976 b, c)
177	*P. fairieanum*	Shoot tip	Heller's	Stewart and Button (1975, 1976 b, c)
178	*P insignae*	Shoot tip	Heller's	Stewart and Button (1975, 1976 b, c)
179	*Paphiopedilum callosum*	Shoot and Leaf tip	GD-D	Allenberg (1976)
180	*Paphiopedilum*	Protoplast (isolation)	Not cultured	Yasugi *et al.*, (1986)
181	*Paphopedilum*	Shoot tip	MS	Huang (1988)
182	*Pipilionanthe teres*	Leaf	Mitra, MS and KC	Vij and Pathak
183	*Phalaenopsis*	Flower stalks	KC	Rotor (1949)
184	*Phalaenopsis*	Floral buds	KC and MS	Tse *et al.* (1971)
185	*Phalaenopsis*	Flower stalk nodes	VW	Intuwong and Sagawa (1974)
186	*Phalaenopsis ambalis*	Vegetative shoots	VW	Intuwong and Sagawa (1974)
187	*P.* X Star of Santacruz	Vegetative shoots	VW	Intuwong and Sagawa (1974)
188	*P.* X Surfrider	Vegetative shoots	VW	Intuwong and Sagawa (1974)
189	*P.* X Arcadia	Vegetative shoots	VW	Intuwong and Sagawa (1974)
190	*P.* X Cochlearis	Vegetative shoots	VW	Intuwong and Sagawa (1974)
191	*Phalaenopsis*	Leaf	KC	Koch (1974 a)
192	*Phalaenopsis*	Leaf	VW and MS	Tanaka *et al.* (1975)

Sl. No.	Orchids Studied	Explants Used	Media Employed	References
193	*Phalaenopsis* Zada	Flower stalk nodes	KC	Zimmer and Pieper (1976, 1978)
194	*P.* Lipperose	Flower stalk nodes	KC	Zimmer and Pieper (1976, 1978)
195	*P.* Zauberrose	Flower stalk nodes	KC	Zimmer and Pieper (1976, 1978)
196	*Phalaenopsis ambalis*	Root tip	MS	Tanaka *et al.* (1976)
197	*Phalaenopsis*	Flower stalk nodes	Knop's solution	Reisinger *et al.* (1976)
198	*Phalaenopsis ambalis*	Leaf	MS	Tanaka and Sakanishi (1977)
199	*Phalaenopsis*	Flower stalk	MS	Tanaka and Sakanishi (1977)
200	*P.* Betty Hausermann	Flower stalk	MS (half strength)	Griesbach, (1983)
201	*Phalaenopsis*	Flower stalk nodes		KC Ernst, (1984)
202	*Phalaenopsis*	Internodes	Thomale's (Macro) + Ringe and Nitsch's (microelements)	Homma and Asahira (1985)
203	*Phalaenopsis*	Leaf	MS and KC	Tanaka Sakamishi (1985)
204	*Phalaenopsis* No. 1603 and *P.* No. 2723	Flower stalk	VW	Yoneda (1986)
205	*Phalaenopsis*	Intermodes	VW	Lin (1986)
206	*Phalaenopsis*	Internodes	VW	Lin (1987)
207	*Phalaenopsis* Celebration	Flower stalk and shoot tip	Kyoto and MS	Wang (1989)
208	*Phalaenopsis* Michelle X Michelle	Protocorm	XE	Lam *et al.* (1991)
209	*Phalaenopsis*	Floral buds	MS	Ichihashi (1992)
210	*Phalaenopsis*	Shoot tip	ND	Tokuhara and Mii (1993)
211	*Phalaenopsis* Pink Leopard 'Petra'	Flower Stalk	VW and Hyponex	Duan and Yazawa (1994)
212	*Phalaenopsis*	Flower stalk	VW	Chen and Piluek (1995)
213	*Phalaenopsis*	Flower stalk	Hyponex, VW and MS	Park *et al.* (1996)
214	*Phalaenopsis*	Floral buds	P	Ichihashi and Hiraiwa (1996)

Sl. No.	Orchids Studied	Explants Used	Media Employed	References
215	*Phalaenopsis* Richard Shaffer 'Santa Cruz'	Leaf	VW	Ishii *et al.* (1998)
216	*Phalaenopsis* Santa Cruz	PLB	VW	Ishii *et al.* (1998)
217	*P.* Wedding Promenade	Floral buds	NP	Islam *et al.* (1998)
218	*P.* Hanboushi *XP* Equestrix 'Ilocos'	Floral buds	NP	Islam *et al.* (1998)
219	*Phalaenopsis*	Flower stalk	MS	Chen *et al.* (1998)
220	*Phalaenopsis*	Callus	NP	Ichihashi and Islam (1999)
221	*Pholidota Chinensis*	Root tips	MS and KC	Yam (1989)
222	*Phragmipedium sedeni*	Floral buds	MS + Nitsch's Microelements	Fast (1973 b)
223	*Phragmipedium caricinum*	Buds	RM	Strokes *et al.* (1975)
224	*Phragmipedium sedenii* 'Candidulum'	Buds	RM	Strokes *et al.* (1975)
225	*Pleione*	Shoot tip	KC	Weatherhead and Harberd (1980)
226	*Ranatanda*	Protoplast	Not cultured	Teo and Neumann (1978 b)
227	*Ranantanda* Sanderi	Leaf	VW	Tan (1982)
228	*Rananthera imschootiana*	Leaf	Mitra *et al.*	Seeni and Latha (1992)
229	*Rananthera imschootiana*	Shoot tip, floral bud and leaf	VW	Sharma (1994)
230	*Rynchostylis gigantea*	Shoot tip	Vajrabhaya and Vajrabhaya	Vajrabhaya and vajrabhaya (1970)
231	*Rynchostylis retusa*	Root	MS	Sood and Vij (1986)
232	*Rynchostylis retusa*	Leaf	Mitra, MS and KC	Vij and Pathak (1990)
233	*Rynchostylis retusa*	Dermal peels	Mitra *et al.*	Vij (1994)
234	*Saccolabium calceolare*	Flower stalk	Mitra *et al.*	Vij *et al.* (1986)
235	*Satyrium nepalense*	Leaf	Mitra, MS and KC	Vij and pathak (1990)
236	*Spathoglottis plicata*	Leaf and Root	MS	Bapat and Narayana swami (1977)

Sl. No.	Orchids Studied	Explants Used	Media Employed	References
237	*Spathoglottis plicata*	Flower stalk	MS	Teng *et al.* (1977)
238	*Spathoglottis plicata*	Leaf and Stem	MS	Teng *et al.* (1994)
239	*Stanthopea intermedia*	Protocorm	VW and KC	Villalobos *et al.* (1994)
240	*Thunia*	Flower stalk	VW	Singh and Prakash (1984)
241	*Vanda* Miss Joaquim	Nodes	VW	Sagawa and Sehgal (1967)
242	*V.* Miss Joaquim	Root, Bud and Shoot tip	White's	Goh (1970)
243	*V.* Miss Joaquim	Shoot tip	VW	Kumsaki *et al.* (1972)
244	*V. insignis X V. tessellata*	Shoot tip	VW	Teo *et at.* (1973)
245	*Vanda* TMA *X.V.* Miss Joaquim	Shoot tip	Mathews and Rao	Mathews and Rao (1985 b)
246	*Vanda*	Root and Leaf	VW	Chaturvedi and Sharma (1986)
247	*Vanda*	Floral buds	KC	Valmayor *et al.* (1986)
248	*Vanda coerulea*	Leaf bases	MPR	Anonymous (1987)
249	*Vanda* TMA *X V. teres*	Root and Leaf	VW	Sharma and Chaturvedi (1988)
250	*Vanda coerulea*	Leaf bases	Mitra *et al.*	Seeni (1988)
251	*Vanda crisata and Vanda testacea*	Leaf	Mitra, MS and KC	Vij and Pathlak (1990)
252	*Vanda coerulea*	Dermal peels	Mitra *et al.*	Vij (1994)
253	*Vanilla planifolia*	Nodes	MS	Kononowicz and Janick (1984)
254	*Vanilla planifolia*	Roots	KC	Philip and Nainar 1986
255	*Vanilla planifolia*	Buds	Linsmaier-Skoog	Gu *et al.* (1987 a)
256	*Vanilla planifolia*	Nodes	MS, WPM and GD	Ganesh *et al.* (1996)
257	*Vanilla planifolia*	Buds	MS	George and Ravishankar (1997)
258	*Vascotylis*	Floral buds	VW	Intuwong and Sagawa (1973)
259	*Vuylsketeara cambria*	Shoot tip	MS	Kukukzanka *et al.* (1989)

MICROELEMENTS

Traces of certain mineral elements are required by all plants. Since these are required in exceedingly small quantities, they are referred to as micro or minor elements (nutrients); thus, the prefix used here is based on the amounts used but not on their importance in the medium (Arditti and Ernst, 1993). Micronutrients essentially include iron, manganese, zinc, boron, copper, molybdenum and chlorine (Marschner 1986).

IRON

Until the advent of chelating agents (EDTA, Na_2EDTA, Na_2 EDTA, $2H_2O$), the incorporation of iron, in an available form in culture media, posed a problem. MS medium includes 37.3 mgl^{-1} Na_2EDTA and 27.8 mgl^{-1} Fe $SO_4 . 7H_2O$, the amount used was assumed to be equimolar until careful recalculations showed otherwise (Singh and Krikorian, 1980). However, the actual concentrations turned out to be 100m mol l^{-1} (100 mm solution) of the iron salt and 111m mol l^{-1} (111 mm solution) of the chelating agent. This left an excess of 11 m mole of chelating agent per litre of medium (Arditti and Ernst, 1993). EDTA too, may have side effects on certain enzyme systems and on morphogenesis in cultures (Bonga, 1982). Growth response of several orchids to Fe EDTA concentration was also studied (Paek *et al.*,1993).

GROWTH REGULATORS

Compounds occurring naturally, which have a regulatory rather than a nutritional role in growth and development, and which are generally active at very low concentrations, are known as plant growth regulators or plant hormones (George and Sherington, 1984). The term 'hormone' should be reserved for naturally occurring plant growth regulators (Dodds and Roberts, 1995). Cytokinin and gibberellin regulation of shoot development was studied in tobacco (Engelke *et al.*, 1973).

The most commonly used auxins in orchid tissue culture media are the naturally occurring IAA and the three synthetics, NAA, IBA and 2, 4-D (Arditti and Ernst, 1993). Some cultures require no exogenous auxins (Street, 1966). An auxin is generally required by most plant cells for division and root initiation (Smith, 2000). Though most synthetic auxins are not destroyed by autoclaving at 110 to 120°C, for 50 to 60 minutes, IAA may be destroyed in such a scenario (Posthumus, 1971). Auxins generally induce cell elongation, swelling of tissues, callus formation, adventitious root formation, inhibition of adventinous and axillary shoot formation and embryogenesis in suspension culture (Pierik, 1987). 2, 4-D, though an effective auxin for callus proliferation, is not found favourable in media involving root and shoot initiation (Gamborg *et al.*, 1976). IBA, 4 CPA, 2, 4, 5-T are also effective auxins. IBA particularly is an effective rooting agent. IAA is rapidly degraded by light (Nissen and Sutter, 1990; Stasinopolis and Hangarter, 1990). NAA was tested by Gautheret in 1942 (Gautheret, 1942) and he also used 2, 4-D (Krikorian, 1982). Auxins are extensively used in orchid tissue culture media (Arditti, 1977a, Zimmer, 1978; Kushnir, 1986).

Cytokinins are often used exogenously to stimulate growth and development. They usually promote cell division, adventitious shoot formation and inhibit root formation (Pierik, 1987). The synthetic compounds, kinetin BAP, BA, DMAP, and naturally occurring, Zeatin are used most commonly in orchid culture media (Arditti and Ernst, 1993). Another chemical with potent cytokinin activity used in orchid tissue culture is TDZ (Chen and Piluek, 1995). Experiments with aqueous solutions of kinetin, zeatin and IPA have shown that they are not broken down when autoclaved for 1 hour (Dekhuijzen, 1971).

Gibberellins are seldom used in culture media for orchids. Autoclaving reduces gibberellin activity by more than 90 percent (Van Bragt and Pierik, 1971). ABA, a growth inhibiting hormone, is generally not used in orchid culture media (Arditti and Ernst, 1993).

An appropriate ration of auxins and cytokinins is essential for accurate morphogenisis (Fig. 12.1 and 12.2)

AMINO ACIDS AND OTHER NITROGENOUS ADDITIVES

With the exception of glycine, which is a component of several media, amino acids are not generally added to plant nutrient media. A medium can be enriched by casein hydrolysate. Casein (bovine milk protein) has eighteen amino acids (Klein and Klein, 1970). Different amino acids used in nutrient media are listed by Huang and Murashige (1977).

VITAMINS

Vitamins have catalytic functions. Whereas nicotinic acid and pyridoxine may stimulate growth (Gamborg *et al.*, 1976) thiamin may be the only essential vitamin used in all plant tissue culture media. However, the high concentrations of ascorbic acid do not imply that the plant has such a large requirement. It is used as an anti-oxidant (Pierik, 1987).

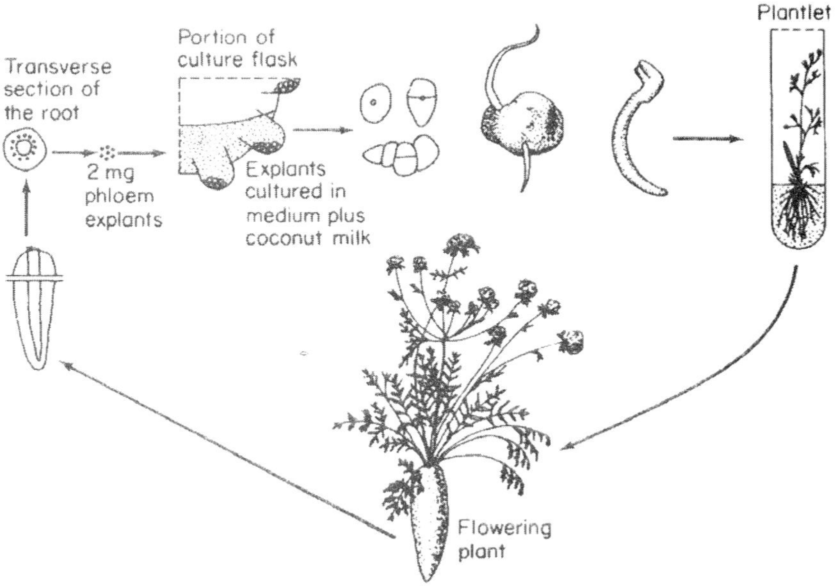

Fig.12.1 Schematic outline of plant root culture.

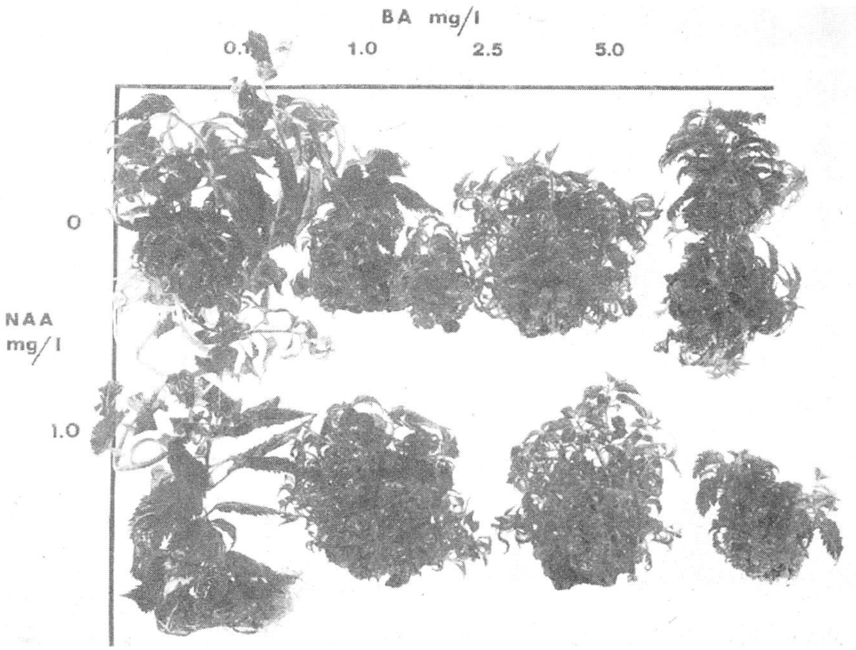

Fig.12.2 Effect of NAA: BA ratio on morphogenisis.

CARBON SOURCES

Since all plant tissue culture media require carbon energy, sucrose is usually added (Dodds and Roberts, 1995). Glucose, maltose, raffinose, fructose, galactose, mannose, and lactose are used as alternatives to sucrose. But sucrose has almost invariably been found to be the best carbon source (George and Sherington, 1984).

CHARCOAL

Establishment of *in vitro* cultures of many plants pose a great difficulty due to lethal browning of the medium (Kumar and Kumar,1998). Activated charcoal can absorb many organic and inorganic molecules from a medium. Although the precise effects of activated charcoal are unknown, there are several possible attributes. It may remove contaminants from agar (Kohlenbach and Wernicke, 1978), and secondary products secreted by the cultured tissues (Wang and Huang, 1976; Fridborg *et al.,* 1978). Activated charcoal has been reported to stimulate embryogenesis (Kohlenbach and Wernicke, 1978). Activated vegetable charcoal improved the growth of *Paphiopedilum* (Ernst, 1974) and *Phalaenopsis* (Ernst, 1975, 1976) in cultures. Similar studies were carried out on other orchids (Wang and Huang,1976; Fridborg *et al.,*1978; Waes,1987).

COMPOSITE ORGANIC SUPPLEMENTS

Natural extracts are often used in culture media when chemically defined substances fail to produce the desired results (Dodds and Roberts, 1995). Liquid endosperm of coconut is widely used in culture media. It has been found to be beneficial for inducing growth of the callus and morphogenesis. The components of coconut water have been identified (George and Sherington, 1984). Johannes van Overbeek was responsible for the discovery of the uses of coconut water (van Overbeek *et al.,* 1941, 1942). A simple procedure for the preparation of coconut water for use in culture media is given by Dodds and Roberts (1995). Besides coconut water, various other adjuvants are also effectively used in the media (Rao, 1977); lower concentrations of tomato juice were recommended for orchid growers (Arditti, 1966b). Apple juice and peptone were effective for the growth of *Cymbidium* and *Paphiopedilum* (Tsmamoto *et al.,* 1963). Extracts of banana were used to culture *Phalaenopsis* (Ernst, 1970). Banana extract was also suitable for the growth of *Cattleya aurantiaca* (Ernst, 1967). Use of CW in *Dendrobium* shoot tip culture was also studied (Soediono, 1983b).

MEDIUM MATRIX

The culture is either suspended in an aqueous medium or it is grown on a semisolid gel or solid matrix (Dodds and Roberts, 1995). Agar is used almost exclusively as the gelling agent for preparing semisolid and solid media (George and Sherington, 1984). Since agar may contain some unknown contaminants (Pierik, 1971), relatively pure gelling agents are preferred. Several substitutes for agar are available, but they have not gained wide acceptance because of high expenses (Arditti and Ernst, 1993).

pH OF THE MEDIUM

It is recommended that a pH in the range of 5.0 - 6.5 is suitable for growth of plants, since pH lower than 4.5 and higher than 7.0 generally stops growth and development (Pierik, 1987).

EXPLANTS USED IN ORCHID TISSUE CULTURE

Clonal propagation methods using several explants have been developed for some orchids by various researchers (Pieper and Zimmer,1976; Arditti,1977; Arditti *et al.,* 1977a; Lavrentyeva, 1980; Kusumoto, 1981; Kim and Kako,1982; Kim,1984; Griesbach,1984; Arditti and Arditti, 1985; Inamura and Saiton, 1987; Chen *et al.,* 1990; Singh, 1991; Hermann and Sell, 1991; Sharma and Tandon, 1992).

MERISTEM

The credit for the initiation of meristem culture technique goes to Morel (1960). He attempted orchid meristem culture technique after the successful results obtained in carnations, dahlias and potatoes (Morel and Martin, 1952, 1955). In a lecture delivered to the *Cymbidium* society, Morel (1965) discussed meristem culture in detail. Leaf primordia, measuring 0.5-1mm were obtained from green bulbs of *Cymbidium*. Meristems isolated from these leaf primordia, regenerated 2-3 protocorms, even on simple Knop's solution. Meristems were taken from the buds of *Cattleya* and cultured on MS medium. About 12 protocorms from each explant were obtained and IAA or NAA were found to be beneficial. Successful results were also obtained in *Dendrobium, Lycaste, Miltonia, Odontoglossum* and *Phaius*. New *Cymbidium* meristems differentiated sub-apically, whereas in *Cattleya,* the wound tissue near the leaf base proliferated and gave rise to meristematic regions (Champagnat *et al.,* 1966; Champagnat and Morel, 1969). The techniques and details involved in *in vitro* meristem tissue propagation have been explained in detail by Isley (1965) and Russon (1965). Meristems were isolated from buds arising from the rhizomes of *Cattleya*. The rhizomes were then sterilised by clorox, CaOCl and 70% ethanol. Meristems were cut with the surrounding rhizome tissue, 1 to 3 mm^3 in size, and cultured on Reinert and Mohr solid medium to obtain plantlets (Reinert and Mohr, 1967). Details regarding the nature of propagation, materials, method of obtaining sterile meristems, size of the explant, pH of the medium and other cultural conditions have been discussed (Bertsch, 1966; Marston, 1966, 1967, 1969, Marston and Voraurai, 1967; Goh, 1971). Thus, a foundation was laid by painstaking research done from 1960 to 1971. Fernando (1979) employed meristem culture for the propagation of *Dendrobium*. Shoots were surface-sterilised by NaOCl before obtaining meristems. Modified KC medium was used for initial culture. Differentiation of PLBs into plantlets occurred on KC solid medium. Meristems were derived from both shoot apices and axillary buds of sixty orchid species and eighty hybrids, and cultured on different media to obtain plantlets at Singapore Botanical Gardens (Lim-Ho, 1981). Shoot meristems of *Cymbidium ensifolium* and *C. goeringii* were cultured on MS or White's solid media, supplemented with NAA and coconut water. The explants formed PLBs within 4-6 months. Plantlets were formed from PLB on MS or Vacin and Went media (Wang, 1988). When shoot meristems of *Dendrobium* **Joannie Ostenhault** were cultured on Vacin and Went medium with 15% coconut water, PLBs formed in 4 weeks. They differentiated into plantlets, on Vacin and Went solid medium with auxins, within 14 weeks (Sharon and Vasundhara, 1990). Meristem culture of *Cymbidium* hybrid **No. 611** was done on MS medium with human bioenergy, to study the effects of human bioenergy on the development of plants (Ardelean, 1995). *Dendrobium* meristems were used for micropropagation (Ye,1995).

Shoot Tips

Wimber (cf. Arditti and Ernst, 1993) concieved the idea of shoot tip cultures. *Cymbidium* shoot tips were cut either vertically or transversely into sectors, cultured in liquid medium for a month and then transferred to fresh agar medium. The sectors finally developed into PLBs (Wilfret, 1966). Similar results were obtained by Sagawa *et al.,* (1966) in six week old cultures. Shoot apices of *Rynchostylis* were cultured on agar medium with NAA and coconut water to obtain callus, which was subcultured to produce plantlets (Vajrabhaya and Vajrabhaya, 1970). Shoot tips of *Dendrobium* were cultured on modified Vacin and Went medium. Explants produced green callus within 45 days, and PLBs developed within three months (Kim *et al.*, 1998). Kunisaki *et al.*, (1972) cultured shoot tips of *Vanda*. Shoot tip explants of *Cymbidium* were sterilised with clorox and cultured on KC or VW medium. Proliferation was induced on Fonnesbech liquid medium with NAA but plantlets formed on solid version of the Fonnesbech medium (Fonnesbech, 1972). Shoot tip culture in orchids was discussed by Bergman (1972).

An earlier successful method for shoot tip culture of *Phalaenopsis* was reported from the University of Hawaii (Intuwong and Sagawa, 1974). *Phalaenopsis ambalis*, *Phalaenopsis* **Star of Santa Cruz**, and *Phalaenopsis* **Surfinder** were used to obtain explants. Surface sterilisation was done using clorox and NaOCl. Solid and liquid modified Vacin and Went medium were used. PLBs formed, were transferred on to sucrose free–solid media to produce plantlets. Shoot tips of *Dendrobium* **Ng Eng Cheow** were cultured on VW medium. All explants were placed on the agar medium, with "right end up" and pushed halfway into the medium. Buds started to grow and plantlets developed within 45 days (Singh, 1976). 1-1.5 mm long shoot tips from 8 months old seedlings of *Dendrobium phalaenopsis* were used as explants. MS medium was used for initial culture and multiple shoot formation. A second modification was used for root formation (Gandawijaja, 1980).

A shoot tip culture method for mass rapid propagation of *Dendrobium* **Jaquelyn Thomas White** was developed. Explants were cultured, using modified VW medium for initial stage, and modified MS medium for rooting stage (Soediono, 1983b). Effects of anti-contaminants on shoot tip cultures of *Cymbidium* on Tsuchiya medium were studied. All but amphotericin-B, nystatin and sodium omadine were found to be deleterious (Vitamic and Arditti, 1984). Shoot tips of *Aranda* **cv. Christine 130** were cultured on VW medium. All shoot tips produced leaves and 44% of the explants produced PLBs (Widiastoety *et al.*, 1986). Shoot tip culture was done in oriental orchids (Wu *et al.*,1987). Shoot tips of *Cymbidium goeringii* and *Cymbidium Kanran* were cultured on different media. Explants on Linsmaier and Skoog media produced rhizomes (Hasegawa and Goi, 1987). Shoot tip explants of *Dendrobium wardianum* were used to study the effects of nitrogenous adjuvants *in vitro* (Sharma *et al.*, 1992). Shoot tips of *Arundina bambosifolia* were cultured on different media to obtain plantlets (Nagaraju and Parthasarathy, 1995). Axenic seedlings of *Dendrobium moschatum* and *Cymbidium aloifolium* were used to obtain the shoot apices. Explants produced PLBs initially and subsequently, plantlets on Nitsch medium

(Devi *et al.,* 1997). Shoot tips of *Cymbidium forrestii* **cv Sojub,** *C. forresti* **cv. Nokwoon** and *C. Kanran* **cv. Soshim** were used to produce rhizomes. The rhizomes were cultured on different media, with varied concentrations of NAA and BA to induce multiple shoot formation (Jae Dong *et al.,* 1998).

Inflorescence

Gavino Rotor (Rotor,1949) first attempted inflorescence culture, while he was a graduate student at Cornell University. He used sections of flower stalks of *Phalaenopsis*, each containing a lateral bud. He surface sterilised them, using household bleach, (NaOCl) and inoculated them on KC agar medium. The swollen buds produced leaves after two weeks. Roots appeared after 2 or 3 leaves had been produced (Rotor, 1949). Tse *et al.,* (1971) recorded poor growth in *Phalaenopsis* as compared with *Cattleya* or *Cymbidium,* by using either of the following techniques (1) 2/3 of the bud were removed or (2) the bud was cut vertically or (3) the bud was punctured longitudinally from the apex to the base. Two weeks later, the intact buds developed into shoot primordia, whereas the injured buds showed callus growth. On further growth, each callus mass developed 3-4 plantlets. MS medium with NAA was more effective than KC medium. Inflorescence primorida of 1.5cm length from *Ascofinetia, Neostylis* and *Vascocystis* were inoculated on a liquid medium. In a few weeks, several proliferations were formed from hypodermal tissues of the axis. Plantlets were formed when the callus was transferred on to an agar medium (Intuwong and Sagawa, 1973). Inflorescences of *Dendrobium* and *Phalaenopsis* were cultured to obtain plantlets (Intuwong *et al.,* 1972; Intuwong and Sagawa, 1974). Tanaka and Sagawa (1978) and Fu (1979 a) achieved clonal propagation of *Phalaenopsis* by lateral buds of inflorescence. Inflorescence segments of *Cymbidium* X **Sazanami cv. Harunoumi** were cultured on MS medium with NAA and BA to produce plantlets (Kim and Kako, 1984). Flower stalk sections of *Phalaenopsis* were cultured on modified VWmedium with BA. While, PLBs formed within 30 days, adventitious buds were formed and grew into plantlets within 60 days of culture (Lin, 1986); 7mm long tips of primary inflorescences of two *Phalaenopsis* hybrids **No.1603** and **No. 2723** were cultured on VW medium with BA to produce plantlets (Yoneda, 1986). Flower stalk segments of *Phalaenopsis* and *Doritis* were cultured to produce plantlets (Tanaka, 1987). Vegetative buds were produced from lateral buds, and apices excised from flower stems and cultured on Kyoto medium with BA, to initiate PLB formation. Subculture of the PLBs on MS medium with BA to initiate PLB formation resulted in the production of plantlets. Plantlets of the white flowered clone *Phalaenopsis* **(Phyllis Key x Bandleader) x Celebration** have been produced on a large scale using this method (Wang, 1989). Floral buds of *Dendrobium crepidatum* and *D. pieradii* were cultured on Mitra medium. NAA and BA favoured shoot development from the explants (Vij *et al.,* 1991). Shoot tips from flower stalks of *Phalaenopsis* and *Deritaenopsis* were cultured on New Dogashima medium with NAA and BA. More than 10,000 PLBs were obtained from a few floral buds on a single flower stalk within one year. Subsequently, plantlets were produced from PLBs (Tokuhara nad Mii, 1993). Adventitious buds formed from the vegetative buds of the flower stalks of *Phalaenopsis* on VW medium with 15 percent CW and TDZ developed into shoots (Chen and Piluek, 1995). PLBs were initiated from inflorescence apices of *Aranda* **cv. Tay Swee Eng,** *Makara* **Khaw Phaik Suan** and *Dendrobium* **Sonia** on VW medium with BA and NAA. Young buds showed best yield of PLBs (Lam-chan and Lee, 1996). 0.8-1cm long segments from *Rynchostylis gigantea* inflorescence axes were also cultured. Segments from the proximal region, with well-differentiated buds failed to regenerate whereas the those from the distal region with more or less undifferentiated buds produced PLBs. As many as 10 plantlets with 2-3 leaves and 1-2 roots were harvested within 20 weeks (Vij *et al.,* 1997). Dormant buds from flower stalks of *Phalaenopsis* **True Lady 'B79-19'** were isolated and sterilised using 70% ethanol and NaOCl. Explants were cultured on half-strength solid MS medium to obtain adventitious shoots. Shoots were then sub-cultured on the same medium with different concentration of NAA and BAP for the growth of plantlets (Chen *et al.,* 1998).

Nodal Buds

Stem segments *Phalaenopsis* (3-5 cm long) with a node have been used as explants. Buds enlarged in two weeks followed by leaf and root development in 6-20 weeks (Sagawa, 1961; Scully, 1966). Axillary buds from pseudobulbs of *Cattleya,* first cultured in liquid medium and then transferred to agar medium after 2-5 weeks showed an increase in size in 4-10 weeks. The axillary buds were subdivided into smaller units and inoculated in a fresh medium, two months later, small plantlets formed (Scully, 1967). Nodal buds were cultured to obtain plantlets in *Vanda* **Miss Joaquim** within 2-3 months (Sagawa and Sehgal, 1967). Buds of back bulbs of *Cattleya* were suface-sterilised using clorox and cultured on KC medium with microelements of Nitsch. Explants turned green after two weeks of culture. PLBs formed from the explants and plantlets were produced eventually (Vajrabhaya, 1978). Axillary bud explants from shoots were surface-sterilised and cultured on Reinert-Mohr medium. One to many shoots developed from each bud. Even PLBs were formed, which later developed into plantlets (Kukulczanka and Wojciechowska, 1983). Scape nodes of *Oncidium* **cv. Gower Ramsey**, *Dendrobium* **cv. Miss Hawaii** and *Phalaenopsis* hybrids were cultured on VW and MS media. Shoot formation was higher in younger scapes (Nurani and Mohammed, 1992). Regeneration potential of uninodal explants of *Cymbidium pendulum* was examined. PLBs were formed within six months (Vij *et al.*, 1994). *In vitro* shoot proliferation was obtained from cultured nodal explants of *Vanilla planifolia*. When MS, Woody Plant Medium (WPM) and Gresshoff and Doy medium were tested, WPM induced maximum shoot length. However multiple shoots were obtained on the medium with BAP (Ganesh *et al.*, 1996).

Leaf

The first unambiguous and well documented report of the production of PLBs was made in cultures derived from *Cymbidium* shoot tips (Wimber, 1965). After the successful culture of palisade tissue of *Arachis hypogea* (Joshi and Ball, 1968a, b) similar attempts were made to culture leaf tips of orchids, which also turned out to be successful. *Epidendrum* and *Laeliocattleya* were used to obtain the explants (Churchill,1971,1973; Churchill *et al.*, 1970, 1971). Production and proliferation of PLBs on the base of thickened leaf sections of *Phalaenopsis* was demonstrated by Koch (1974a). When leaves of *Dendrobium* **Alice Spalding** of *in vitro* origin were cultured on modified VW medium. A single plantlet is formed at the base of each leaf (Fu, 1978). Leaf segments were succesfully cultured by Tanaka and Sakanishi (1985). Leaves from *Dendrobium crumenatum* were transferred to solid medium. Callus formed after 40 days in culture and PLBs appeared within three months (Manorama *et al.*, 1986). Leaf explants of *Doritis pulcherrima* were used for clonal propagation (Tanaka, 1987). An *in vitro* method was developed to regenerate plants from the basal parts of the leaves using Mitra et. al., medium (Seeni and Latha, 1992). The dermal peels from the leaves of *Rynchostylis retusa* and *Vanda coerulea* were inoculated on solid Mitra medium with various combinations of NAA, IAA, IBA and BAP. PLBs and plantlets were formed from the dermal tissues (Vij, 1994).

Internodes

Internodal segments have been rarely used as explants in orchid culture. Lack of meristematic tissue perhaps is one of the reasons for most of the workers ignoring this source. Alternatively, several advantages of using internodes such as relatively more amount of tissue for use as explant source, easier procedure for isolation, sterilization, etc., have been listed by a few authors who have used them as explants.

Internodal segments, measuring 5 mm in length were obtained from several portions on the inflorescence axis of *Phalaenopsis*. Explants were cultured on the medium with Thomales medium's macroelements and Ringe and Nitsch's microelements, 10% CW, NAA and BA. PLBs formed within 100 days of culture and plantlets were obtained successfully in about 380 days. About 400 PLBs were obtained from one stalk in a year (Homma and Asahira, 1985). Internodal segments of *Phalaenopsis* and *Doritaenopsis* were also cultured to obtain plantlets (Lin, 1986).

Root

The idea of orchid root tip culture originated originated from a theoretical article of Beechey (1970). *Epidendrium* roots cultured on a modified Ojima and Fujiwara medium, grew longer, but after two years they lost their chlorophyll (Churchill *et al.*, 1972). *Phalaenopsis* roots, which sometimes produce plantlets spontaneouly (Anonymous, 1887; Reichenbach, 1883), were cultured by Tanaka *et al.*, (1976), successfully. Later, roots of *Catasetum* were cultured to obtain plantlets (Kerbauy, 1984a). Roots of *Rynchostylis* were successfully cultured and plantlets were produced (Vij *et al.*, 1984; Sood and Vij, 1986). Regeneration of plants from the root tips of *Catasetum pileatum* was also successfully achieved (Kraus and Kerbauy, 1987). Root segments of *Rynchostylis retusa* were cultured on MS medium with NAA and kinetin or BA and plantlets were produced (Vij *et al.*, 1987). Sanchez (1988) succeeded in producing plants from root explants of *Cyrtopodium*. Root segments of *Vanda testacea* were cultured to obtain plantlets (Vij, 1994a). Plantlets of *Phalaenopsis* were produced *in vitro* from root tip cultures (Yoneda and Momose, 1988). *In vitro* transformation of root meristems to shoots and plantlets was done in *Vanilla planifolia* (Philip and Padikkala, 1989). Callus was induced from the root tips of *Bletilla striata, Cleisostoma fordii, Pholidota chinensis* on MS medium, with NAA and BA to obtain plantlets (Yam and Weatherhead, 1991). *In vitro* conservation of the endangered red orchid *Renanthera imschootiana* was achieved by utilising root tip explants (Sharma, 1994). Root segments of *Ascocenda, Aerides multiflorum, Cattleya* **Almakee**, *Cymbidium alorifolium, C. mastersis, C. pendulum, Dendrobium denudans, D.* **Emma White**, *Gastrochilus calceolaris, Phaius tankervilliae, Rhynchostylis retusa, Vanda bicolor*

V. coerulea, V. cristata, V. **Kasem**, *V. teres* and *V. testacea,* were washed and surface sterilised with NaOCl, and cultured on Mitra and MS medium with various combinations of NAA, IAA, IBA, 2,4-D, Kinetin and BAP. Out of the 21 orchids tested, 12 responded positively and produced plantlets (Vij, 1994) (Fig. 12.3).

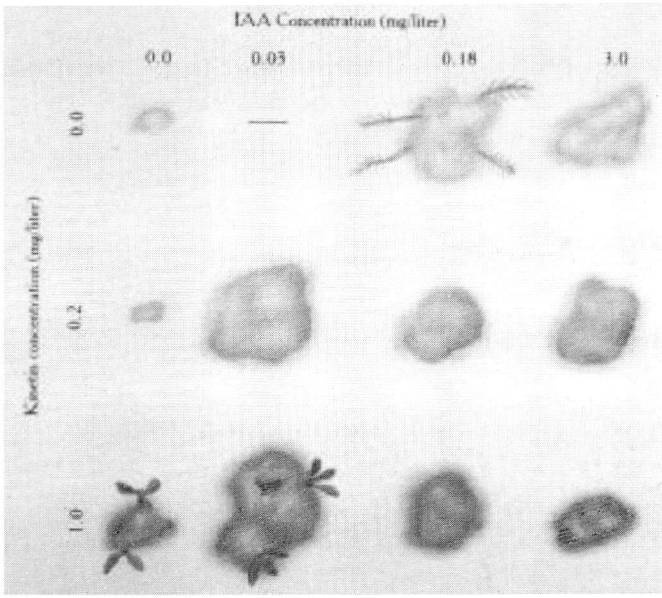

Fig.12.3 Effect of IAA: kinetin ratio on morphogenisis.

Rhizome, Tuber and Pseudobulbs

Rhizome tips of *Cymbidium goeringii* and *Cymbidium pumilum* were inoculated on agar medium, supplemented with kinetin. They gave rise to shoots (Ueda and Torikata, 1972). *Nervilia aragoana* was successfully propagated from tuber explants on MS medium with BA, NAA and activated charcoal (Chow, 1986). Rhizome explants of *Cymbidium Kanran* were inoculated on MS agar medium to obtain plantlets (Kim *et al.,* 1998). *Cymbidium forrestii* was propagated using rhizomes as explants. The rhizomes were cultured and proliferated on MS medium to produce plantlets (Paek and Yeung, 1991).

To study the effects of sucrose gradient and changes of free sugar content during organogenesis through rhizome, the rhizome explants were grown on the Hyponex medium. *Cymbidium ensifolium, C.* **Donghui**, *C. Kanran*

Tofa and *C. Kanran* **Jeyu** × *C. goeringii* were used in this investigation (Paek *et al.*, 1993). Rhizome explants of *C. niveomarginatum* were cultured on MS medium to study the effect of sucrose concentration on rhizome growth and shoot formation. Uninodal pseudobulb segments of *Bletilla striata* were cultured on four different media. Modified Ichihashi medium favoured the development of multiple shoots (Vij and Dhiman, 1997).

Cell and Protoplast

Haberlandt made the first attempt to culture mechanically isolated cells, as early as 1898 (Krikorian and Berquam, 1969), but he failed because the media used did not contain plant growth regulators. Half a century later, La Rue (1953) attempted to obtain cell colonies from pollen grains. A year later, Muir *et al.* (1954) failed in an attempt to produce callus. Muir placed single cells of *Nicotiana tabaccum* and *Tegetes erecta* on filter papers that were in contact with callus but without any success. The first convincing report that single cells can divide, was published by Vasil and Hilderbrandt (1965a, b). The first report of orchid cell suspension culture was on *Cymbidium* (Steward and Mapes, 1971); Production and fusion of orchid protoplasts were first reported by the Malaysian orchid scientist, Teo, and German botanist, Neumann (Teo,1978 b; Teo and Neumann,1977, 1978a, b). Subsequently, orchid protoplasts were isolated by several others (Seeni and Abraham, 1986; Hew and Yip, 1986; Gopalakrishnan and Seeni, 1987 Koh *et al.*, 1988; Yasugi, 1990). The American Orchid society with the funds provided by British orchid grower, Eric Young Fiske, (Fiske,1982) has declared a prize for the first successfull report of a somatic orchid hybrid through protoplast fusion. No scientist was able to succeed in this challenging endeavour until 2001 (Kanchanpoom *et al.*, 2001).

To improve the efficiency of regeneration without sacrificing the Genetic stability, thin section culture system developed in *Aranda, Cattleya, Dendrobium and* other orchids (Prakash *et al.*, 1996).

SOMACLONAL VARIATIONS

Stability of plant genomes under *in vitro* conditions is an important factor, determining the successful applications of any culture system, either for micropropagation, selection of mutant types, or for genetic transformation. In some plant systems, lack of such stability might result in increased variability, which is commonly described as "somaclonal variation" (Mythili *et al.*, 1995). There are very few reports on the utilisation of somaclonal variations in orchids (Devi and Deka, 2001). This has been extensively reviewed (Larkin and Scowcroft, 1981; Larkin *et al.,* 1985). Interest in measuring the genetic variations derived from tissue culture regenerates has been reported in many plant species (Bajaj, 1990). A few investigations have been conducted to verify the phenotypic and genetic variations in somaclones of *Phalaenopsis* (Tokuhara, 1992). A 'mericlone mutant' *Brassolaeliocattleya* **Greenwich 'Irish fair'** was obtained by Green (1983). The morphological and genetic variations were studied in the somaclones of *Phalanopsis* **True Lady 'B79-19'** which were derived from tissue culture. Laneri (1990, cf. Devi and Deki, 2001) observed two somaclones in *Cymbidium* which produced variant flowers. The random amplified polymorphic DNA (RAPD) data indicated that normal and variant somaclones were not genetically identical (Chen *et al.,* 1998).

Plate 1. *Dendrobium* **Queen Sonia**. A single flower.

Plate 2. Differentiation of PLBs and plantlets from callus segments of *Dendrobium* **Queen Sonia** on VW medium with 0.5 mgl^{-1} NAA and 1 mgl^{-1} BAP after 10 weeks in culture (see Table 20).

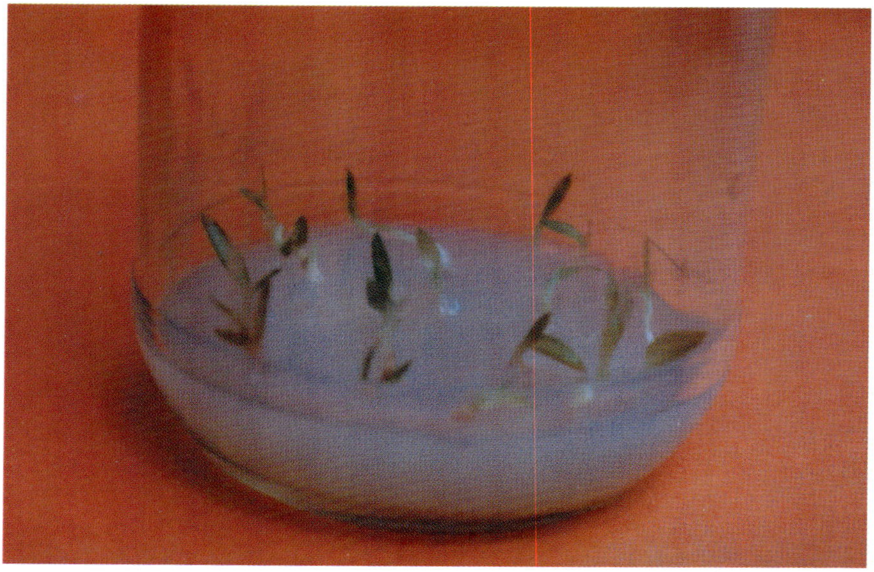

Plate 3. Sub-Culturing of Plantlets of *Dendrobium* **Queen Sonia** on VW medium.

Fig 4. *Dendrobium* **Queen Sonia** plantlets on KC medium after 12 weeks of culture showing slow growth (see Table 41).

Plate 5. Plantlets *of Dendrobium* **Queen Sonia** on VW medium with 2.5 mgl^{-1} kinetin and 5 mgl^{-1} NAA after 10 weeks of culture (see Table 52).

Plate 6. Plantlets of *Dendrobium* **Queen Sonia** on MS medium with 5 mgl^{-1} NAA and without BAP, showing unusually long leaves in 12 weeks culture (see Table 39).

Plate 7. *Dendrobium* **Queen Sonia** plantlets showing slow stunted growth and no multiplication on VW medium with 2 mgl^{-1} NAA and without BAP in 12 weeks culture (see Table 60).

Plate 8. Plantlets of *Dendrobium* **Queen Sonia** on VW medium with 4 mgl^{-1} 2,4-D and 1 mgl^{-1} BAP in 10 weeks of culture showing unusual narrow leaves (see Table 60).

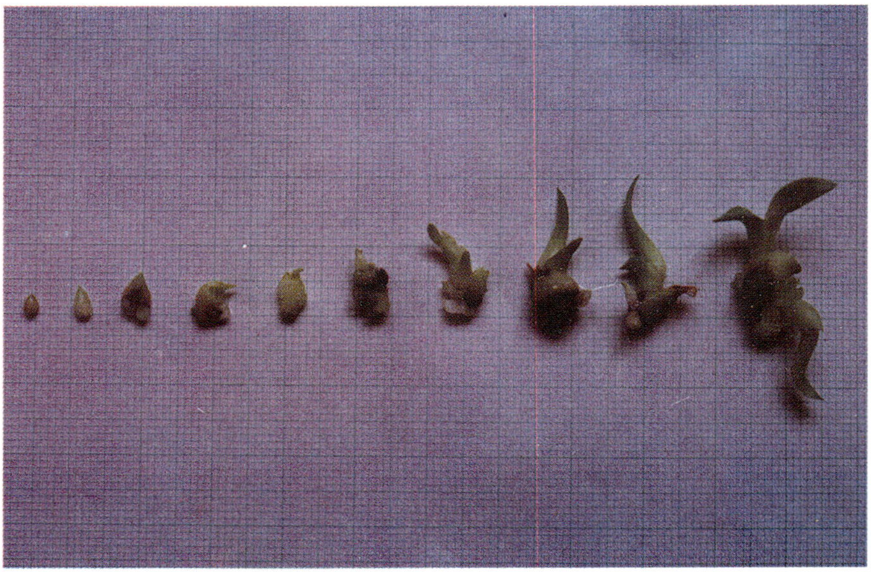

Plate 9. *Dendrobium* **Queen Sonia** plantlets showing different stage of growth.

Plate 10. *Dendrobium* **Queen Sonia** plantlets showing high degree of multiplication ration in VW medium with 1 mgl^{-1} NAA and 4mgl^{-1} BAP (see Table 60).

Plate 11. *Dendrobium* **Emma White;** A potted flowering plant.

Plate 12. *Dendrobium* **Emma White**—an young infloresence.

Plate 13. *Dendrobium* **Emma White** showing multiplication on MS medium (see Table 51).

Plate 14. *Dendrobium* **Emma White** plants on KC medium not showing multiplication (see Table 51).

Plate 15. *Dendrobium* **Emma White** plants on VW medium showing multiplication and Uniform growth (see Table 51).

Plate 16. Root exaplants of *Dendrobium* **Emma White** on MS medium with 5mgl^{-1} NAA and 2 mgl^{-1} BAP after 16 weeks of culture (not shown in Tables).

Plate 17. Plantlets *of Dendrobium* **Emma White** on VW medium with 10 mgl^{-1} NAA and 2 mgl^{-1}BAP showing unusual growth of roots. After 16 weeks of culture (not shown in Tables).

Fig 18. *Dendrobium* **Emma White** plants on VW medium with 1 mgl^{-1} BAP and 10 mgl^{-1} NAA (see Table 57).

Plate 19. *Dendrobium* **Emma White,** showing no multiplication on VW medium with 1mgl^{-1} NAA and without BAP (see Table 61).

Plate 20. *Cattleya* **Naomi Kerns**—A flowering plant.

Plate 21. *Cattleya* **Naomi Kerns** plantlets showing high degree of multiplication on MS medium with 2 mgl^{-1} BAP and 1 mgl^{-1} NAA (see Table 62).

Plate 22. Plantlets of *Cattleya* **Naomi Kerns** on MS medium with 1 mgl^{-1} BAP and 4 mgl^{-1} 2,4-D, showing abnormal growth (see Table 58).

Plate 23. *Cattleya* **Naomi Kerns** plantlets on MS medium with 150 ml 1^{-1} CW showing healthy growth with normal rooting (see Table 45).

Plate 24. Plantlets of *Cattleya* **Naomi Kerns** on MS medium with 4 mgl^{-1} NAA and 1 mgl^{-1} BAP (see Table 58).

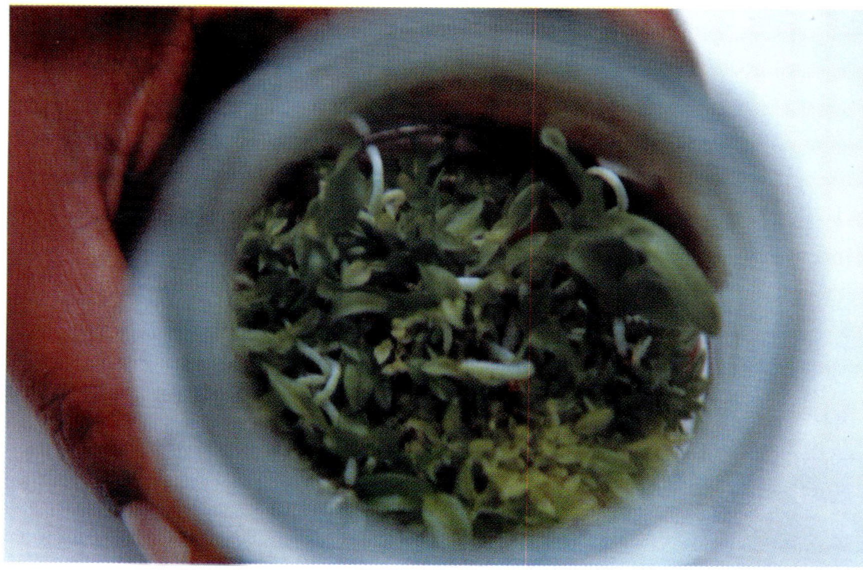

Plate 25. Plantlets of *Cattleya* **Naomi Kerns** on MS medium with 4 mgl^{-1} NAA and 1 mgl^{-1} BAP.

Plate 26. Ex-agar *Cattleya* **Naomi Kerns** plantlets in community pots.

Plate 27. Flowers of *Phalaenopsis* **Queen Emma.**

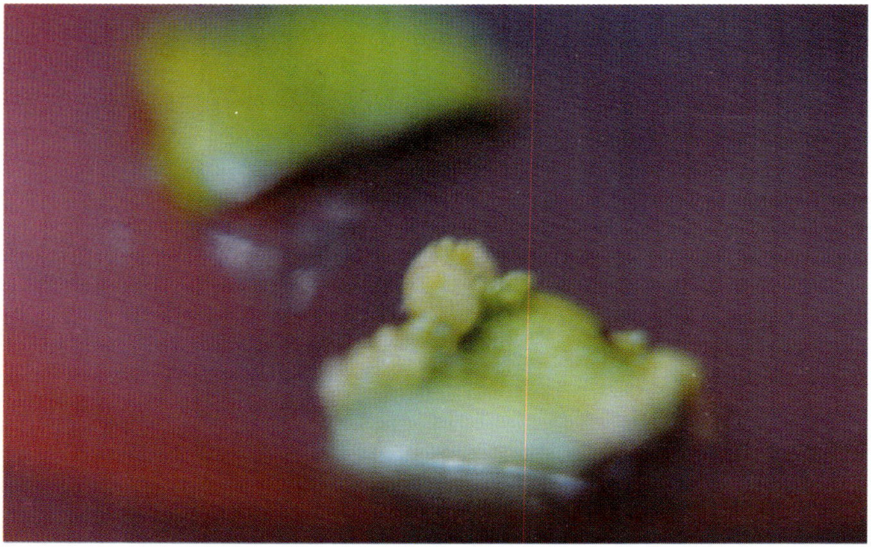

Plate 28. Leaf explants *Phalaenopsis* **Queen Emma** on MS medium with 1 mgl^{-1} BAP and 5 mgl^{-1} NAA (see Table 28).

Plate 29. Leaf explants of *Phalaenopsis* **Queen Emma** on VW medium with 1 mgl^{-1} BAP and 4 mgl^{-1} NAA (see Table 29).

Fig 30. Leaf explants of *Phalaenopsis* **Queen Emma** on VW medium with 5 mgl^{-1} NAA and 1 mgl^{-1} BAP (see table 29).

Fig 31. Leaf explants of *Phalaenopsis* **Queen Emma** on KC medium with 5 mgl^{-1} NAA and 1 mgl^{-1} BAP (see Table 30).

Plate 32. Differentiation of *Phalaenopsis* **Queen Emma** plantelets directly from leaf explants on MS medium with 1 mgl^{-1} BAP and 5 mgl^{-1} NAA. After 16 weeks of inoculation (not shown in tables).

Plate 33. Differentiation of *Phalaenopsis* **Queen Emma** plantlets directly from nodal explant. bud explant on VW medium with 5 mgl^{-1} NAA and 1 mgl^{-1} BAP (see Table 32).

Plate 34. Differentiation of PLBs and plantlets from leaf explants *of Phalaenopsis* **Queen Emma** on MS medium with 5 mgl^{-1} NAA and 1 mgl^{-1} BAP (see Table 28).

Plate 35. Subculturing of *Phalaenopsis* **Queen Emma** plantlets VW medium with 7 mgl^{-1} NAA and 1 mgl^{-1} BAP (see Table 49).

Plate 36. Multiplication of *Phalaenopsis* **Queen Emma** plantlets on MS medium with 8 mgl^{-1} BAP and 2 mgl^{-1} NAA (see Table 63).

13

PROTOCOLS FOR THE *IN VITRO* PROPAGATION OF SOME ORCHID HYBRIDS

OVERVIEW

The protocols were developed by the author by conducting research work at the Post Graduate Department of Botany, St. Joseph's College, Bangalore, India and the Orchid Laboratory, Indian Institute of Horticultural Research (IIHR), Hessaraghatta, Bangalore. Plants collected for the experiments, as well as the plants produced during experimentation, were maintained in the greenhouse attached to the Orchid Laboratory at IIHR. Standard procedures were used for maintaining the experimental plants.

PLANT MATERIAL

For this study, a few commercially important hybrids were purchased from different sources. Healthy flowering plants were selected. Initially, the following cultivars(cv) were brought for selecting a suitable plant material.

After screening several hybrids, the following hybrids were selected. Two from the genus *Dendrobium* and one hybrid each from *Cattleya* and *Phalaenopsis* were selected. The details are as follows:

1. *Dendrobium* **Queen Sonia**
2. *Dendrobium* **Emma white**
3. *Cattleya* **Naomi Kerns** and
4. *Phalaenopsis* **Queen Emma**

Dendrobium *Schwartz*

The genus is named after the Greek word for 'tree' and 'life' (Black, 1974). It comprises 1000-1400 species and is distributed in the Himalayas,

Thailand, Burma, Malaysia, Indonesia, New Guinea extending to Australia. About 80 species are reported in general cultivation, together with several hybrids (Cullen,1992).

Dendrobiums are epiphytic sympodial herbs, stems are nodose or of pseudobulbs, leaves are sessile and are never plicate. Leaves are variable with or without sheathing leaf bases. Flowers are in recemes borne on leafy or leafless stems. Upper sepals are free, lateral ones are attached to the elongate column foot, forming a conspicuous, variously shaped abellum which is sometimes spur-like. Petals are smaller than sepals. Lip is entire or more frequently three lobed, attached to the apex and variable in shape. Column is short with an extended foot (Cullen,1992).

There are several interspecific and intergeneric hybrids. *Dendrobium* hybrids play a very important role in the tropical cut-flower trade. White, biocolour (white and pink combination) and purple types are in much demand. The species *Dendrobium phalaenopsis* has been responsible for the development of a good number of hybrids for the cut-flower market, especially in Thailand and Singapore. Based on the involvement of *D. phalaenopsis* parentage, these hybrids are classified into phalaenopsis, semi-phalaenopsis and intermediate types.

Dendrobium plants develop long sturdy canes which may bear one or two inflorescence spikes and about 18 - 20 flowers, compactly arranged on each flower stalk. *Dendrobium* **Queen Sonia** is a Phalaenopsis type hybrid. The flowers are bicoloured (pink and white) and measure about 2½ inches across. *Dendrobium* **Emma White** is a semi-phalaenopsis type hybrid. The flowers are white in colour and measure about 2¼ inches across.

Cattleya *Lindley*

The genus, named after the famous botanist, William Cattley (Black, 1977). It includes 40 species from Central and South America, and is often used in bi- and tri-generic hybrids (Cullen,1992). They are distributed from Mexico to tropical South America and West Indies.

Cattleya is epiphytic or terrestrial with prominent pseudobulbs.

The rhizomes are creeping, bearing erect, swollen or cane-like stems with several nodes. Leaves 1-4 are borne on top of the stems. The flowers are solitary or in racemes. Sepals and petals are free. Lip is 3-lobed or entire. Lateral lobes or basal part are rolled around the column. Column is free from lip, slightly flattened, often winged (Cullen,1992).

Cattleya **Naomi Kerns** is a beautiful hybrid, producing large fascinating flowers which are used for making corsages. The flowers are bicoloured with red and orange combinations. Petals and sepals are with fringed edges. The flowers are produced in pairs on a shorter stem, and measure about 3-4 inches in diameter.

Phalaenopsis *Blume*

Phalaenopsis literally means reminiscent of butterfly (Teuscher, 1977). It is a genus including 40-55 species, occurring in the Himalayas, South-East Asia and North Australia. Their classification is based on the complex lip structure (Cullen, 1992).

It is a monopodial epiphyte without pseudobulbs. The leaves form a basal rosette, dorsally keeled. Inflorescence is an unbranched receme. Floral bracts are concave, oppressed and green. Sepals are almost equal and free, usually spreading. Petals are similar or larger than sepals. Lip is relatively small and often more brightly coloured than the sepals and petals. It is 3-lobed and continuous with the column foot or inserted at a right angle. Lateral lobes are erect with cushion-like swellings in the middle; central lobe is fleshy, usually firmly attached to the lateral lobes. The column is often constricted in the middle, usually wingless, and produced into a short foot at the base (Cullen, 1992).

Phalaenopsis **Queen Emma** is a compact prolifically flowering pot plant. The flowering stem is 15-20 inches long and bears 25-30 flowers. Flowers are pinkish white and splashed with bluish pink dotted stripes. The lip is pink. The flowers measure about 3 inches across and stay for about 60-90 days on the plant.

EXPLANTS USED

After the first attempt of Haberlandt (1902) to culture isolated leaf palisade and mesophyll cells, many workers tried to culture parts of different plants, including orchids. Over the years, almost all living parts of orchid plant body have been cultured, but certain explants are used extensively. The selection of the source of the explant is primarily based on the choice of the researcher. The other basis for the selection of the explant is its ability to respond to the nutrient media according to the designs of the experiments. Of late, the focus has shifted towards the commercialisation and thus, explants of all types have become useful, but with an emphasis on the retention of the mother plant as the main criterion. This is very significant in view of the fact that many orchids are becoming endangered due to the very nature of their habit like being monopodial or environmental threats.

The first successful indefinite culture of root tips was achieved by White (1934a) in tomato. Orchid roots were cultured first by Tanaka *et al.* (1976). [See also reviews by Churchill (1973) and Vij (1994). Orchid roots are a very economical source of explants as damage to the plant is minimal and not lethal. In the present study, roots of all the orchids studied have been successfully utilised for micropropagation. Thus, orchid root culture holds enormous prospects, particularly for commercial exploitation of the ornamental ones.

Some of the early attempts to culture plants were made with leaf explants. These attempts failed initially, but experiments with mature leaves in it differentiated palisade parenchyma of some plants were successful (Joshi and Ball, 1968). In orchids, the tendency of juvenile leaves or protocorms to produce PLBs led to the development of micropropagation methods for the culture of leaf bases (Churchill *et al.*, 1970).

The first unambiguous and well-documented report that leaves can produce PLBs was made in cultures derived from *Cymbidium* shoot tips (Wimber, 1965). Since then, several orchids have been cultured using leaf explants. Formation of PLBs from cultured leaf segments has been reported in *Laeliocattleya* (Ball *et al.*, 1971), *Vanda, Aranda* (Fu, 1979) *Ranantanda Rhynchostylis* (Vij *et al.*, 1997) and *Phalaenopsis* (Tanaka, 1990). Like roots, even leaves are a potential source of explants. All orchids studied in the current project have been cultured using different parts of leaves.

Shoot tips are commonly used as explants in the clonal propagation of orchids. They are either inoculated directly or they are utilised to isolate the meristems. Mayer (1956) reported the growth and organogenesis of *Pelargonium zonale* and *Cyclamen persicum,* using shoot explants. Using Mayer's method, Thomale (1957) cultured buds containing explants from tubers of *Orchis maculata* and obtained shoots and roots. This work of Thomale was published 3 years before the first report on *Cymbidium* through the culture of the generally virus free meristems. Morel did not suggest that this method could be used for mass propagation, although he stated that "Very often the PLB divided into clumps of four to five identical structures, each of them producing a new plant". The production of virus free *Cymbidiums* was an important advancement in itself. It also led to the clonal propagation of orchids through tissue cultures in general and shoot tips in particular. The first clearly described and reproducible method for mass rapid clonal propagation of orchids was published by Wimber (1963). Later, Morel extended this technique to other orchids, *Odontoglossum, Miltonia,* and *Phaius* (Morel, 1963).

With regard to shoot tip culture of other plant, Robbins (1922) attempted to culture cotton and pea stem tips, but obtained only limited growth. White (1933) was also unsuccessful in his attempts to culture bud meristems of *Stellaria media*. The first successful culture of shoot tips *in vitro* was obtained by Loo (1945) in *Asparagus* (Loo, 1945).

In the present study, shoot tips and nodal buds of all the orchids are utilised as the source of explants. But it was uneconomical except in the case of the monopodial orchid, *Phalaenopsis*. Hence, inflorescence axis was used instead of shoot tips, and buds were used both directly and as a source for meristems.

Ovaries were the first orchid flower segments to be cultured (Ito, 1960). Since then, immature seeds of many orchids have been cultured. In some orchids, this is the preferred method of propagation since it saves time and facilitates the germination of several species (for reviews, see Withner, 1959; Rao, 1977; Czerevczenko and Kushnir, 1986 and Sagarik, 1994). Rotor (1949) cultured flower stalks of *Phalaenopsis* on KC medium. This was the first attempt to asexually propagate an orchid, probably any plant. But credit has been assigned to others for this invention. Rotor's name is usually not even mentioned, for unknown reasons; perhaps due to a desire not to rock the historical boat, strong convictions, vested interests, and may be even national pride (Gautheret, 1983). Later, several others utilised inflorescence axes for clonal propagation. Tse *et al.* (1971) have cultured *Ascofinetia, Neostylis* and *Vascostylis*. Similarly, inflorescence explants of *Dendrobium* and *Phalaenopsis* were used for clonal propagation by Intuwong *et al.* (1972) and recently, *Phalaenopsis* **True Lady "B79 - 19"** was cultured using inflorescence (Chen *et al.*, 1998). Inflorescence explants of all the four orchids have been utilized in the present study. The response was good in all the cases.

Stem sections without meristem have rarely been used in orchid tissue culture. The culture of stem sections was first reported at the First World Orchid Conference at California in 1966 (Bertsch, 1966). The same orchid was cultured by Mitra (1971) using stem discs. Internodal segments of *Phalaenopsis* were successfully cultured to obtain plantlets (Homma and

Asahira, 1985). Similarly *Phalaenopsis* and *Doritaenopsis* were cultured using internodes as the source of explants (Lin, 1986).

Internodes have not been exploited as the source of explants as they do not contain meristems. But if the techniques are developed to elicit response from these explants they will turn out to be one of the most economical source of explants. In the present study internodal segments of all the orchids were used successfully and plantlets were produced. But the yield of plantlets from internodes is less as compared to shoot tips, nodal buds, and leaf explants.

ORGANISATION OF THE CULTURE PROTOCOL

Generally, micropropogation involves four stages and by introducing a '0' (zero) stage, the protocol has been made a five-stage process (Debergh and Maene, 1981, cf. Bhojwani and Razdan, 1993). To make the study more comprehensive, the present study is discussed under a 6 stage protocol, allowing more time for analysis and documentation of the morphogenetic development of orchids studied *in vitro*. The different stages of the protocol viz., (1) an *in vivo* (stage '0') preparatory stage (2) initiation of culture (3) multiplication of callus (4) regeneration of plantlets (5) multiplication of plants, and 6) growth of plants. But generally micropropagation involves only three *in vitro* stages and an *in vivo* preparatory stage, described briefly, below.

Stage '0' is a preparative stage, involving the preparation of plants to provide quality explants for better establishment of aseptic cultures in stage 1.

Stage 1 is an initiation of the cultures stage. This stage is mainly concerned with the production of either callus or plantlets obtained directly from explants, excision of parts to be inoculated, sterilisation and monitoring cultures till plantlet formation. Depending on the type of explant used and the purpose of study, this stage needs to be further divided. For example, in orchids, there is an intermediate PLBs formation stage between callus formation and plantlet formation stages. There is also a need for subdivision of this stage so that the plantlets may either follow direct or indirect organogenesis.

Stage 2 is the crucial multiplication stage. It is at this point where most of the failures in micropropagation occur (Bhojwani and Razdan, 1993), because it is in this stage, that the morphogensis of organs and structures that are the precursors of new intact plants occurs. Shoots produced in this stage can also be called propagules as they can be subcultured to increase their number. Thus, the addition of one more stage in the protocol may be necessary in view of the commercial production of a large number of *in vitro* products.

In Stage 3, somatic embryos may develop directly into plantlets as in most other plants, but in orchids PLBs are formed instead of somatic embryos. The plantlets derived in stage 2 are very small and not capable of self-supporting growth. Hence, they have to be induced to produce roots *in vitro* (George and Sherringtone 1984), and they require an additional step of maturation to acquire the capability to grow independently in the outside environment.

Stage 4, is the stage of transplantation. The ultimate success of commercial micropropagation depends on the ability to transfer plantlets out of culture vessels, with a high rate of survival. This however is not an *in vitro* stage. But different researchers have employed protocols of their own choice, which is discussed below:

Over the years, in the micropropagation of orchids researchers have used different number of stages. In leaf segment culture of *Phalaenopsis ambalis,* only one *in vitro* stage was employed. The explants were sterilised and inoculated on a MS medium with auxins and cytokinins. But, out of 20 explants inoculated, only five showed PLB formation (Tanaka and Sakanishi, 1976). Shimasaki and Uemota (1991) employed two stages in the micropropagation of *Cymbidium goeringii*. Flower stalk sections were initially inoculated on MS medium with NAA and BA to obtain rhizome mass, which was subcultured under different conditions. Niimi *et al.* (1995) used MS medium for obtaining callus from *Neofinetia falcata* and different revised Hyponex media for organogenesis from the callus. The purpose here was to select the best media for organogenesis. Park *et al.* (1996) employed only one stage for the propagation of PLBs from the flower stalk explants of *Phalaenopsis*. They employed VW liquid medium and obtained 80% multiplication of PLBs. For obtaining the plantlets from leaf explants of *Phalaenopsis hybrid* and for studying somaclonal variations, five stage micropropagation was used by Chen *et al.* (1998). They cultured young leaf explants of *in vitro* origin on modified VW medium. After ten days of culture, they were transferred to another modification of the same medium. The derivatives were then subcultured on modified KC medium for the induction of PLBs. Then the PLBs were cut into half once a week and cultured on the same medium for further development. Plantlets finally were obtained by further subculturing in L8 medium. To make histological observations in the PLBs the outer tissue of PLBs derived from *Cymbidium* hybrids were cultured on MS media, which is only a one stage process. (Ishii *et al.* 1998). A complete study of the orchid, *Dendrobium* including micropropagation and hardening, was carried out by Lim *et al.* (1993).

The above examples clearly indicate that the number of *in vitro* steps vary based on the purpose of the study. To carry out a comprehensive morphogenetic study of orchids, it was considered necessary to include two more stages as mentioned above to the conventional 3 stages, thus making it *in vitro* five stage study. The five stages are as follows:

(1) Induction of callus or PLBs

(2) Multiplication of callus and PLBs

(3) Regeneration of plantlets

(4) Multiplication of plantlets

(5) Growth of plantlets

In these five stages the stages of multiplication of callus and PLBs and multiplication of plantlets are the additional stages. Multiplication of callus and PLBs are always necessary as the amount of callus or number of PLBs formed initially will generally be less. If the same source is used directly in the next stage, the yield of plants would be considerably low. So, in order to increase the number of plants to be regenerated, this stage was introduced. Similarly, the stage of multiplication of plantlets was also added to increase the number of plantlets produced.

EXPLANT SELECTION, ISOLATION AND STERILISATION

Initially, all explants used were obtained from plants brought from nurseries and grown in greenhouses. Explants such as stem tips, inflorescence axes, root tips, nodal buds, internodal segments, leaves, etc., were isolated from healthy plants. Subsequently, plants were produced *in vitro* and they were used as the source of explants for experimental purposes.

Generally, meristems are located at the tips of the shoots and buds, where they are protected by sheathing petioles, leaves or scales. The shoots tips and buds were used directly as explants and were also used to obtain the meristems for culture. Shoot tip and meristem cultures were found to be economical in *Cattleya* and *Dendrobium* hybrids, but in case of *Phalaenopsis,* hybrid excision of shoot tips meant destruction of a complete plant, since it is a monopodial orchid. Hence, they were obtained from inflorescence axes and young scape nodes, when explants were obtained from greenhouse grown plants. However, shoot tips and axillary buds were used from the plants produced *in vitro*.

ISOLATION AND STERILIZATION OF SHOOT TIPS AND MERISTEMS

Shoot tips and inflorescence tips were isolated from *in vivo* plants, washed with 1% (v/v) Teepol detergent (Glaxo India Pvt. Ltd., Bombay) and scrubbed gently with a soft brush under running water, and finally kept in autoclaved distilled water. Later, they were transferred into an inoculation chamber to bottles containing 0.1% (v/v) $HgCl_2$, where they were kept for 5 minutes and cleaned five times with autoclaved distilled water. (for shoot tip and bud cultures), while shoot tips and inflorescence tips were, directly used, meristems were dissected in the inoculation room under laminar airflow by carefully removing the leaves and using a sharp edged forceps without damaging the young internal tissues. Only a small terminal region was isolated and mounted on a piece of autoclaved cork and the meristem was isolated by microexcision under a dissection microscope.

ISOLATION AND STERILIZATION OF NODAL AND AXILLARY BUDS

Buds are located at the nodal regions of the stems as well as in the inflorescence stalks in the axil of leaves. Generally, buds are covered by scales or leaves. They are clearly visible and appear as raised domes. Buds are usually 0.25 to 0.5cm in length.

Buds were isolated along with the adjacent regions of the stem and inflorescence stalks. Newly grown young stems of new growth and inflorescence stalks on which a few flowers had already bloomed were ideal for obtaining buds. Inflorescence stalks containing a few dormant buds at the basal regions were also used. The stem was cut on either sides of the buds, using sharp blade with a sliding action to avoid peeling off of the tissue. The nodal buds were sterilised like the shoot apices.

A perusal of literature indicates that orchids belonging to more than 90 genera have been successfully cultured *in vitro*. Researchers have used various explants for the purpose of micropropagation. Different media with a variety of combinations of growth regulators have been used. In the present study, an attempt is made to investigate some commercially important orchids. Almost all sources of explants, except cells and protoplasts have been used and cultured on commonly used media.

Experiments were planned economically by using low cost bottles as culture vessels and utilizing explants with an emphasis to protect the main plant specimens, used as source materials. The effects of different synthetic auxins and cytokinins were studied, as most of the orchids require auxins and/or cytokinins for *in vitro* growth and development (Arditti and Ernst, 1993), the combinations, concentrations and the ratios between them being critically important (Teng *et al.,* 1997).

Presently, micropropagation of monopodial orchids is considered difficult (Fu, 1978; Vajrabhaya and Vajrabhaya, 1970) and there has been no report on the induction of embryogenic callus in the monopodial orchid *Phalaenopsis* (Ishii *et al.,* 1998). But in the present study, *Phalaenopsis* hybrid has been micropropagated successfully using explants from different sources, including leaf segments. However, there are reports indicating that callus induction from leaf segments is difficult (Ishii *et al.,* 1998).

ISOLATION AND STERILIZATION OF LEAF SEGMENTS

All regions of the leaves were used to obtain explants. 2-4 week old young leaves were better, as compared to old leaves. Leaves of *in vitro* origin were segmented into leaf bases, leaf tips and laminar regions, inoculated directly by segregating them from one another, and cultured separately.

Explants from plants grown in greenhouses were cleaned with water and washed with 1% (v/v) Teepol, scrubbed gently with a soft brush under running water, finally, placed in autoclaved distilled water, and taken into an inoculation chamber. They were surface sterilized using 0.1% (v/v) $HgCl_2$ for 5 minutes, followed by washing five times with autoclaved distilled water. Margins of leaf segments exposed to $HgCl_2$ were removed, the remaining part of the leaf was cut into small pieces measuring 0.25 to 0.5 cm in diameter, and then inoculated.

ISOLATION AND STERILIZATION OF ROOT SEGMENTS

Roots were collected from the greenhouse grown plants as well as from the plants grown *in vitro*. However, only the terminal 1 cm of the root tips were used. Both aerial roots and the normal roots, grown in the potting media, were used. *In vitro* roots were segmented and inoculated without sterilisation, whereas the roots of *vivo* origin had to be sterilised.

Aerial roots were first cleaned with water and washed with 1% (v/v) Teepol. Explants were then washed under running water tap 4-5 times. Finally, they were kept in autoclaved distilled water and taken inside the inoculation chamber. Normal roots, which had been in direct contact with the potting medium, were cleaned with water, washed with 1% (v/v) detergent (Teepol), scrubbed with a soft brush under tap water and soaked in 5% (v/v) fungicide (Bovestein) for 30 minutes. After 30 minutes, the explants were cleaned with water 4 to 5 times, kept in autoclaved distilled water and taken into the inoculation chamber. Sterilisation of root explants with $HgCl_2$ was carried out in the same way as for shoot tips (described earlier).

EXPLANT STERILISATION METHODS

Explants of *in vitro* origin were found to be better for micropropagation, as compared to those obtained from greenhouse grown plants. They were also easy for sterilisation, as the chance of contamination was almost nil. Chow *et al.* (1982) cultured lateral bud explants of *in vitro* origin and found that there was no need for surface sterilization. They obtained seedlings by germinating the seeds of *Anoectochilus roxburghii* and used these seedlings as the source of lateral buds. Similarly, leaf tip explants of *Aranda sp.* were obtained from axenically grown plants and cultured on VW medium without sterilization, for the purpose of micropropagation (Loh *et al.*, 1975). Similarly, pseudobulb segments from 40 week old seedlings of *Dendrobium crysanthum*, grown *in vitro*, were used as explants without sterilisation and cultured to obtain plantlets (Vij and Pathak, 1989).

Explants of *in vivo* origin, however, require sterilization as indicated by the works of others. The process varies with the kind of explant used in the culture and the choice of the researcher. Roots and older portions of the plants require elaborate sterilisation procedures, as compared to younger portions because of their relatively extra exposure to nature or the environment.

The surface sterilization of plant material may be accomplished with an aqueous solution of either sodium hypochlorite(NaOCl) or calcium hypochlorite (CaOCl). Most workers use a common household bleach such as clorox, which contains 5% active NaOCl (Dodds and Roberts, 1995). Sodium ions, in NaOCl can induce abnormal development in some seedlings. Moreover, hypochlorite can cause severe bronchial irritation, and skin contact can be harmful (Windholz, 1983).

Ethanol is used for sterilisation of plant material because rectified spirit (90% alcohol) dehydrates too much. Mercuric chloride ($HgCl_2$) is also rarely used for this purpose. But $Hgcl_2$ is used in a concentration of 0.01-05% (v/v) for 2-12 minutes. Rinsing must be thorough, as it is toxic

(Peirik, 1987). Although antibiotics are used routinely in animal cell cultures, they have not been widely used in plant tissue cultures (Eichholtz *et al.,* 1982). Falkiner (1990) has reviewed the criteria for the selection of appropriate antibiotics for the control of bacteria in plant tissue culture. Cheah and Sagawa (1978) removed all leaf portions, except the sheaths, from the upper nodes to expose the axillary buds of *Aranthera* **James Storei** and immersed them in 10% clorox for 10 minutes. The sheaths were removed and then the sections were immersed in 5% clorox for 5 minutes. Buds excised from such sections were used for culture.

The goal in surface sterilisation is to remove all the microorganisms with minimum damage to the plant tissues. Some explants pose a problem especially if they contain microorganisms within the tissue sample. Surface treatments will obviously be completely ineffective in such cases. In some cases the achievement of this goal is empirical and the worker must be selective in his approach to the problem (de Fossord, 1976). In the present study, such plant parts were discarded to avoid the problem of internal pathogens.

In the meristem culture in *Cattleya,* all the dead scales were removed from the axillary shoots and their surface was swabbed with 95% ethyl alcohol, rinsed with distilled water, soaked for 20 to 30 minutes in 0.4 to 0.5% CaOCl, and finally rinsed in a large volume of sterile distilled water, before being used in the culture.

Shoot tips from buds on pseudobulbs of *Cymbidium* were dipped in 75% ethyl alcohol, placed in CaOCl for 20 minutes, rinsed with sterile distilled water and cultured for the production of PLBs and plantlets (Morel and Champaganath, 1966a). For the purpose of micropropagation of *Cymbidium ensifolium,* shoot meristems and axillary buds were scrubbed with a soft brush and sprayed with 70% ethanol. Then the explants were dipped in 70% ethanol for 20 seconds and immersed in NaOCl for 30 minutes. Finally, they were rinsed six times with sterile distilled water before being inoculated (Wang *et al.,* 1981). Shoot tips of *Dendrobium sp.* were first wiped several times with a cheese cloth, then soaked in 95% ethanol and finally immersed in 10% clorox for 15 minutes. Following removal of the bracts, the stems were placed in 10% clorox for 10 minutes and then dipped in sterile distilled water for 3 minutes before inoculation.

Elimination of contaminants from contaminated cultures of *Dendrobium sp.* was achieved by dipping callus segments, PLBs and plantlets, in 70% alcohol for 1 to 2 seconds, and then rinsing three times with sterile distilled water (Soediono, 1983a). For the purpose of clonal propagation of *Mormodes histrio,* using sections of pseudobulbs, the explants were decontaminated with NaOCl and active chlorine. The best results were obtained with 25 to 35 gl^{-1} active chlorine. The explants were exposed to chlorine for 5 minutes (Holters and Zimmer, 1990a).

Shoot tips were used as a source for meristem culture of 11 taxa belonging to *Paphiopedilum*. Shoot tips were sterilised by placing them in a solution of 3-4 drops of Tween 20, 100g carbowax 4000, and 50 ml clorox in one litre of water. Tween 20 and carbowax 4000 were dissolved in water and autoclaved before adding clorox. The shoots were immersed in the solution for only 2 minutes, rinsed in sterile distilled water and drained on sterile paper towels (Bubeck, 1973).

Shoot apices were isolated from emerging laterals of plants in the initial stage of flowering of the *Paphiopedilum* and were used as explants. Lateral branches were excised from the plant and washed with pHisoHex solution. This was followed by rinsing with distilled water and removing of all leaves, except the smaller upper ones that encase the shoot apex. To retard browning, the trimmed shoots were stored in a mixture of 100 ml ascorbic acid and 150 mg citric acid in a one litre solution made with distilled water. The trimmed shoot apices were surface sterilised by immersing them in 0.5% sodium hypochlorite for 15 minutes. At the end of this period, the solution was decanted and replaced with 0.005% NaOCl. However, after this stage, distilled water rinsing was not necessary.

In most of the cases, NaOCl and CaOCl solutions have been used, followed by clorox and 70% ethanol. But the use of $Hgcl_2$ as a surface sterilant has been very rare. Yao and Krikorian (1981; cf. George and Sherington, 1984) excised nodal segments and cultured them successfully, using 0.2% $HgCl_2$ with exposure of 3 minutes.

In the present research, 0.1% of $HgCl_2$ solution has been used with 5 minutes of exposure. Explants were cleaned with Teepol before being treated with $HgCl_2$ and they were rinsed 3 to 5 times in autoclaved distilled water before being inoculated.

NUTRIENT MEDIA

The knowledge that we have about the nutritional requirements of the individual organs and tissues of plants is obviously less than what we have about the whole plant. Biosynthetic activity of isolated tissues and organs is limited; at the same time they may exhibit changes in their metabolic pathways over a period of time. The requirement for a particular organic supplement may change due to a shift metabolism under *in vitro* conditions. In order to elicit a response from explants *in vitro,* the nutrient media are to be supplemented with macronutrients, micronutrients and growth regulators. The exact requirements will be known only after a thorough research in each instance.

White (1934; cf. Dodds and Roberts, 1995) discovered that isolated tomato roots had the potential for unlimited growth, if they were provided with a liquid medium containing a mixture of inorganic salts, sucrose, thiamin, pyridoxine, nicotinic acid, and glycine. Since then, many media have been used for *in vitro* studies. It is necessary to take all aspects into consideration while selecting a medium for culturing the plant in question. Each plant requires a different composition and there is no single universal medium that can be used for culturing all plants/plant parts.

In orchid tissue culture, MS, VW and KC media have been used extensively. MS medium includes macronutrients, ammonium nitrate, calcium chloride, magnesium sulphate, potassium nitrate, monopotassium phosphate, and seven micronutrients, boric acid, cobalt chloride, manganese sulphate, potassium iodide, sodium molybdate, and zinc sulphate. Besides, it contains chelated iron, glycine, myoinositol, nicotinic acid, pyridoxine and thiamin. Hence, it includes a large number of nutrients as compared to VW and KC media. VW medium contains only five macronutrients—ammonium sulphate, calcium phosphate, magnesium sulphate, monopotassium phosphate and potassium nitrate; only one micronutrient—manganese sulphate; and chelated iron. KC medium has only four macronutrients — ammonium sulphate, calcium nitrate, magnesium sulphate, monopotassium phosphate; one micronutrient—manganese sulphate; and chelated iron.

It is evident that MS medium is the most complex among these three and hence, it is the costliest. In micropropagation, selecting a medium is a crucial factor. Even though MS medium is better in terms of initiated response, growth and multiplication, one has to explore the possibilities of obtaining almost the same degree of success from alternative media, so as to make the endeavour a cost effective one. Thus, researchers always try to modify the media, without altering the basal nutrients, in order to get better results. In fact, it is sometimes necessary to make substitutions and modifications in 'recipes', due to shortages or response of the taxa. For example, 100mg of K_2PO_4 (dipotassium phosphate, MW 17.18; 44.89% potassium, 17.79% phosphorus) contains 1.56 times as much potassium as an equal weight of KH_2PO_4 (monopotassium phosphate, MW 136.09;

28.73% potassium, 22.76% phosphorus). Using 50 mg K_2PO_4 as a substitute for 100 mg KH_2PO_4, will result in the same amount of potassium by only 0.78 times as much phosphorus (Arditti and Ernst, 1993).

But, substituting one salt for another is not as simple as it seems and it even leads to complexities sometimes. For example, if KCl is used to replace one of the potassium phosphate salts, phosphorus will be eliminated entirely and the chloride content may become supraoptimal. In such cases, another salt many have to be added to supply phosphorus. But this could introduce other complications. Some substitutions can be relatively simple. For example, $MgSO_4$ (magnesium sulphate) can be replaced by $MgCl_2$ (magnesium chloride) and $FeCl_36H_2O$ (ferric chloride), and Fe $(NO_3)_3$ $9H_2O$ (ferric nitrate) can be used as replacements for $FeSO_4$ $7H_2O$ (ferrous sulfphate).

Culture media vary widely in the use and content of microelements. Their concentrations, and even presence or absence, do not seem to be critical. Single shoot tips and other explants from the same genus can be cultured on media with different concentrations and formulations of microelements (Arditti and Ernst, 1993).

Growth regulator requirements for most callus cultures have the same concentrations / combinations of auxin and cytokinin. The most commonly used auxins in orchid culture media are the naturally occurring IAA and their synthetics, NAA, IBA and 2, 4-D. The relative stability of IAA and IBA in nutrient media has been studied and it has been found that IBA is more stable (Nisser and Sutter, 1990). Both auxin and cytokinin concentrations are usually a result of trial and error method and it is best not to make unnecessary changes in substitutions. If any changes are made, it is necessary to keep in mind that auxins and cytokinins may differ from each other, both qualitatively and quantitatively. Thus, the effects of one auxin on a certain species may be different from those of other auxins on other orchids. It also means that different levels of an auxin may have different effects on a given species and that these differences may not be the same for other orchids. For example, for the explants from the leaves of *Dendrobium crumenatum,* MS medium with 2 mgl^{-1} each of NAA and 2, 4-D were used along with 5 mgl^{-1} BAP (Manorama *et al.,* 1986). In a similar experiment, leaves of *Dendrobium* **Alice Spalding** were inoculated on VW medium without any growth regulators but they still initiated the cultures successfully (Fu, 1978). There are several such examples. It is obvious that the choice of auxins and formulation of concentrations are entirely at the discretion of the researchers.

Cytokinins are very important for the regulation of growth and morphogenesis in tissue cultures (George and Sherrigton, 1984). They promote cell division in plant tissues. In addition to callus initiation, cytokinins stimulate bud proliferation and inhibit rooting (Dodds and Roberts, 1995). As with auxins, the choice of cytokinins and their concentrations are based on empirical findings (trial and error). The effects of different cytokinins and their concentrations differ like those of auxins. Vij *et al.* (1989) have used 1 mgl^{-1} kinetin in MPR medium for the induction of shoot buds as rhizome segments from seedlings of *Eulophia hormusjii*. The same combination and concentration of growth regulators were used for the plantlet regeneration of rhizome segments of the above orchid. For the culture of pseudobulbs of *Mormodes histrio,* 0.5 mgl^{-1} BAP was used in KC medium (Holters and Zimmer, 1990a). The root explants of the above orchid was cultured on KC medium, using 2 mgl^{-1} BAP along with 0.2 mgl^{-1} NAA (Holters and Zimmer, 1990b).

Since the time Skoog and Miller (1957) showed that shoot formation could be induced predictably from tobacco callus, auxin and cytokinin interaction has been established as a criterion for the induction of morphogenetic response. Using relatively low levels of auxins and a high level of cytokinin in the growth medium have controlled many aspects of cellular differentiation and organogenesis in tissue culture, and even organ cultures have been found to be controlled by an interaction between cytokinin and auxin concentrations. Gibberellins have been rarely used in orchid tissue culture. When tobacco pith was cultured in media containing gibberellic acid, shoot formation was observed. The formation of stem structures was strikingly suppressed by concentrations of 0.5 mgl^{-1} and above (Murashige, 1961).

Many media with different combinations of growth regulators have been devised and tried for the tissue culture of orchids. However, the exact relation between the composition of a medium and the development of an orchid is not clear. Only a few of the media have been systematically investigated for the role of each organic element. In the methods extensively used for the study of nutrient compositions, some mineral salts were kept in constant concentration while the others varied. The disadvantage of this method is that the total concentration of the nutrient solutions varies or often, an unrequired ion is present. However, the effects of mineral ions in nutrient media can be studied by following the triangulation method. But there are very few systematic investigations about mineral nutrient requirements of orchids. Miller (1965) has reported some aspects of inorganic nutrition in plant tissue cultures. Total ionic concentration of some nutrients in about 8 media used for orchid tissue, was studied by Tokato *et al.* (1990) and he recommended the respecific composition of macroinorganic nutrients required for orchid micropropagation.

Complex additives, such as coconut water, tomato juice, banana extract, and fish emulsion, etc., have been used in orchid tissue culture (Rao, 1977). In the present work also, both coconut water and banana pulp have been used. Coconut water contains minerals, nitrogenous compounds, amino acids, organic acids, vitamins, sugar, sugar alcohol, auxin, cytokinin, gibberellins, etc. whereas, banana pulp is found to contain minerals, carbohydrates, pectic substances, organic acids, lipids, tannins, and growth promoters like zeatin, purine, etc. (Arditti and Ernst, 1993). In the current research, coconut water has been used in many experiments, and banana pulp has been used for the induction of rooting.

PREPARATION OF THE NUTRIENT MEDIA

Knudson C (Knudson, 1946), Vacin and Went (Vacin and Went, 1949) and Murashige and Skoog (Murashige and Skoog, 1962) media were prepared by carefully following the instructions given in the original publications. But it was found necessary to make substitutions and modifications in the recipes due to shortages or preferences by orchid cultivars selected for the study. Culture media vary widely in the use and content of microelements. It was simpler to substitute microelements than macroelements, because the concentrations of the former are lower. Iron was provided in the form of $Na_2EDTA\ 2H_2O$ (disodium ethylene diamine tetraacetic acid MW 272.24). A solution of Na_2 EDTA was prepared by dissolving Na_2 EDTA and Fe SO_4 $7H_2O$ in distilled water. Chemicals were supplied by Hi-Media (Hi-Media Laboratories, Private Limited, Bombay).

COMPONENTS OF MEDIUM

Growth Regulators

To elicit specific responses, such as callus induction, callus multiplication and maintenance, formation of PLBs and formation of plantlets, different auxins and cytokinins were tested at various concentrations, either individually or in combination.

Vitamins

Niacin (nicotinic acid) pyridoxine (vitamin B_6) and thiamine (vitamin – B_1) were used. They were used for preparing nutrient media without autoclaving as they are not heat resistant. Vitamins were supplied by Hi-Media.

Complex Additives

Banana pulp and coconut water were used in different concentrations, either individually or in combination. Fresh tender coconut (*Cocos nucifera*) was brought from the local market. The outer fibrous fruit wall was scraped at one end till the shell was exposed, then a hole was made in the shell, and the water was collected. Coconut water (the liquid endosperm) was prepared freshly, as and when necessary, and used in the medium immediately. Ripe banana fruits (Scavindish variety) were brought from the local market. The skin was removed and the inner fleshy portion was ground without adding water. Banana pulp was also prepared, as and when necessary.

Sugar

Sucrose, also called kitchen sugar, (B-D fructofuranosyl-a-D glucopyranoside, MW 342.20) was used in different concentrations.

Antibiotics

Streptocyclin was added to the media meant for root culture in different concentrations.

Charcoal

Activated Charcoal (AC) was used for the absorption of substances. It leached while being prepared for removing contaminants. AC was added at different suitable concentrations.

Solvent

Double Distilled Water (DDW), prepared in the laboratory, using quartz double distillation setup (Borosil) was used as a solvent.

Solidifier

Agar (Hi-Media) was used in different concentrations as the gelling agent or solidifier.

Maintenance of the work area and instruments

Cleanliness of the work area was given utmost priority. Floors of the preparation room, inoculation room and growth room were cleaned and swabbed with antibiotics, [Dettol [Reckitt and Colman (India) Limited, Mumbai] everyday. Working tables were first washed, with Teepol and water and then rubbed with 70% ethanol (Analar Lab. Grade). All culture bottles, caps, beakers, funnels, measuring cylinders dissection instruments etc., were washed with Teepol and water, before and after use, and rubbed with 70% ethanol, and allowed to dry in clean and dust free areas.

pH Maintanance

pH maintenance was important to the medium as it affected the availability of nutrients to growing tissues and plants. If the pH was either below 4 or above 8, the solid medium did not solidify. pH of the medium was adjusted as per the requirements by using 1N NaOH or 1N NaCl which was measured by using a digital pH meter.

PREPARATION OF STOCK SOLUTIONS

Stock solutions were prepared for most of the media components. Two separate stock solutions were prepared, one each for macro and micronutrients. Separate stock solutions were prepared for each of the growth regulators. Though a separate stock solution was for chelated iron complex prepared, but stock solutions were not prepared for vitamins. All stock solutions were prepared at least every two weeks to avoid contamination. The stock solutions prepared were stored in bottles and properly labelled with details. A magnetic stirrer was used for dissolving some chemicals. Growth regulators were initially dissolved in a little quantity of NaOH, and the final volume was made up with DDW.

CULTURE BOTTLES

Glass bottles and screw caps were supplied by Indo-American Hybrid Seeds Company, Bangalore. They were cleaned with water and soaked in the

Teepol for 12 hours. They were then washed under tap water and dried. Finally, they were swabbed with 70% alcohol and autoclaved for 30 minutes at 106 kg/cm^2. The bottles were very economical and could be reused after cleaning and autoclaving. The mouth of the bottles was sealed using autoclavable foodgrade tape, *Klinwrap* (Anu Marketing Company, Bangalore).

PREPARATION OF THE MEDIUM

Measured quantities of concentrated stock solutions of macro and micronutrients, growth regulators and chelated iron complex were taken in a stainless steel container as per the recipe, and sucrose with freshly prepared complex additives (if needed) was added as per the requirements. Finally, the volume of the medium was adjusted to one litre and the pH was adjusted between 5.6 to 5.8. The whole mixture was placed on an electric heater and when the mixture just started simmering, the required quantity of agar was added slowly and stirred. The hot medium was then dispensed into 500 ml glass culture bottles. Each bottle was filled to 1/5 of its capacity and closed with autoclavable caps. The mouth region of the bottles was then sealed with *Klinwrap* tape.

AUTOCLAVING

All bottles containing nutrient medium were autoclaved at a pressure of 1.06 kg/cm^2 and temperature of 121°C for 15 minutes. Distilled water was filled to ½ the capacity of each bottle, closed with autoclavable cap, and the mouth was sealed with *Klinwrap*. They were autoclaved at 1.06 kg/cm^2 pressure at 121°C for 30 minutes. Media bottles were taken out after autoclaving and allowed to cool down. Then, they were stored in the culture room with controlled temperature, maintained at 25°C ± 2°C.

INOCULATION

Inoculation was carried out under rigid aseptic conditions of an inoculation room under laminar airflow. All the instruments used for inoculation, such as surgical blade holders, forceps, stainless steel dissection plates, were also cleaned with 70% ethanol and heat sterilised by being kept in the hot air oven at 150°C for 60 minutes, and exposed to UV light for 30 minutes. Inoculation was initiated after washing the hands with soap [Dettol, Rockitt and Colman (India) Ltd., Mumbai] and scrubbed with 70% ethanol. Forceps and blade holders with blades were placed in a container with absolute alcohol before use. Then both the forceps and blades were heated over the flame and kept on a sterilised stainless steel plate to allow them to cool down. Cleaned explants kept in autoclaved distilled water were used within 30 minutes of excision.

All explants were transferred into 0.1% $HgCl_2$ inside the inoculation chamber and allowed to remain there for 2-5 minutes; and were then transferred to the autoclaved distilled water and rinsed five times by being kept them for 3-5 minutes in each bottle of autoclaved distilled water. These explants were placed on dissection plates and further dissected and carefully inoculated on to the medium, using a long forceps. About 10 segments were inoculated into each bottle. The bottles were closed using autoclaved caps and sealed using *Klinwrap*. The bottles were then labelled and stored in the growth room, with a temperature of 25°C ± 2°C, with light intensity of 2000 lux (Philips fluorescent lamps 40 W, Philips India Ltd., Mumbai) for 16 hours per day.

MORPHOGENESIS *IN VITRO*

Since orchids require different nutritional conditions for various stages of morphogenisis, various studies have been conducted to know the requirements of orchids (Chia *et al.*,1988; Kanase *et al.*,1993; Fu *et al.*,1997). They generally follow different pathways of morphogenesis during *in vitro* the formation of plantlets from explants.

(a) Explants may directly give rise to plantlets.

(b) Explants first produce PLBs, which in turn give rise to plantlets.

(c) Callus is formed from the explants, which then produce PLBs, and finally plantlets are formed from PLBs.

PLBs formation is a necessary and crucial step in the *in vitro* morphogenetic process. The reason for these differences in the behaviour of different explants depends on the nature of the explant used, the composition of the nutrient media, and the growth regulators used. Different developmental sequences have been reported by different researchers. All these fall under three general categories of developmental sequences:

(a) Plantlets formed directly from the explants in the initial stage, itself (direct organogenesis).

(b) Initiation of cultures begins with the formation of PLBs (organogenesis through PLBs).

(c) Initiation by producing callus and redifferentiation of the callus cells into plantlets (indirect organogensis).

(a) Direct Organogenesis

Nodal explants obtained from the inflorescence of *Phalaenopsis* were inoculated on a modified VW medium without growth regulators. Shoots with roots were produced within two months (Intuwong *et al.,* 1972). The main advantage of this method is that the plant is not damaged or endangered. A disadvantage is the lack of callus formation. This limits propagation to the available flower stalk buds.

Nodal sections of flower stalks of *Phalaenopsis* hybrids were inoculated on a modified KC medium with 0.3 mgl^{-1} NAA and 2 mgl^{-1} BAP. Plantlets developed in 2 months (Koch, 1974). Young shoots of *Dendrobium* hybrid, measuring 8 cm long, were inoculated on VW medium with 0.2 mgl^{-1} NAA. After 3 months of culture the explants developed single shoots and cluster of shoots (Singh, 1976).

Leaves taken from plants grown *in vitro* were inoculated on a VW medium with 500 mgl^{-1} CW, but without growth regulators. A single plant formed at the base of each leaf (Fu, 1978). Apical and axillary buds of *Dendrobium* **Caesar Red Lip** were inoculated on modified VW medium with 4 mgl^{-1} NAA. The explants did well, but only one shoot developed from each bud (Fernando, 1979). Axiallary buds from the shoots of *Dendrobium antennatum* and *Dendrobium phalaenopsis* were induced to produce plantlets on RM medium with 1.75 mgl^{-1} each of IBA and NAA and 5 mgl^{-1} BAP (Kukulczanka and Wojciechowska, 1983). Shoot apex explants of *Cattleya* on a modified MS medium with 0.1 mgl^{-1} NAA, 10 mgl^{-1} adenine sulphate and 1 mgl^{-1} BA elongated and formed leaves (Huang, 1984). In another experiment, flower stalk explants, Himalayan orchid of *Thunia alba,* were cultured on a modified VW medium with 100 mgl^{-1} TCA, an antiauxin, 10 mgl^{-1} BA and 150 mgl^{-1} CW. Initially, buds started growing and rooted plantlets were obtained within 67 days in the same medium with 5 mgl^{-1} IBA (Singh and Prakash, 1984). Stem nodes of *Bletilla striata* were placed on Knop's medium with 70 mgl^{-1} TCA without any growth promoters. Plantlets formed directly from the buds of nodes (Yam, 1989).

In the present study, plantlets were produced through direct organogenisis (see Fig. 87) in all the plants used for the study. But the number of plants produced was very low and all explants did not respond to direct organogenisis, for example root explants. When micropropagation is to be carried out on a commercial scale, this method is not at all a viable procedure. However, chances of variations is minimum in this method, as compared with indirect organogenesis.

(b) Organogenesis Through PLBs

In a pioneering experiment, Morel (1960) inoculated shoot tips from buds on pseudobulbs of *Cymbidium* on a modified KC medium with 150 mgl^{-1} ripe banana pulp. Within 2 months, the explants produced PLBs of 1-2 mm in size. These bodies were colourless at first, but soon turned green and produced leaves and roots. This method provided a means of quick *in vitro* propagation by which about 4 millions plantlets could be produced from a single bud in a year. Leaves of *Cattleya* obtained from *in vitro* grown seedlings were inoculated on Heller's medium with 1 mgl^{-1} kinetin. When axillary meristem is retained, it starts to grow and inhibits the formation of adventitious buds. But if it is removed, numerous PLBs form at the cut end of the leafs blade (Champagnath *et al.*, 1970), indicating the significance of axillary buds in the *in vitro* morphogenetic pathway that the explant follows. Shoot tips of *Dendrobium* were inoculated on a modified VW medium without any growth regulators. Explants produced PLBs after 3 months in culture. The highest number of PLBs were yielded when PLBs were subdivided and subcultured (Sagawa and Kunisaki, 1982).

The last emerging leaf (youngest) from *Phalaenopsis ambalis* was removed and inoculated on MS medium containing 1 mgl^{-1} NAA and 10 mgl^{-1} BAP. Only a few explants survived and PLBs were produced after 173 days, which subsequently gave rise to plants on a KC medium (Tanaka and Sakanishi, 1985). Leaf tips from seedlings of *in vitro* origin were cultured on modified Heller's medium with 0.2 mgl^{-1} NAA and 1 mgl^{-1} BA. PLBs formed after about one month of culture (Yam and Weatherhead 1989).

Formation of PLBs during the morphogenesis of orchids is a natural phenomenon. PLBs are formed before the plantlets are produced from the explants, callus, or seeds. But in some instances plantlets are produced without the formation of PLBs.

In the present study, in most of the cases, the PLBs were formed, which gradually transformed into plantlets. Thus, formation of plantlets from PLBs is a routine process; and this method is commercially viable as a large number of plantlets can be produced. Morel (1965) produced several million plantlets from one shoot tip, using this method.

(c) Indirect Organogenesis

In the third category of *in vitro* developmental pathway, explants produced callus, an undifferentiated mass of tissue of various textures and colours, which gradually differentiated to give rise to PLBs, which in turn was subsequently transformed into plantlets. Callus formation is very useful because more number of plantlets can be produced from less number of explants and at the same time, the resultant callus can also be multiplied.

In a classic experiment, Vajrabhaya and Vajrabhaya (1976) developed a method for tissue culture of the monopodial orchid, *Arachnis*. They inoculated lateral buds on MS medium with 0.5 mgl^{-1} 2, 4-D. Callus masses formed within a month. When subcultured plantlets formed on suitable media. Steward and Mapes (1971b) produced callus from the shoot tip explants of *Cymbidium sp.* on White's medium with 5 mgl^{-1} NAA and 100 mgl^{-1} CW. Subsequently, PLB and plantlets were produced by subculturing on suitable media. Seedlings of *Phalaenopsis ambalis* hybrids were used to obtain root tip explants measuring 3mm. The explants were inoculated on MS medium without any growth regulators. About 3 months later, an increase in cellular activity was observed. Small "bumps" formed at the tip or lateral sides of the root tip explant which differentiated into a callus, from which PLBs developed gradually (Tanaka *et al.,* 1976).

Tubers of 6 month old seedlings of *Pachystoma senile* grown *in vitro* were sliced into 3 to 5mm thick segments, and inoculated on MPR medium with 1 mgl^{-1} 2,4-D to produce callus (Vij *et al.,* 1983). Internodal sections of flower stalks of *Phalaenopsis* were cultured on modified VW medium with 5 mgl^{-1} BAP. Yellowish callus formed on some sections with a frequency of 53% (Lin, 1986).

Production of plantlets through callus is very economical, as with a large scale production of plants can be achieved with a limited amount of source material. In the present study, the experiments were designed to first produce callus from the explants and to regenerate PLBs, and subsequently produce plantlets from the callus. Several modifications of MS, VW and KC media were made by varying the compositions and concentrations of growth regulators and complex additives.

PRODUCTION OF CALLUS

Shoot tip explants from all four orchids used in the current study were inoculated on all the three media with varied concentrations of BAP and 2,4-D. Explants from *Phalaenopsis* formed callus in the shortest period of time, i.e., 18 days in MS medium with 5 mgl^{-1} BAP and 2 mgl^{-1} 2, 4-D. On the same medium *Cattleya* explants, however, showed callusing in 26 days, whereas *Dendrobium* **Queen Emma** required 35 days, and *D.* **Emma White** required 34 days. *Phalaenopsis* explants in KC medium with 1 mgl^{-1}

BAP and 2 mgl^{-1} 2, 4-D showed callusing after 68 days in culture. On MS medium, callus was formed after 48 days in culture of *Cattleya D.* **Emma White** while *D.* **Queen Sonia** showed callusing after 62 and 65 days. It was observed that while 2, 4-D helps in the initiation of callus, at the same time BAP or a cytokinin, was also found to be necessary for callus formation. In the above experiments, the concentration of 2, 4-D was kept constant and only the concentration of BAP was altered. The number of days taken for callus initiation varied between 18 to 68 in *Phalaenopsis,* 24 to 48 days in *Cattleya,* and 22 to 65 days in *Dendrobium*. Though initiation of callus occurred in all the combinations of auxin and cytokinin, the extent and amount of callus formed varied considerably. When leaf tips of *Epidendrum* were inoculated in modified MS medium with 0.5 mgl^{-1} BA and 1 mgl^{-1} 2, 4-D, the explants remained without any apparent changes or growth for nearly 60 days. But about 7% of formed calli when transferred to Linsmaier-Skoog medium with 10 mgl^{-1} Kinetin developed to form plantlets. (Arditti, 1967).

Buds and shoot tips of *Rhynchostylis gigantea* were inoculated on a specially modified semisolid medium with 0.1 mgl^{-1} NAA. During the first week of culture, most explants enlarged and some portions turned green. Within 2 to 4 weeks, a few produced new tissue, but some of these died. At the end of 2 months, only one in 30 explants survived. The survivors formed calli (Vajrabhaya and Vajrabhaya, 1970). Shoot tips of *Paphiopedilum* were cultured on modified Heller's medium with 1 mgl^{-1} 2, 4-D. They took 3 months for the initiation of callus (Stewart and Button, 1975). Leaf tips, measuring approximately 5mm in length were removed and inoculated on VW medium with 200 mgl^{-1} CW. The leaf tips remained green and formed cream or green coloured callus after 4 to 6 weeks. Callus formation was greater on the upper epidermis (Loh *et al.,* 1975).

Root tips of *Bletilla striata* were inoculated on MS medium with 1 mgl^{-1} NAA and 0.2 mgl^{-1} BA. The explants showed swelling and formed callus, which produced multiple shoots on KC medium after 7 months of culture (Yam and Weatherhead, 1991).

Higher concentration of auxins and cytokinins are found to be necessary for the production of good amounts of callus. In *Rhynchostylis,* explants cultured with 0.1 mgl^{-1} NAA remained unchanged for four months and only one of the 30 survived. The survived ones produced a little amount of callus (Vajrabhaya and Vajrabhaya, 1970). Similarly, in an earlier experiment conducted by Arditti (1967) on *Epidendrum,* 7% of the explants produced a little amount of callus while others did not produce any at

0.5 mgl^{-1} BA, but at 10 mgl^{-1} Kinetin large quantities of callus was produced. Internodal segments of *Phalaenopsis* showed 26% callusing on VW medium with 1 mgl^{-1} BAP, but at 5 mgl^{-1} BAP callusing was 53% (Lin, 1986). Though 2, 4-D is most frequently used for inducing callus, it can cause genetic variations, if used repeatedly in subculturing. The callus has to be transferred to media containing NAA or IBA for further development (George and Sherrington, 1984).

In the present study, various explants such as meristems, shoot tips, nodal buds, leaf segments, internodal segments and root segments were inoculated on all the three media with 2 mgl^{-1} BAP as constant and various concentrations (0.5 mgl^{-1} to 5 mgl^{-1}) of NAA. This experiment was conducted to mainly study the degree of callusing and percent of callusing.

It was found that all the explants showed callusing at some or the other concentrations. But some explants showed callusing at all concentrations tried. However, the degree and percentage of callusing varied from explant to explant and orchid to orchid, indicating that each explant type and each orchid has its specific requirements. The percentage of callusing was high in case of meristem explants. It was maximum 82% in the case of *D.* **Queen Sonia** at 2 mgl^{-1} NAA, in MS medium, 91% in VW, and 70% in KC medium. There was a high rate of survival of explants in VW medium. Percent of callusing initially increased with the increase in the concentration of NAA from 0.5 mgl^{-1} to 2 mgl^{-1}, but later it declined. Shoot apex explants of the above orchid showed a maximum of 83% callusing in VW medium, 77% in MS medium and it was maximum 69% in KC medium. Nodal buds of this orchid also showed high percent of callusing in VW medium only, as compared to other media. Similarly all other explants, including roots, also showed higher percentage of callusing in VW medium. *Dendrobium* **Emma White** also showed a similar response as the other *Dendrobium* hybrid, but with some differences. This orchid showed 90% callusing in MS medium at 1 mgl^{-1} NAA concentration. At the same concentration of NAA, D. **Queen Sonia** showed 70% callusing. In VW medium, *D.* **Emma White** showed 92% of callusing. Shoot apices, followed by nodal buds, internodal segments and leaf segments showed lesser percent of callusing.

Root explants in both the *Dendrobium* hybrids responded very well and showed good percent of callusing. For *Dendrobium* hybrids, VW medium was found to be very suitable through MS medium was also good, but not as good as VW medium. KC medium was not found to be good for these orchids, especially for the purpose of producing callus.

Cattleya **Naomi Kerns** showed better percent of callusing in MS medium than in VW medium and KC medium. But the difference in the percent of callusing between MS and VW media was not much. However, since MS medium contains a large number of nutrients than VW medium, it is expensive. So for commercial purposes VW medium is better as *Cattleya* hybrids showed 90% callusing. Meristem explants of *Cattleya* callused at 1 mgl^{-1} NAA in MS medium, but at the same concentration of NAA, shoot apices showed 88% callusing. While meristem showed 89% and shoot apex showed 86% callusing in VW medium at 1 mgl^{-1} NAA nodal buds in this hybrid showed almost equal percent of callusing. 86% callusing was recorded at 1 mgl^{-1} NAA in MS medium and even the internodal and root segments showed good percent of callusing. But leaf segments showed only a maximum of 40% callusing at 1 mgl^{-1} NAA in MS medium.

In *Phalaenopsis,* a monopodial orchid, all explants except internodal and root segments showed good percent of callusing. Though monopodial orchids are considered difficult to culture (Vajrabhaya and Vajrabhaya, 1970; Arditti and Ernst, 1993), callus formation was first reported by Sagawa (1990 a,b) and Ishii *et al.* (1998), who reported that callus induction from leaf segments is difficult. In the present study, meristem explants showed 90% callusing at 2 mgl^{-1} NAA in MS medium, but the same explants showed 75% callusing in VW medium. In VW medium, the same explant showed 90% callusing at 1 mgl^{-1} NAA. Unlike *Dendrobium,* both meristem and shoot apices showed almost similar results in *Phalaenopsis*. However, for nodal buds, VW medium was better. At 2 mgl^{-1} NAA, 82% callusing was observed in VW medium while in MS medium, 78% callusing was recorded at the same concentration of NAA.

Leaf segments showed good percent of callusing, both in VW and MS media. But in KC, the response was relatively low. Internodal segments showed a maximum of 79% callusing at 2 mgl^{-1} NAA in MS medium. The percentage decreased with the further increase in the concentration of NAA. Lin (1986) noticed increase in the percent of callusing with the increase in Kinetin.

Root segments also showed good percentage of callusing on MS medium. In the present study, a maximum of 78% callusing was recorded in MS medium at 2 mgl^{-1} NAA. In VW medium, root segments showed 69% callusing at the same concentration of NAA. There was further decrease in the percent of callusing with the increase in the NAA concentration. In VW and KC media, root segments did not show any callusing at 0.5 mgl^{-1} NAA concentrations. AT 5 mgl^{-1} NAA concentration, root segments showed 39% callusing in KC medium and 50% callusing in VW medium.

Though meristem was found to be a very good source of explants in all the orchids used in this study, it was difficult to isolate a meristem. Shoot tip on the other hand is a very good explant that can be collected easily, and at the same time it shows an equally good response. The reason for the difference in the percentage of callusing between the meristem and shoot apex is that the meristem is just the meristematic region isolated from the remaining non-meristematic tissues, whereas in the shoot apex the meristem is conjoined with differentiated leaf primordia. Shoot apices may show a little higher percent of callusing if they are maintained for some more days in the media, but in the current study, the results were recorded upto 10 weeks in culture.

Nodal buds are also comparable to shoot apices in their response. Nodal buds showed, lesser percent of callusing as they received a lesser quantity of auxins the terminal shoot apices.

Leaf explants of *D.* **Queen Sonia** showed a maximum of 75% callusing in VW medium at 3 mgl^{-1} NAA, which is exceptional, since generally the percent of callusing was below 40 on all media and concentrations.

Internodal segments are rarely used in orchid tissue culture. Clonal multiplication of *Phalaenopsis,* obtained by culturing thin discs of young internodal sections, was reported by Lin (1986), who recorded 26% callusing on VW medium with 1 mgl^{-1} BAP. But the percentage of callusing was 53 at 5 mgl^{-1} BAP.

A perusal of literature reveals that there have been quite a few attempts at *in vitro* culture of orchid roots with moderate success. The explants developed into thin etiolated structures in *Vanda* (Goh, 1970) and *Epidendrum* (Churchill 1972). In *Epidendrum o'brienianum*, they callused, and differentiated normal roots formed in most of the cultures; the callus however, developed into a plantlets through PLBs mediated organogenesis. In one of the cultures (Steward and Button, 1978), callus mediated PLBs development was also reported in the root tip explants from *in vitro* grown *Oncidium varicosum* seedlings, the PLBs multiplied in liquid medium and differentiated plantlets when cultured in a semisolid medium (Kerbauy, 1984b). The *in vitro* regeneration response was tested in 21 orchid taxa, out of which 12 of them responded positively, by generating plants (Vij, 1994).

Generally in all orchids, root explants showed moderate percent of callusing except in certain concentrations. It is known that roots have the ability to spontaneously develop shoot primordia (Churchill *et al.,* 1971). Several attempts have been made so far to keep the *in vitro* regeneration potential of excised roots (Goh 1990; Arditti and Ernst, 1993). In the present study, all the explants showed callusing at some or the other concentrations, but the amount of callus produced was always low. At the same time, the regenerative capacity of the callus produced from the root was also low.

The degree of callusing was studied in all the explants and generally, the degree of callusing increased with the increase in the concentration of NAA. The increase ranged between 1 to 4.9. However, the number of explants producing callus may also decrease with the increase in NAA level, as at higher concentrations of NAA, the percent of callusing and survival of explants is low. Thus, the fewer number of explants that have survived and produced callus would produce more amount of callus, leading to increased degree of callusing.

MULTIPLICATION OF CALLUS

A callus consists of a mass of loosely arranged, thin-walled parenchyma cells arising from the proliferation of cells of the cultured explant. Some calli are heavily lignified and hard in texture, whereas others may be easily separated into small units (fragile callus). Thus, there is considerable variability in the anatomy of callus (Dodds and Roberts, 1995). After the callus has grown for a while in association with the original tissue, it becomes necessary to subculture the callus on a fresh medium. Growth on the same medium for an extended period will lead to a depletion of essential nutrients and to gradual desiccation of the gelling agent. Metabolites secreted by the growing callus may accumulate to toxic levels in the medium. The other purpose for the subculturing of calluses is to multiply the callus. In the medium where the callus has formed, the nutrients may not be sufficient for a healthy growth of callus and different combinations and concentrations of growth regulators may be required for the growth of the callus. If the transferred callus segment is too small, it may exhibit a very slow rate of growth or none at all (Street, 1969). Successive subcultures are usually performed every 4 to 6 weeks with cultures maintained on agar medium. Timing of the transfer may also be important, because non-embryogenic callus from the original explant at the expense of the competent tissue (George and Sherrington, 1984).

Subculturing is imperative to maintain the culture for a long time or to increase the volume of cultures. The duration of one subculture is sometimes called a passage. The risk of obtaining abnormal variations in plants has deterred laboratories from using the regeneration of plants from callus, but genetically abnormal plants are produced by other methods of micropropagation also. Some genetic variations have been reported in mericloned *Vanda* (Teo and Teo, 1974) and *Dendrobium* (Vajrabhaya and Vajrabhaya, 1974). Hence, in the present study an additional *in vitro* stage has been included to allow multiplication of the culture before it is actually utilised for regenerative purposes. *Dendrobium* hybrids showed good rate of multiplication in MS medium at 2 mgl^{-1} NAA and 1 mgl^{-1} BAP concentrations. Though *Cattleya* also showed a preference for MS medium for multiplying callus, but it required higher concentration of BAP (3 mgl^{-1}). MS medium was better even for the multiplication of *Phalaenopsis* callus. High degree of callus multiplication was observed at 3 mgl^{-1} NAA and 2 mgl^{-1} BAP.

REGENERATION OF PLBs AND PLANTLETS

The success of tissue culture technique lies not only in the ability to produce large number of plants, which can be grown later on in the outside environment but in producing callus and regenerating only a few plants. Hence, inducing callus to produce plants is a very crucial stage.

The creation of a new form and organisation, where previously it was lacking, is termed morphogenesis. Generally there are two methods of morphogenesis, direct organogenesis and indirect organogenesis. Many interacting factors like the nature of the explant and composition of the nutrient media influence organogenesis *in vitro*.

Depending upon the plant species and the origin of callus, the formation of adventitious organs, somatic embryos or PLBs can take place. The regenerative capacity of cells and callus tissue may reduce or be completely lost if growth is allowed for too long. As to the reason why the calluses of some plants are only capable of forming adventitious organs and no embryos whereas others behave entirely opposite. Some plants are capable of regenerating either by adventitious shoots or adventitious embryos (Pierik, 1987).

Orchids show a different morphogenetic pathway under *in vitro* conditions, prior to the formation plantlets and PLBs. Morel (1960) obtained PLBs in meristem cultures of *Cymbidium,* which were extremely similar to those from seed germination. A protocorm that has developed after germination from an orchid seed has a morphological state that lies between an undifferentiated embryo and a shoot. PLBs are also formed by vegetative propagation or by meristem culture, which means that a shoot tip of an adult plant returns from the adult into the juvenile phase. Rice protocorms have many close similarities with PLBs. In the current study, the callus produced was induced *in vitro* by altering the composition of the media, especially by altering the ratio of auxin and cytokinin, to produce PLBs and plantlets. In certain experiments however, plantlets developed without the intervening callus stage.

In this study, the nutritional requirements of different explants and callus segments were examined, for producing PLBs, by inoculating all explants and callus segments of all the orchids selected for study on all the three media. It was found that *Dendrobium* and *Cattleya,* hybrids showed a good rate of PLB formation in VW medium while *Phalaenopsis* showed a preference for MS medium for PLB formation. *Cattleya* and *Phalaenopsis* hybrids, on the other hand, showed a good rate of differentiation on MS medium, but a moderate rate of differentiation on VW medium with the rate of differentiation being the least in the case of KC medium.

Dendrobium **Caesar Red Lip** was induced to produce PLBs on modified VW medium, by Fernando (1979). Lim-Ho (1981) has reported to have used modified VW medium for the purpose of differentiation of about 13 taxa belonging to the genus *Dendrobium*. Fu (1978, 1979b) concluded that VW medium is the most suitable for *Dendrobium*. Similarly, in case of *Dendrobium crumenatum* (Monorama *et al.,* 1986), differentiation of PLBs was achieved by inoculating callus segments on modified VW medium. In *Dendrobium,* differentiation was successfully done, using other media also such as RM medium (Kukulczanka and Wojciechowska, 1983). Morel (1970) used modified MS medium, with both NAA and kinetin, for inducing differentiation in the meristem culture of *Cattleya*.

Three media—all based on MS medium—were used for the purpose of initiation of culture, differentiation, shoot multiplication and rooting in the case of *Cattleya* (Huang, 1984). In *Cattleya,* induction of PLB has been achieved even on other media like KC medium (Scully, 1967; Vajrabhaya, 1978). For the induction of PLBs in *Phalaenopsis,* modified MS medium was used (Tanaka and Sakanishi, 1977, 1980). Leaf segments of *Phalaenopsis ambalis* were induced to produce PLBs on Ms medium (Tanaka and Sakanishi, 1985). Hass-Von-Schmude (1985) also used Ms medium for the purpose of differentiation of PLBs in 4 hybrids of *Phalaenopsis*.

Other media, such as Knop's (Ball *et al.*, 1974), VW (Intuwong and Sagawa, 1974), and KC (Koch, 1974) were also used to induce PLB formation in *Phalaenopsis*.

Callus segments, shoot tips and inflorescence apex have showed good rate of PLB formation in the media used in the present study. Moderate and low rate of PLB formation was observed in other explants such as leaf buds, internodal buds and root segments. The growth regulator requirements for these explants were different for different explants. For example, root segments of *D.* **Queen Sonia** showed good differentiation in PLBs in VW medium at 1 mgl^{-1} BAP and 2 mgl^{-1} NAA. But the shoot apex of this orchid required 0.5 mgl^{-1} NAA for good rate of differentiation. Shoot apex of *Cattleya* **Naomi Kerns** showed good differentiation in MS medium at 1 mgl^{-1} BAP and 1 mgl^{-1} NAA. But root segments showed requirement of 2 mgl^{-1} NAA in the same medium with 1 mgl^{-1} BAP.

Phalaenopsis showed the requirement of higher concentrations of auxin in the case of all explants. Through inflorescence apex of *Phalaenopsis* showed good rate of PLB formation in MS medium with 5 mgl^{-1} NAA and 1 mgl^{-1} BAP, but callus segments showed good differentiation in the same medium with 3 mgl^{-1} NAA and 1 mgl^{-1} BAP. While all explants failed to show any differentiation at 0.5 mgl^{-1} NAA concentration in MS medium, callus showed a little differentiation at this level.

It appeared through the experiments that more than the kind of media, the concentrations of auxins and cytokinins determined the rate of differentiation. In both the *Dendrobium* hybrids, shoot apex explants showed the same rate of differentiation at 0.5 mgl^{-1} NAA and 1 mgl^{-1} BAP in both MS and VW media. Shoot apex explants of *Cattleya* also showed same rate of differentiation at 1 mgl^{-1} each of NAA and BAP in MS, as well as VW media.

A correlation between survival rate of explants, wet weight of the callus and rate of PLB formation was established by selecting only the shoot tip of all the orchids studied. The explants were inoculated on the selected, media by observing the results in the previous experiments. Explants of each orchid were then cultured on the selected media in which they had given best results in the previous experiments. For instance, *Dendrobium* hybrids were cultured in VW medium, whereas *Cattleya* and *Phalaenopsis* hybrids were cultured on MS medium. All the media contained different concentration of BAP (0.1, 1 and 2 mgl^{-1}) and 2,4-D (0.1, 1, 2, 3, 4, and 5 mgl^{-1}).

On the basis of the results obtained, three comparative studies were made. Although, each of these parameters has been studied independently for some orchids, no comparative/correlative studies are available in the literature.

a) Survival rate and wet weight of the callus.

b) Survival rate and rate of PLBs formation.

c) Wet weight of the callus and rate of PLBs formation.

Correlation between percentage of survival and wet weight of the callus:

In case of *Dendrobium* hybrids, survival rate was indirectly proportionate to the wet weight of the callus. At 0.1 mgl^{-1} each of BAP and 2,4-D, the percentage of survival was 95 and wet weight of the callus was 236 mg. At 5 mgl^{-1} 2, 4-D and 0.1 mgl^{-1} BAP, percentage of survival was 0.5 and wet weight was 326 mg. While the percentage of survival of explants decreased with the increase in 2, 4-D concentration, wet weight of the callus segments increased with the increase in concentration of 2, 4-D. To be specific here, 2, 4-D prove to be inihibitory at higher concentration. The auxin, most frequently used to initiate callus growth is 2, 4-D, but the cells of callus cultures maintained on 2, 4-D are particularly liable to become genetically variable (George and Sherrington, 1984).

The reason for the increase in the wet weight with the increase in the 2, 4-D concentration may be that the number of surviving explants is reduced by the increasing 2, 4-D level, and the ones that have survived have more amount of nutrition at their disposal. In this experiment, weight of the callus was calculated by considering only the surviving explants, hence the weight was always high. But the utility of these large masses of calli was not good in terms of PLB formation. Even in *Cattleya* hybrid, the survival rate decreased and the weight of the callus decreased with the increase in the level of 2, 4-D. But *Phalaenopsis* showed a requirement for higher concentrations of 2, 4-D than *Dendrobium* for better survival of explants. 85% survival was observed at 0.1 mgl^{-1} 2, 4-D and 90% survival was seen at 1 mgl^{-1} 2, 4-D. But as there was further increase in 2, 4-D, the rate of survival decreased. Percentage of survival was almost constant even when there was a change in the concentration of BAP. It was 85%, 80%, and 80% at 0.1, 1 and 2 mgl^{-1} BAP, respectively. Except at the initial levels, the decreased percentage of survival increased the wet weight of the callus. Increase in BAP did not increase the weight much, as cytokinins are not often used to induce callus (George and Sherrington, 1984).

Correlation between percentages of survival of explants and PLB formation

In *Dendrobium* hybrids, percentage of survival of explants and PLB formation were closely related. With the increase in the 2,4-D level there was a decrease in the percentage of survival as well as % formation of PLB. Even though there was increase in wet weight, the rate of PLB formation was low with the increase in PLB formation as higher concentration of 2,4-D can induce mutations and inhibit photosynthesis (Pierik, 1987). At 4 mgl^{-1} and above concentrations of 2, 4-D, there was no PLB formation at all. In *Cattleya* also, similar results were observed. In this case, the rate of PLBs formation reduced even with the increase in BAP. At 1 mgl^{-1} BAP and 0.1 mgl^{-1} 2, 4-D, the percentage of survival was 90, and the percentage of PLB formation was 85 and 45, respectively and at 3 mgl^{-1} and above concentration of 2, 4-D, there was no PLB formation. This indicates that not only auxins but cytokinins are also required for PLB induction, but only upto certain levels (Reinert and Mohr, 1967).

In *Phalaenopsis,* the results were different; increase in 2, 4-D decreased the rate of survival, but increased PLB formation upto 2 mgl^{-1} concentration. With further increase in 2,4-D, the number of PLBs decreased. It indicated that a certain combination of cytokinins and auxins is needed for PLB formation. It also showed the higher requirement of 2, 4-D for PLB formation. Even at 5 mgl^{-1} level, there was PLB formation, when the concentration of BAP was 0.1 mgl^{-1}. But as the concentration of BAP was increased to 5 mgl^{-1} 2,4-D, there was no PLB formation.

Correlation between the wet weight of the callus and percentage of PLB formation

In *Dendrobium,* increase in the wet weight decreased the percentage of PLBs formation in *Cattleya* and *Phalaenopsis*. Increase in wet weight showed increase in PLB formation, although there was a gradual decrease later. This is probably because of the higher concentration of 2,4-D. Increased concentration of 2,4-D, no doubt increased the amount of callus, but it did not show any regeneration in this case. Reduction in chlorophyll content by auxins was noted in source callus cultures (Hilderbrandt *et al.,* 1963, cf. George and Sherrington, 1984). At high concentrations, most synthetic auxins are phytotoxic (George and Sherrington, 1984). Though, initially 2, 4-D was used for callus induction for regenerative purposes, later cytokinins with other auxins have been used (Churchill *et al.,* 1971; Fonnesbech, 1972a; Tran Thanh Van 1974; Huang, 1984).

Comparisons between rate of survival of explants, rate of PLB formation, colour of PLB and rate of plantlet formation were made by subculturing callus segments on selected media with varied concentrations of NAA and BAP. Auxins and cytokinins were used in higher concentrations (1, 5, 10 and 15 mgl^{-1}).

Dendrobium hybrids did not survive at NAA and BAP concentrations above 10 mgl^{-1}. Rate of survival was very high (100% in *D.* **Queen Sonia** and 98% in *D.* **Emma White**) at lower concentrations of NAA and BAP. But the rate decreased to 48% at 1 mgl^{-1} NAA and 5 mgl^{-1} BAP, and 36% at lower levels and very low at higher levels of auxins and cytokinins. At very high concentrations all the explants died. There was a strong correlation between percentage of survival and rate of PLBs formation, as expected. At 1 mgl^{-1} each NAA and BAP, the rates of survival and PLBs formation were 100% and 90%, respectively. At these levels, the PLBs formed were all green in colour with a high rate of plantlet formation. At higher levels of auxins and cytokinins there was a low rate of survival (16% at 10 mgl^{-1} NAA and 1 mgl^{-1} BAP; 4% at 15 mgl^{-1} NAA and 1 mgl^{-1} BAP). The PLB formation was almost nil. At 15 mgl^{-1} NAA and 1 mgl^{-1} BAP, only a few pale yellow PLBs formed, but they all eventually died.

Several experiments were carried out to determine the effects of various factors on the survival of shoot tip explants. It was found that either NAA or 2,4-D was essential for initiating the growth of explants, but the concentrations should be low (Ichihashi and Kako, 1973). A similar study was also made on the effects of growth regulators and organic supplement concentrations on PLBs and plantlets of *Cattleya* (Kusumoto 1979a, b). Shoot formation was enhanced by 1 mgl^{-1} 2, 4-D or 0.1 mgl^{-1} BA and 6.1 mgl^{-1} 2, 4-D, and proliferation of protocorms was enhanced by 5 mgl^{-1} BA and 0.1 mgl^{-1} NAA.

Phalaenopsis, unlike *Dendrobium* and *Cattleya,* requires higher cytokinin concentrations. Ball *et al.,* (1974,1975) have used 20 mgl^{-1} BAP in Knop's medium to successfully produce plants from the inflorescence stalk nodes of *Phalaenopsis*. Similarly, for the propagation of leaf sections of *Phalaenopsis ambalis,* Tanaka and Sakanishi (1977) have used 1 mgl^{-1} NAA and 10 mgl^{-1} BAP in MS medium. In the present study, 80% survival and 40% PLBs formation was recorded at 5 mgl^{-1} of each NAA and BAP. At 5 mgl^{-1} NAA and 10 mgl^{-1} BAP 62% survival and 24% PLB formation was noticed. In both the cases, the PLBs formed were green and they developed into plantlets, both on the same medium as well as on subculturing medium.

MULTIPLICATION OF PLANTLETS

Generally micropropagation of orchids includes three phases (a) transformation of explants into PLBs (b) propagation of PLBs by cutting them into pieces and (c) development of these PLBs into rooted shoots (Teo, 1978; cf, Pierik, 1987). In the present study, two more *in vitro* phases have been added to the list: (a) production of callus, and (b) multiplication of plantlets.

In most of the experiments conducted in the present study, an intervening callus stage was observed and from this callus the PLBs were formed. There are innumerable instances in literature where callus was reported to be formed prior to the formation of PLBs (Vajrabhaya and Vajrabhaya, 1976; Loh *et al.,* 1975; Kukulczanka and Wojciechowska, 1983; Tanaka and Sakanishi, 1985). Hence, production of callus has been identified as an important phase in the micropropagation of orchids, at least in the cases where it is reported. Similarly, multiplication of plantlets is important to increase the yield, especially, when micropropagation is carried out on a commercial basis and the orchids selected for the present study are all commercial varieties. Hence the plantlets produced *in vitro* were allowed to multiply before they were inoculated on a rooting medium. At this stage, the bases of the small plantlets were injured and cultured on selected media with varied concentrations of auxins and cytokinins. Side shoots started appearing from the bases of plants, each of which developed into a plant, resulting in the multiplication of plants. Rate of multiplication of plantlets was calculated, as mentioned earlier. Effects of medium composition for multiple shooting and growth of plants in *Cymbidium* were studied Chung (1998).

Dendrobium hybrids showed high rate of multiplication in VW media with 4 mgl^{-1} BAP and 1 mgl^{-1} NAA. *Cattleya* showed maximum rate of multiplication in MS medium with 2 mgl^{-1} BAP and 1 mgl^{-1} NAA, and *Phalaenopsis* showed good rate of multiplication at 8 mgl^{-1} BAP and 4 mgl^{-1} NAA in MS medium. It was clearly shown that not only auxins but even cytokinins are important for the multiplication of plantlets. Plantlet multiplication occurred by the formation of side shoots arising from the base of the plants. Even *Phalaenopsis* which is a monopodial orchid, produced side shoots. Such studies are rare in literature as most researchers tend to develop and grow plants. But there have been reports about the formation of plantlets. Lim-Ho (1981) has reported the successful production of plantlets in about 13 orchid taxa. There have been some reports on the multiplication of callus (Lindemann *et al.*, 1970) and PLBs (Tanaka and Sakanishi, 1977, 1980).

DEVELOPMENT AND ROOTING OF PLANTLETS

Generally, once the plantlets are formed on a medium they are allowed to grow on the same medium till they develop into a stage from where they can be used for hardening. However, some workers have employed a separate medium for the development and rooting of plantlets (Kusumoto, 1979a, b). All the media used in the present study were also supplemented with activated charcoal, banana pulp and coconut water, at various concentrations. VW media was found to be suitable for the development of *Dendrobium* hybrids. VW media has been used by several researchers for the propagation of *Dendrobium* taxa (Sagawa and Shoji, 1967; Lim-Ho, 1981; Sagawa and Kunisaki, 1982). MS medium was found to be suitable for the development of *Cattleya* plantlets. This same medium was used for this orchid by other earlier workers, earlier on (Huang, 1984). Similarly, in the current study, even *Phalaenopsis* showed better development on MS medium, and many earlier workers had also reported it (Haas-Von-Schmude, 1983,1984, 1985). The effects of growth regulators were examined by using different auxins (NAA, IBA and 2,4-D) and cytokinins (Kinetin and 2, 4-D). It was found that different auxins and cytokinins produced different results.

For the induction of caulogenesis and rhizogenesis, usually an adjustment of the levels of auxins and cytokinins is usually necessary (George and Sherrington,1984). Different orchids require different levels of growth regulators, which can be understood only by meticulous study. Initially, the concentration of NAA was kept constant and concentrations of kinetin and 2iP were altered. It was found that 2iP was slightly better than kinetin in both the hybrids of *Dendrobium*. But in *Cattleya,* though kinetin was better for an increase in the length of plantlets and the formation of leaves, 2iP was better for the formation of roots. In case of *Phalaenopsis* also, 2iP was better than kinetin in all respects. When the concentration of BAP was kept constant and the concentration of auxins (NAA, IBA 2, 4-D) was altered, NAA was found to be the best auxin for good growth in case of all the hybrids. 6 mgl^{-1} NAA was best for *Dendrobium*, 4 mgl^{-1} NAA was suitable for *Cattleya* and *Phalaenopsis* showed good development at 10 mgl^{-1} NAA. Parameters such as length of the plant, number of leaves formed and number of roots produced were considered. As such, at certain higher levels of auxins as well as cytokinins, growth appears to be excellent. But, at higher levels of auxins and cytokinins the plants showed a number of abnormal features and the plantlets also showed side shoot formation.

In case of *Dendrobium* **Queen Sonia** and *Dendrobium* **Emma White,** increase in NAA enhanced growth upto certain levels but with the increase in BAP the growth of the plant was less, however, there was side shoot formation. *D.* **Emma White** plantlets were comparatively smaller than those of *D.* **Queen Sonia**. As there was a more than 5 mgl^{-1} increase in NAA the plantlets showed unusual growth and elongation of internodes with narrower and sparingly arranged leaves. *Cattleya* **Naomi Kerns** and *Phalaenopsis* **Queen Emma** also showed unusual growth at higher levels of growth regulators.

Higher levels of cytokinins in all the cases induced unusual and non functional root formation. The roots were small, thin and abnormal as compared to the healthy roots produced at lower levels of cytokinins. In fact high concentrations of cytokinin (10 mgl^{-1}) generally inhibit root formation (Schrandolf and Reinert, 1959; Harris and Hart, 1964, cf. George and Sherrington 1984).

In the early days, plants produced through tissue culture, with *in vitro* roots were directly grown *in vivo by* considering the *in vitro* formed roots as sufficient. This method is still widely used and cited in current tissue culture literature, but there is an increasing trend towards inducing extra vitrums rooting procedures in commercial propagation laboratories (George and Sherrington, 1984). Debergh and Maene (1981) emphasise that the *in vitro* rooting of shoots is expensive and ineffective, and has some drawbacks. An *in vitro* method for rooting, developed by McComb and Newton (1981, cf. George and Sherrington, 1984) does overcome some of the disadvantages. In the present study, plantlets of all selected orchid taxa were cultured in all three media, with varied concentrations of banana pulp to study its effect on the rate of rooting. All orchids reached saturation point with the requirement for banana pulp at 150 mgl^{-1}. *Dendrobium* and *Cattleya* showed good rooting in all the three media but *Phalaenopsis* showed good rooting in MS and VW media.

HARDENING OF TISSUE CULTURED PLANTS

A biological approach towards increasing the rates of survival of micropropagated plants has been reviewed by Sahay and Varma (2000). Generally the plants ready for hardening are taken out, cleaned and put in the potting medium (Kumar and Kumar 1992). But in the current study, the bottles containing plants to be hardened were as such kept in the greenhouse for 3-4 days, and only after that they were taken out, cleaned and potted. A universal medium for culturing orchid explants is still elusive. Sheehan (1958), summarising the history of orchid media, concluded that a wide variety of materials have been successfully used by orchid growers. Mc Lellan (1956) suggested use of fir bark for epiphytic orchids. Baryley (1956) proposed use of tree fern fibre which lasted 4-5 years under natural conditions. It has water retention capacity similar to *Osmunda.* Though there are numerous media and fertiliser recommendations for growing orchids, the various media have neither been compared scientifically, nor have the effects of microelements been studied (Poole and Sheehan, 1977).

Brazilian *Cattleya labiata* was successfully grown on charcoal or fir bark under constantly humid conditions (Pessoa and Pessoa, 1985). A *Phalaenopsis* hybrid was grown on five *foam* substances, and *peat* and foam substances without peat proved suitable for picking out plants, but after 18 months, plants in foam showed chlorosis due to nutrient deficiency. Kumar and Kumar (1992) transplanted *in vitro* grown plants directly onto 10 different potting media comprising of broken tiles, charcoal, coconut husks, coconut shells, fern roots, grass roots, *granel* moss, rubber seed husks and wood shavings. They recommended that rubber *seed husks*, gravel and coconut shells be used as cheap and abundant substitutes. 'Oasis' was used as a substratum for *Phalaenopsis* culture *in vivo* (Cocozza *et al.*, 1994). *Dendrobium* **cv. Linnapa** was grown on different media such as (a) Perlite, Netromix 700 and Charcoal (PMC) (b) coarse fir bark (c) medium fir bark, peat moss perlite, vermiculate and Par Gro rockwool (RM). Vegetative growth was good on PMC but flowering was better on RM. Influence of medium on the growth of *Dendrobium* was studied by Paul *et al.,* (1992).

Some media might have given good results. But it is also necessary to look into aspects of feasibility, economy and availability of these exotic potting media. Hence, in the current study, easily and locally available cost effective materials have been used to obtain good growth and flowering. Charcoal pieces, broken burnt clay pieces, brick pieces, moss, coconut husk, rock wool, clay balls, fern blocks, wooden blocks, split coconut pericarp and coir were used in different combinations. *Cattleya* and *Dendrobium* hybrids showed good growth on clay balls and coconut husk pieces, and *Phalaenopsis* grew well on the medium containing fern blocks.

Supplying nutrients is also important. The effect of time and supply of nitrogen was important in the flowering of *Phalaenopsis* (Kuboto and

Yoneda, 1994) There are several ratios and combinations suggested in different reports. Sakai *et al.,* (1982) suggested the supply of 48 mg of nitrogen per pot per month, as it increases the number of flowers. Wang and Wang (1995) recommended the use of 1g nuturient mixture (20-N, 8.6-P and 16.6-K), at each and every third irrigation per plant. In the current study, a mixture of NPK (30gl^{-1} N, 20 gl^{-1} P, 20 gl^{-1} K) and microsalts of MS medium are dissolved in 1 litre of water. 0.1% of the mixture is diluted and sprayed on the plant once a week. Most of the workers have suggested the supply of nutrients to the potting media, but it has been found that supplying nutrients through foliar spray is better.

Bottles containing fully grown plants ready for hardening were kept in the greenhouse for 3-4 days, before removing the plants out of the nutrient medium. This procedure helped in allowing the *in vitro* grown plants to acclimatise to the greenhouse environment. The cleaned ex-agar plants were treated with 1% (V/V) antibiotics for 5 minutes by dipping the basal portions in an antibiotic solution.

Ex-agar plants were placed on a bed of potting medium for 2 - 3 months. Once they developed roots, they were planted in individual pots containing a suitable potting medium.

RECORD OF DATA

Cultures were observed twice a week and the results were recorded. Morphogenetic stages such as initiation of callus, multiplication of callus, formation of PLBs, formation of plantlets, growth and multiplication of plantlets were observed, and photographs and the results were recorded.

STATISTICAL ANALYSIS

Statistical analysis has become an integral part of agricultural and biological experiments. Analysis of variances is a commonly used statistical analysis in agriculture experiments. Its interpretation is valid when the random errors are independently distributed according to a normal distribution, with zero mean and an unknown but fixed variance (Steel and Terrie, 1980). Strong deviations from one or more of the assumptions must be checked and corrected before statistical analysis and interpretation of the results (Fernandez, 1992). Data transformations are done to make them conform more closely to the assumptions underlying the analysis of variance, with three main objectives —(a) to make the error variance nearly more homogeneous, (b) to improve additivity, and (c) to produce a nearly normal error distribution (Finney, 1989). Such analysis have rarely been carried out in orchid tissue cultural studies. Duncan's multiple range test was done by Tong *et al.* (1997) in a publication on the micropropagation of *Spathoglottis*. Ishii (1998) has calculated the standard error micropropagation of *Phalaenopsis*, but the results have not been discussed.

In the current study, either standard deviation (SD) or standard error (SE) have been calculated for most of the results. These are mainly done to know the mean deviation and to make the errors homogenous. All experiments have been conducted using two sets of experiments to estimate the standard error and to understand the deviation from the standard error. For example, in an experiment to study the number of days taken for the formation of callus from shoot tip explants of *D.* **Queen Sonia**, two sets of identical experiments were conducted. Each final result was considered by taking the average of two independent results of identical experiments. At 2 mgl^{-1} 2, 4-D and 1 mgl^{-1} BAP, the explants took 41 days for the initiation of callus. The SD was $\pm$ 1.4, meaning that the expected variation is less. At 5 mgl^{-1} BAP and 2 mgl^{-1} 2,4-D, the time taken was 24 days with SD $\pm$ 9.19. More the standard deviation higher will be the rate of variability. Similarly, SE was also calculated as it automatically gives an idea about the relation rate of expected errors that could possibly arise in the experiment. One table of results with standard errors or standard deviation is as good as two tables with a hint about possible errors that could arise. In the experiment done by Ishii *et al.* (1998), SD is almost equivalent to the result itself (23.5 $\pm$ 23.5), thus giving an idea about the chances of succeeding in the objectives.

CALCULATIONS

(a) The number of days taken for the initiation of callusing was calculated as follows:

Each bottle with about 10 segments was considered as a unit. The value was considered by taking the average of all the segments. For each experiment, two such units (replicants) with similar treatments were considered and their average data was taken as the final value.

(b) Percent callusing was calculated as following:

$$\text{Percent callusing} = \frac{\text{No. of explants callused}}{\text{No. of explants inoculated}} \times 100$$

To obtain a value, two replicants were considered (each bottle was considered as a replicant) and their average was taken as the final value.

(c) Weight of the callus was calculated by volumetric method by measuring the net weight. Average of all segments in the two replicants was considered as the final value.

(d) Degree of callusing (DC) was calculated by non-parametric method. In this method, calculations were made by visual observations and the grading was done as follows:

	Nature of callusing	Points	Grade
1.	Excellent	5	+++++
2.	Very good	4	++++
3.	Good	3	+++
4.	Moderate	2	++
5.	Poor	1	+

Degree of callusing for a particular treatment (protocol):

Replicants	Points
R1	3
R2	5
R3	4
R4	3
R5	4
R6	5

Therefore,

$$DC = \frac{R1 + R2 + R3 + R4 + R5 + R6}{n} = \frac{3 + 5 + 4 + 3 + 4 + 5}{6} = \frac{24}{6} = 4$$

where, 'n' is the number of replicants

In the above case, six bottles, having similar conditions such as the nature of the explant, and composition of the nutrient medium were considered to obtain a single final value. Hence, it is an average of six replicants. Therefore, the degree of callusing in the above example is equal to 4 i.e., very good (++++).

(e) Rate and degree of multiplication of callus, degree of induction of PLBs from callus and explants, rate of formation of plantlets, growth of plantlets and degree of rooting were also calculated by non-parametric method, as described earlier.

Percentage of PLB and plantlet formation from callus or explants directly, was calculated as follows:

Percentage of plantlet formation

$$= \frac{\text{No. of segments giving palntlets or PLBs}}{\text{No. of segments cultured}} \times 100$$

Similarly, percentage of survival of callus segments was also calculated. To obtain each value, two replicants of similar treatments were considered with each value being the average of two independent experiments.

(f) Plantlet growth was calculated by parametric method by measuring the stem length and counting the number of leaves and roots formed. Each value obtained is the average of all the plants in a bottle; each bottle is a replicant and the final value is the average of two independent replicants of similar treatments.

(g) To measure the rate of multiplication, multiplication ratio was calculated as:

$$\text{Multiplication Ratio (MR)} = \frac{\text{No. of plants produced}}{\text{No. of plants inoculated}}$$

STATISTICAL ANALYSIS

Two replicants were considered for each treatment with identical cultural conditions. There was considerable difference between the results obtained from the replicants maintained under similar conditions. The variations observed made it difficult to make conclusive remarks and it was found necessary to carry out statistical analysis. Hence, for most of the results, standard deviation and standard error were calculated.

Standard Deviation (SD) is the positive square root of the variance. It is the value of the deviation from the mean of the two replicants. SD was calculated as follows:

Number of days taken for the initiation of callus.

$$SD = \Sigma = \sqrt{\frac{\Sigma(X - \overline{X})^2}{n-1}}$$

	Replicants	No. of days taken for the initiation of the callus
1.	R1 = X1	41
2.	R2 = X2	45

Number of replicants n = 2

Mean difference $= \frac{86}{n}$

$= 84/2 = 43$

$$\Sigma (X - \quad)^2 = (X_1 - \overline{X})^2 + (X_2 - \overline{X})^2$$
$$= (41 - 43)^2 + (45 - 43)^2$$
$$= 2^2 + 2^2$$
$$= 4 + 4$$
$$= 8$$
$$\Sigma = \frac{8}{2-1}$$
$$\Sigma = 2.8284271$$
$$SD \underset{(SE)}{\Sigma} = 2.8$$

Standard Error of the mean was calculated as follows.

R1 = 41 and R2 = 45

$$SE = \frac{\Sigma}{n} = \frac{2.8}{2}$$

$$SE = \frac{2.8}{1.4142135}$$

$$SE = 1.97$$

RESULT AND DISCUSSION

For successful micropropagation, it is well known that orchids require at least three different stages of explant / callus / PLB / plantlet transfers to different media. Each medium with its own distinctive composition of nutrients, growth regulators, etc., has a characteristic influence on the morphogenetic process, thus reducing the time to yield the required transplantable plantlets. In this study, the generally employed three step process is extended to a five step process, i.e., five different media combinations were used from the explant stage to the regenerated transplantable plantlet stage. Further, to evaluate the relative merits of commonly used media in orchid tissue culture, these five different stages were repeated with VW, MS and KC media. These three media were selected for the study after screening the literature on *in vitro* study of about 260 orchids, which showed that of the thirty four media compositions used, more than 65% of the orchids had been studied using these three media.

In vitro morphogenesis was monitored at regular intervals (twice a week) and recorded. *In vitro* behaviour of different explants used in the study was recorded separately for each of the orchids. In all, six explant sources have been used, in this study viz., shoot apex, inflorescence, leaf, root, node and internode. In *Phalaenopsis,* however, shoot apex was not used in view of the monopodial nature of the orchid and the fact that excision of a shoot tip means death of the plant itself.

Before initiating the *in vitro* studies, careful explant screening and selection method was followed to ensure a higher rate of regeneration. Plants from both commercial nurseries as well as regenerated sources were used to obtain the required explants. It was found that the plants from the nurseries needed an in-house maintenance for at least fifteen days, for yielding good quality explants.

The following are the five stages under which morphogenesis of each explant, under appropriate media conditions, was studied:

1. Induction of callus or PLBs
2. Multiplication of callus or PLBs
3. Regeneration of plantlets
4. Multiplication of plantlets and
5. Growth and rooting of plantlets

INDUCTION OF CALLUS: Time Taken for the Visual Appearance of Callus

Callus was induced from explants in all the three media, mentioned earlier, with varied concentration of growth regulators. The data on the callus generated was studied with regard to requirement of time for callus initiation, percentage of callusing, degree of callusing, and degree of multiplication of callus.

For purposes of callus initiation, the duration required for the visual identification of the whole or part of the explant used for dedifferentiation was taken into account. Thus, the number of cells dedifferentiated could vary at this initial stage. In general, for all the four orchids, the initiation time of 18-69 days was required. The minimum time was recorded in the case of *Phalaenopsis* on MS medium and the maximum was recorded in the case of *Cattleya* on KC medium. Similarly, percentage of callusing was recorded in terms of the success rate of explants producing callus and the degree of callusing in terms of the amount of callus produced. Both these were studied by visual observation, as a precise qualitative estimate would disturb the *in vitro* process and also terminate the experiment itself.

To study the time required for the initiation of callus, shoot apices of all the experimental orchids were inoculated onto all the three basal media, with 8 gl^{-1} agar, 15 gl^{-1} sucrose, 100 mgl^{-1} citric acid, 150 mlL^{-1} CW, 2 mgl^{-1} 2,4-D and varied concentrations (1-5 mgl^{-1}) of BAP, and observations were made daily (see Tables 2-5).

Shoot tips of *Dendrobium* **Queen Sonia** (DQS) started producing callus in VW medium with 5 mgl^{-1} BAP within 24 days of inoculation. They initiated callus after 42 days of inoculation, when the concentration of BAP was 1 mgl^{-1} in the same medium. In MS medium, the initiation of callus was comparatively slow. It occurred in 35 days with 5 mgl^{-1} BAP. Initiation of callus occurred in 41 days with 1 mgl^{-1} BAP. Initiation of callus was slowest in KC medium as it took nearly 62 days at 1 mgl^{-1} BAP concentration. But callusing started in about 39 days in the same medium with 3 mgl^{-1} BAP. Though the concentration of 2,4-D was constant (2 mgl^{-1}), the increase in the concentration of BAP from 1 mgl^{-1} to 5 mgl^{-1} in all the medium considerably hastened the speed of callus induction upto certain levels. VM medium with 2 mgl^{-1} 2, 4-D and 3 mgl^{-1} BAP was found to be the best combination for callus induction from shoot apices of DQS, where callus was initiated within 21 days and only this callus was used for further studies (see Table 2).

Shoot apices of *Dendrobium* **Emma white** (DEW) also showed a similar response. But it was slightly faster in response. In VWM with 5 mgl^{-1} BAP, callusing started in about 23 days and it took 41 days to respond in the same medium with 1 mgl^{-1} BAP. In MS medium, initiation of callusing started in 34 days with 5 mgl^{-1} BAP. Callusing initiated after 40 days with 1 mgl^{-1} BAP. While KC medium with 1 mgl^{-1} BAP induced callusing in 65 days, but 1 mgl^{-1} BAP in the same, induced callus in 47 days. VW medium supplemented with 3 mgl^{-1} BAP was found to be the best for callus induction from the shoot apices of DEM, where callus initiation started in 20 days (see Table 3) and only this callus was used for further studies.

Shoot apex of *Cattleya* **Naomi Kerns** (CNK) showed faster callus induction in MS medium, as compared to VW and KC media. The callus was induced within 26 days of inoculation in MS medium with 5 mgl^{-1} BAP. In the same medium, with 1 mgl^{-1} BAP, callus induction occurred after 36 days. In KC medium, callus induction was a little slower. Shoot apex of *Cattleya* hybrid started producing callus after 39 days of inoculation in KC medium with 5 mgl^{-1} BAP, and callus formation started in 48 days in the same medium with 1 mgl^{-1} BAP. In VW medium, callus induction began slowly. With 5 mgl^{-1} BAP, callus formation started in 39 days and with 1 mgl^{-1} BAP, it started after 46 days of inoculation.

MS medium with 4 mgl^{-1} BAP and 2 mgl^{-1} 2, 4-D was found to be the best combination for inducing callus formation in the shoot apices of *Cattleya* hybrid (see Table 4), and only this callus was used for further studies.

While, shoot apex of *Phalaenopsis* **Queen Emma** (PQE) initiated callus formation within 18 days in MS medium with 5 mgl^{-1} BAP, but callus formation started after 33 days in MS medium with 1 mgl^{-1} BAP. In VW medium with 5 mgl^{-1} BAP, shoot apex of *Phalaenopsis* hybrid started producing callus after 34 days, and it took nearly 46 days in the same medium with 1 mgl^{-1} BAP. Shoot apex of *Phalaenopsis* hybrid initiated callus formation in KC medium supplemented with 1 mgl^{-1} BAP after 68 days of inoculation. The shortest time of only 18 days was taken by the shoot apices of *Phalaenopsis* hybrid which started producing callus in the MS medium with 5 mgl^{-1} BAP (see Table 5), and only this callus was used for further studies.

Callus started forming from the base of the shoot apices. Initially, the callus was white in colour and slowly started turning green. As the formation of callus started, the explants turned brown or black in colour, but sometimes they remained green and even produced very small leaves. Though initiation of callus was observed in all the above combinations, the nature and amount of callus formed, varied considerably depending on the orchid and the medium used.

Percent Callusing (PC) and Degree of Callusing (DC)

Different explants were inoculated in all the three basal media, with $8 gl^{-1}$ agar ,$15 gl^{-1}$ sucrose, 100 mlL^{-1} citric acid, 150 mlL^{-1} coconut water, 2 mgl^{-1} BAP, as constant and varied concentrations of NAA (0.5 to 5 mgl^{-1}) and observations were made in 10 weeks old culture (see Table 6-17).

Callus was induced successfully on all the explants used. It was found that leaf segments and internodal segments were comparatively difficult to induce callus. Meristem was the best explant, but it was difficult to isolate it. Shoot apices and nodal buds were found to be the most viable explants as they showed a good response and required a simpler procedure. Root segments were also equally good for the induction of callus, but they produced some unusual types of callus in certain concentrations, and plants obtained from the root explants are yet to show flowering, whereas the plants obtained from all other explants have already flowered.

Dendrobium Queen Sonia in MS medium with 2 mgl^{-1} BAP as constant and varied concentrations of NAA (see Table 6)

Meristem explants showed 61% callusing at 0.5 mgl^{-1} NAA, and 63% callusing at 5 mgl^{-1} NAA. Maximum percentage of callusing (82%) was obtained at 2 mgl^{-1} NAA. It showed 2.1 degree of callusing (DC) at 0.5 mgl^{-1} NAA and 5.0 DC at 5 mgl^{-1} NAA. DC increased with the increase in the concentration of NAA. Percentage of callusing increased initially, but started declining beyond the optimum level of NAA concentration.

Shoot apex showed 55 and 48 percent callusing (PC) at 0.5 and 5 mgl^{-1} concentrations of NAA, respectively. 77 PC was recorded at 2 mgl^{-1} NAA concentration. DC was 1.8 at 0.5 mgl^{-1} NAA and 5.0 at 5 mgl^{-1} NAA. Nodal buds showed 55 PC at 0.5 mgl^{-1} NAA and 40 PC at 5 mgl^{-1} NAA level. DC was 2.0 at 0.5 mgl^{-1} NAA and 4.6 at 5 mgl–1 NAA. Leaf segments showed low PC and DC. They showed no callusing at 0.5 mgl^{-1} NAA, where they remained green for about three weeks and started turning brown and became black by 6 weeks. At 1 mgl^{-1} NAA, leaf segments showed whitish callus in the margins. Leaf segments showed wrinkling and swelling after about 2 weeks. Callus started appearing on the margins in the fifth week. At this level (2 mgl^{-1}) of NAA, 20% callusing was observed at 5 mgl^{-1} NAA, and 5 PC was recorded.

Internodal segments showed 32 PC at 3 mgl^{-1} NAA, and 23% PC at 0.5 mgl^{-1} NAA and 15 PC at 5 mgl^{-1} NAA level. They recorded 1 DC at 0.5 mgl^{-1} NAA and 2.4 DC at 5 mgl^{-1} NAA concentration.

Root segments showed very good response at all levels of NAA. PC increased with the increase in NAA concentration, but it started declining from 3 mgl^{-1} concentration onwards. It showed 60 and 53 PC at 0.5 and 5 mgl^{-1} NAA levels, respectively. Maximum PC (78) was observed at 2.0 mgl^{-1} NAA level. DC showed a steady increase, as the concentration of NAA increased. 2.0 DC was recorded at 0.5 mgl^{-1} NAA level and 4.5 DC was recorded at 5 mgl^{-1} NAA.

***Dendrobium* Queen Sonia in KC medium with 2 mgl^{-1} BAP and varied concentration of NAA (see Table 8).**

Meristem showed 70 PC at 2 mgl^{-1} NAA and showed 59 PC, both at 0.5 mgl^{-1} and 5 mgl^{-1} levels. 2.0 DC was observed at 0.5 mgl^{-1} NAA, and 4.9 at mgl^{-1} NAA concentration. DC increased with the increase in the level of NAA concentration. Shoot apex showed a maximum of 69 PC at 2 mgl^{-1} NAA concentration. It showed 50 and 48 PC at 0.5 mgl^{-1} and 5 mgl^{-1} NAA concentrations, respectively. 2.0 DC at 0.5 mgl^{-1} NAA increased to 4.4 at 5 mgl^{-1} NAA concentration. Both PC and DC were better in meristem explant as compared to shoot apex. Nodal buds showed 67 PC at 2 mgl–1 NAA. 54 PC was recorded at 0.5 mgl^{-1} NAA and 38 PC at 5 mgl^{-1} NAA level. In this case, overall DC was not as good as in meristem and shoot apex explant. A maximum of 4 PC was recorded at 5.0 mgl^{-1} NAA.

Leaf segments showed response to callus induction within a short range of 1 to 4 mgl^{-1} NAA. 30 PC was recorded at 3 mgl^{-1} NAA and 10 PC was recorded at 1 mgl^{-1} NAA. DC was also very low it was 1.5 at 2.0 mgl^{-1} NAA level and decreased with both increase and decrease in the NAA level.

Internodal segments showed 20 PC at both 0.5 mgl^{-1} and 4.0 mgl^{-1} levels. A maximum of 30 PC was observed at 2.0 mgl^{-1} concentration. DC ranged between 1 and 1.9 and it increased as the concentration of NAA increased.

Root segments showed 65 PC at 3.0 mgl^{-1} NAA which was the optimum level. They showed 49 PC and 45 PC at 0.5 mgl^{-1} and 5 mgl^{-1} levels of NAA, respectively. DC was 1.5 at 0.5 mgl^{-1} NAA and increased to 4.0 at 5 mgl^{-1} NAA.

***Dendrobium* Queen Sonia in VW medium with mgl^{-1} BAP as constant and varied concentrations of NAA (see Table 7).**

VWM was found to be the best for all as maximum PC and DC was recorded in this medium.

91% of the meristems inoculated, showed the formation of callus at 2mgl^{-1} NAA level. By the end of sixth week, yellowish masses of calli were formed at 2.0 mgl^{-1} NAA level. At 5 mgl^{-1} NAA concentration, 68% of the meristems showed callusing and the remaining meristems dried up at 5 mgl^{-1} NAA level. DC was high (4.8) at 5 mgl^{-1} NAA level. 65 PC was observed and 2.2 DC was recorded at 0.5 mgl^{-1} NAA.

Shoot apex showed 83 PC at 2 mgl^{-1} BAP. At 1 mgl^{-1}, it showed 80 PC, even meristem showed a good rate of 89 PC at 1 mgl^{-1}. NAA shoot apices showed 4.6 DC at 5.0 mgl^{-1} NAA and 2.5 DC at 0.5 mgl^{-1} NAA.

Nodal buds showed 75 and 76 PC at 1 mgl^{-1} and 2 mgl^{-1} NAA levels, respectively. At 3 mgl^{-1} and 0.5 mgl^{-1} levels of NAA, they showed 49 and 66 PC, respectively. DC however increased with the increase in the concentration of NAA.

Leaf segments showed no response at 0.5 mgl^{-1} NAA but they showed a good response of 75 PC at 3 mgl^{-1} NAA and they showed 39 PC and 40 PC, at 2 mgl^{-1} and 4 mgl^{-1} levels of NAA, respectively. A maximum of 2.0 DC was recorded at 2.0 mgl^{-1} NAA level.

Internodal segments showed a maximum of 39 PC at 2.0 mgl^{-1} and a maximum of 2.0 DC at 5 mgl^{-1} NAA.

Root segments showed a better response than the nodal buds and internodes. A maximum of 80 PC was observed at 2 mgl^{-1} NAA, and 79 PC at 1 mgl^{-1} NAA. PC ranged between 60 to 50 with a maximum of 80, in the concentrations between 0.5 mgl^{-1} to 5 mgl^{-1} of NAA. 3.0 DC was observed, at 2mgl^{-1} concentration, but it increased to 4.6 at 5 mgl^{-1} NAA level.

***Dendrobium* Emma White in MS medium with 2 mgl^{-1} BAP and varied concentrations of NAA (see Table 38).**

Meristem showed 90 PC at 1 mgl^{-1} NAA level and showed 88 PC at 2 mgl^{-1}. In these levels of NAA, 2.5 and 3.0 DC were recorded. At 5 mgl^{-1} NAA level, 58 PC and 4.9 DC were recorded. At 0.5 mgl^{-1} NAA level, 60 PC and 2.1 DC were recorded.

Shoot apex showed 86 PC at 1.0 mgl^{-1} NAA and 79 PC at 2.0 mgl^{-1} NAA. At 0.5 mgl^{-1} and 5 mgl^{-1} NAA levels, 58 and 50 PC were recorded.

At 5 mgl^{-1} NAA, 5.0 DC was recorded and 2.2 DC was recorded at 0.5 mgl^{-1} NAA concentration.

Nodal buds showed 72 PC at 1 mgl^{-1} NAA and 67 PC at 2 mgl^{-1} NAA. At 0.5 mgl^{-1} and 5 mgl^{-1} NAA concentrations, they showed 49 and 45 PC. DC ranged between 1.9 and 3.8.

Internodal segments showed 68 PC at 2 mgl^{-1} NAA and 58 PC at 1 mgl^{-1} NAA. They showed 1.0 and 2.8 DC at 0.5 mgl^{-1} and 5 mgl^{-1} NAA concentrations, respectively.

It was difficult to induce callus in leaf segments. There was not a much difference as far as the induction of callus was concerned in basal, middle and apical regions of the leaf. While leaf segments showed very slow callus formation, callus formed from the cut regions of the leaf segments. It appeared as a white growth first, turning yellow gradually. At 0.5 mgl^{-1} NAA, the leaf segments turned yellow first and then turned brown and there was no callus formation, and all segments died. At 2.0 mgl^{-1} NAA, about 48% of the leaf segments showed callus formation within six weeks, whereas the others wrinkled and remained green for another two weeks and then turned brown and died. At 2 mgl^{-1} NAA level, 3.1 DC was recorded.

Root segments showed response to callus induction at all levels of NAA studied. At 3 mgl^{-1} NAA level, they showed 70 PC. PC ranged between 45 to 50. DC was maximum (4.5) at 5 mgl^{-1} NAA concentration.

***Dendrobium* Emma White in KC medium with 2 mgl–^{1} BAP as constant and varied concentrations of NAA (see Table 14).**

Meristem showed 86 PC in 2 mgl^{-1} NAA concentration. It showed 60 PC, both at 0.5 mgl^{-1} and 5 mgl^{-1} NAA concentrations. It showed 4.8 DC at 5 mgl^{-1} NAA concentration. Shoot apex showed 84 PC at 2 mgl^{-1} NAA concentration. It recorded 60 62 PC at 0.5 and 5 mgl^{-1} NAA levels. It showed 4.6 DC at 5 mgl^{-1} NAA levels, respectively. It showed 4.6 DC at 5 mgl^{-1} NAA concentration, but at 0.5 mgl^{-1} level it showed 1.6 DC.

Nodal buds showed 72 PC at 2 mgl^{-1} NAA level and showed 48 and 49 PC at 0.5 mgl^{-1} and 5 mgl^{-1} NAA levels, respectively. They showed 3.9 DC at 5 mgl^{-1} NAA level. Leaf segments did not show any response at 0.5 mgl^{-1} and 1 mgl^{-1} NAA levels. At 2.0 mgl^{-1} level, leaf segments showed 22 PC. DC ranged between 0.5 at 5 mgl^{-1} NAA levels to 1.2 at 2 mgl^{-1} level. Internodal segments showed a maximum of 23 PC at 2 mgl^{-1} level and was even lower at 0.5 mgl^{-1} and 5 mgl^{-1}, where it was 10 and 13, respectively. Root segments showed 75 PC at 2 mgl^{-1} NAA concentration and 4.5 DC at 5 mgl^{-1} NAA concentration.

***Dendrobium* Emma White in VW medium with 2 mgl^{-1} BAP as constant and varied concentrations of NAA (see Table 13).**

VWM, was found to be the best medium for this hybrid. Meristem, followed by shoot apex were the best explants.

Meristem showed 92 PC at 2 mgl^{-1} NAA. It showed 65 PC at 0.5 mgl^{-1} level and 66 PC at 5 mgl^{-1} NAA concentrations. This explant showed 5.0 DC at 5 mgl^{-1} NAA concentration and 1 DC at 0.5 mgl^{-1} NAA concentration and 3.2 DC at 2 mgl^{-1} NAA level.

Shoot apex showed 86 PC and 2.5 DC at 2 mgl^{-1} NAA level. It showed 45 PC at 5 mgl^{-1} NAA and 61 PC at 0.5 mgl^{-1} NAA level. DC was maximum at 5 mgl^{-1} NAA where it recorded 4.5 DC.

Nodal buds showed a good response in the above medium. They started producing callus by the end of the fifth week after inoculation. Buds in contact with the medium produced callus, whereas the cut stem region along which the buds were inoculated remained green. Only the cut terminal ends turned black. Buds showed 78 PC at 2 mgl^{-1} NAA level and 69 PC at 0.5 mgl^{-1} NAA level. While, PC was 30 at 5 mgl^{-1} NAA level, DC ranged between 1.0 at 0.5 mgl^{-1} NAA and 4.4 at 5 mgl^{-1} NAA levels.

Leaf segments did not respond at lower levels of NAA concentration. At 3 mgl^{-1} NAA level, a maximum of 28 PC was observed and at 4 mgl^{-1} NAA concentration, 15 PC and 3 DC were recorded.

Internodal segments showed 38 PC and 2.2 DC at 2.0 mgl^{-1} NAA concentration level. This explant showed a maximum of 3.3 DC at 5 mgl^{-1} NAA concentration.

Root segments showed 80 PC and 79 PC at 1mgl^{-1} and 2 mgl^{-1} NAA levels, respectively. DC which was found to be 0.5 at 0.5 mgl^{-1} NAA level, increased to 4.8 at 5 mgl^{-1} NAA level.

***Cattleya* Naomi Kerns in VW medium with 2 mgl^{-1} BAP and varied concentrations of NAA (see Table 16).**

In case of meristem explants 89 PC was recorded at 1.0 mgl^{-1} NAA level while 0.5 mgl^{-1} and 5 mgl^{-1} NAA levels recorded 58 and 48 PC respectively, DC increased from 1.9 at 0.5 mgl^{-1} NAA level to 5.0 at 5 mgl^{-1} NAA level.

Shoot apex showed 86 PC at 1.0 mgl^{-1} NAA and it showed 56 and 45 PC at 0.5 mgl^{-1} and 5 mgl^{-1} NAA levels, respectively; and 5.0 DC was observed at 5 mgl^{-1} NAA concentration.

Nodal buds showed 79% callusing at 1 mgl^{-1} NAA. PC decreased with and increase in the NAA concentration above this level. At 1 mgl^{-1} NAA level they showed 2.3 DC which increased to 3.9 at 5 mgl^{-1} NAA level.

Leaf segments showed callusing at all concentrations ranging from 1 mgl^{-1} to 5 mgl^{-1}. Leaf segments showed swelling and wrinkling within a weeks time. They remained green for nearly two weeks then they turned yellow-green and subsequently produced whitish yellow callus. Leaf bases gave better results as compared to laminar and apical regions of the leaf. At 2.0 mgl^{-1} NAA, leaf segments showed 38 PC and at 1 mgl^{-1} NAA 52 PC was observed, but DC ranged between 1.0 to 1.8. It was maximum (2.5) at 5.0 mgl^{-1} NAA level.

Internodal segments showed 49 PC and 2.2 DC at 1 mgl^{-1} level. Root segments, however, showed a better response as compared to leaf segments, internodal segments and nodal buds. They showed 80 PC at 1.0 mgl^{-1} NAA which decreased as the NAA level increased. At 1.0 mgl^{-1} NAA concentration 2.0 DC was recorded, which increased to 3.9 at 5 mgl^{-1} NAA.

***Cattleya* Naomi Kerns in MS medium with 2 mgl^{-1} BAP and varied concentrations of NAA.**

For callus induction in this hybrid, MS was the best medium. Most of the explants showed good callusing at 1 mgl^{-1} NAA level. Meristem showed 90 PC at 1 mgl^{-1} NAA. It showed 58 and 50 PC at 0.5 $m l^{-1}$ and 5 mgl–^{1} NAA concentrations, respectively and 4.6 DC at 5 mgl^{-1} NAA.

Shoot apex showed 88 PC at 1 mgl^{-1} NAA and it showed 4.9 DC at 5 mgl^{-1} NAA.

Nodal buds showed 86 PC at 1 mgl^{-1} NAA. At 5 mgl^{-1} NAA, they showed 44 PC with a maximum of 4.2 DC being observed at 5 mgl^{-1} NAA.

Leaf segments did not show any response at 0.5 mgl^{-1} NAA. 40 PC was observed at 1 mgl^{-1} NAA and 2.5 DC was recorded at 5 mgl^{-1} NAA.

Internodal segments showed 77 PC at 1 mgl^{-1} NAA. They showed 40 PC both at 0.5 mgl^{-1} NAA and 5.0 mgl^{-1} NAA levels. A maximum of 3.0 DC was observed at 5 mgl^{-1} NAA level.

Root segments showed 79 PC at 2 mgl^{-1} NAA, and at this level they showed 2.6 DC. But at 5 mgl^{-1} NAA concentration, they showed 4.0 DC.

***Cattleya* Naomi Kerns in KC medium with 2 mgl^{-1} BAP as constant and varied concentrations of NAA (see Table 17).**

Meristem explants showed 73 PC and 3.9 DC at 2 mgl^{-1} NAA level. PC decreased with the increase in the NAA concentration. At 5 mgl^{-1} NAA level, 4.5 DC was observed.

Shoot apex showed 69 PC at 2 mgl^{-1} NAA, and at this level it showed 2.1 DC. 4.0 DC was observed at 5 mgl^{-1} NAA level.

Nodal buds showed 68 PC and 1.6 DC at 2.0 mgl^{-1} NAA. At 5 mgl^{-1} NAA, they showed 33 PC and 2.6 DC.

Leaf segments did not show any response, both at 0.5 mgl^{-1} and 5 mgl^{-1} NAA. They showed 25 PC and 1.0 DC at 1 mgl^{-1} NAA concentration which was the best response obtained.

Internodal segments showed 44 PC and 1.8 DC at 2.0 mgl^{-1} NAA, and at 5 mgl^{-1} NAA, they showed 25 PC and 2.2 DC.

Root segments showed 60 PC and 3.0 DC at 2.0 mgl^{-1} NAA and at 5 mgl^{-1} NAA concentration, they showed 30 PC and 3.8 DC.

***Phalaenopsis* Queen Emma in MS Medium with 2 mgl^{-1} BAP and varied concentrations of NAA (see Table 18, Fig. 28).**

In this hybrid, inflorescence stalk was used to obtain most of the explants. For example, instead of shoot apex, inflorescence apex was used; instead of stem nodal buds, inflorescence nodal buds were used; and instead of internodal segments of stem, internodal segments of inflorescence stalk were used. All these segments showed good response.

Meristem gave best results. It showed 90 PC at 2.0 mgl^{-1} NAA concentration. The callus, which was initially whitish, turned green after three weeks of callus formation. Meristem showed 90 PC at 2 mgl^{-1} NAA and 4.9 DC at 5.0 mgl^{-1} NAA level. Inflorescence apex showed 80, 85 and 90 PC at 0.5 mgl^{-1}, 1.0 mgl^{-1} and 2 mgl^{-1} NAA levels, respectively. But maximum DC of 5.0 was recorded at 5 mgl^{-1} NAA.

Nodal buds showed 77 and 78 PC at 1.0 and 2.0 mgl^{-1} NAA levels. But DC was maximum at 4.8 at 5 mgl^{-1} NAA level.

Leaf segments obtained from young plants showed a good response. Basal as well as the apical regions showed response for callusing. In leaf segments, 50 PC was observed at 2.0 mgl^{-1} NAA and 2.8 DC was observed at 5 mgl^{-1} NAA, which was the maximum for this explant. At this concentration, PLBs and plantlets were directly produced from the cut regions of the leaf segments. This occurred without the appearance of callus.

Internodal segments showed 79 PC and 2.7 DC at 2.0 mgl^{-1} NAA. They showed 42 PC and 3.2 DC at 5 mgl^{-1} NAA levels.

Root segments showed 76 and 78 PC at 1.0 mgl^{-1} and 2.0 mgl^{-1} NAA levels, respectively. They showed 41 PC and 4.2 DC at 5 mgl^{-1} NAA level.

***Phalaenopsis* Queen Emma in KC medium with 2 mgl^{-1} BAP as constant and varied concentrations of NAA (see Table 19, Fig. 31).**

Meristem showed 75 PC at 2 mgl^{-1} NAA level, and 2.9 DC at this level of NAA. It showed 60 PC and 4.8 DC at 5 mgl^{-1} NAA concentration. Inflorescence apex showed 70 PC and 2.8 DC at 2 mgl^{-1} NAA. It showed 50 PC and 4.9 DC at 5 mgl^{-1} NAA concentration.

Nodal buds showed 72 PC and 3.0 DC at 2 mgl^{-1} NAA, and 48 PC and 4.8 DC at 5 mgl^{-1} NAA level.

Leaf segments showed 40 PC and 2.0 DC at 2 mgl^{-1} NAA. At 5 mgl^{-1} NAA, they showed 5 PC and 1.0 DC. No response was noted at 0.5 and 1mgl^{-1} NAA.

Internodal segments showed no response at 0.5 mgl^{-1} NAA. At 2.0 mgl^{-1} NAA, they showed 51 PC and 1.8 DC, and at 5 mgl^{-1} NAA, 15 PC and 2.2 DC were observed.

Root segments showed no response at 0.5 mgl^{-1} NAA. But at 2.0 mgl^{-1} NAA, they showed 68 PC and 2.2 DC. DC was maximum (3.9) at 5.0 mgl^{-1} NAA concentration.

***Phalaenopsis* Queen Emma in VW medium with 2 mgl^{-1} BAP and varied concentrations of NAA (see Table 20, Figs. 29 and 30).**

VWM was found to be the best medium for callus induction in almost all explants of this hybrid.

Meristem showed 90 PC at 1 mgl^{-1} NAA and at this level of NAA 2.5 DC was recorded. At 5 mgl^{-1} NAA level, 45 PC and 5.0 DC were observed.

Inflorescence apex showed 90 PC at 2.0 mgl^{-1} and 85 PC at 1.0 mgl^{-1} NAA. It showed 4.5 DC at 4.0 mgl^{-1} NAA and 5.0 DC at 5 mgl^{-1} NAA concentrations.

Nodal buds showed 79 PC at 1.0 mgl^{-1} NAA and 82 PC at 2 mgl^{-1} NAA. They showed 3.9 DC at 4.0 mgl^{-1} NAA and 4.5 DC at 5.0 mgl^{-1} NAA.

Leaf segments showed response at all levels of NAA. PC was lowest (20) at 5.0 mgl^{-1} and highest (50) at 2.0 mgl^{-1} NAA. Maximum DC was recorded (2.5) at 5 mgl^{-1} NAA concentration.

Internodal segments showed 60 PC and 2.8 DC at 3.0 mgl^{-1} NAA and 42 PC and 2.9 DC at 5.0 mgl^{-1} NAA. Root segments showed 69 PC and 3.0 DC at 2.0 mgl^{-1} NAA and 39 PC and 4.1 DC at 5 mgl^{-1} NAA concentration.

Multiplication of Callus

The Degree of Multiplication of callus (DM) was observed visually and catagorised under five observable visual stages, as detailed in methods. These stages represent the relative quantities of callus growth from the callus segments subcultured. Callus obtained from the shoot tip was used in this study (see Table 18).

Callus was segmented into small pieces and inoculated on all the three basal media with 30 mgl^{-1} sucrose, 8gl^{-1} agar, 100 mgl^{-1} ascorbic acid, and varied concentrations of NAA and BAP.

Dendrobium **Queen Sonia** showed 5 DM at 2 mgl^{-1} NAA and 1 mgl^{-1} BAP in MS medium. At the same level of growth regulators in VW medium, it showed 4.6 DM and in KC medium, it showed 3.8 DM. DM increased with the increase in NAA and BAP levels, but after the above mentioned levels it again started decreasing.

Dendrobium **Emma White** also showed similar results at 2 mgl^{-1} NAA and 1 mgl^{-1} BAP, showed at 5 DM, 4.8 DM and 4.0 DM MS, VW and KC media, respectively.

Cattleya hybrid showed a maximum of 5 DM at 2 mgl^{-1} NAA and 3 mgl^{-1} BAP in MS medium; at this level it showed 4.2 and 3.8 DM in VW and KC media, respectively. It showed 4 DM at 1 mgl^{-1} NAA and 3 mgl^{-1} BAP in MS medium.

Phalaenopsis hybrid showed 5 DM, 4 DM and 3.3 DM at 3 mgl^{-1} and 2 mgl^{-1} BAP in MS, VW and KC media, respectively. In MSM at 1 mgl^{-1} NAA and 3 mgl^{-1} BAP, it recorded 4.1 DM, and at 3 mgl^{-1} each of BAP and NAA in MSM, it showed 4.4 DM.

Requirement of NAA and BAP for callus multiplication was found to be low in *Dendrobium* hybrids, moderate in *Cattleya,* and high in *Phalaenopsis*. However, PLBs and occasionally plantlets were also produced in the above treatments.

Regeneration of PLBs and plantelts

For this study, different explants and callus segments obtained from the earlier stages were inoculated on all the three selected media with varied concentrations of growth regulators (see Tables 22-41, Figs. 5-7, 32, 33, 53,54, 56, 58, 59, 86, 90 and 91-94).

As in earlier cases, here also, visual, non-parametric observations were made at regular intervals for 10 weeks and recorded in the Tables mentioned above. The purpose of this study was to identify the best medium suited for the individual orchids selected for study. Explants from shoot apices of all orchids yielded good results. VW medium was found to be suitable for *Dendrobium* hybrids and MS medium was better for *Cattleya* and *Phalaenopsis*. Hence these combinations were further experimented with in this study. Performance in terms of weight of the callus produced, percentage of surviving explants, percentage of explants producing callus and percentage of PLBs formation were studied.

A similar experiment was also conducted with callus segments, by growing them on selected media. In addition to the parameters of percentage of survival and percentage of PLB formation, colour of PLBs and number of plantelets derived from each cultured segment have also been studied. For the purposes of above studies, explants and callus segments were inoculated on all the three basal media with 20 gl^{-1} sucrose, 100 mgl^{-1} ascorbic acid, 100 mgl^{-1} citric acid, 1 mgl^{-1} BAP and varied concentration of NAA.

Dendrobium **Queen Sonia** showed the highest degree of PLB formation in VW medium, while it was moderate in MS medium. In KC medium, the degree of PLB formation was least. In all the cases, an increase in the concentration of NAA decreased the degree of PLB formation beyond 1 mgl^{-1} level, and the best results were obtained at 0.5 mgl^{-1} NAA. Of all the explants used, shoot apex, inflorescence apex, nodal buds and callus segments showed high degree of PLB formation. Root segments showed least degree of PLB formation and they required higher levels of NAA.

Dendrobium **Emma White** showed high degree of PLBs formation in VW medium. In MS medium, it showed a moderate response. The least degree of response was observed in KC medium. 0.5 mgl^{-1} NAA level was found to be most suitable for shoot apex. Inflorescence apex, callus segments and leaf segments responded well, both at 0.5 mgl^{-1} and 1 mgl^{-1} NAA levels, whereas root segments showed the best response at 2 mgl^{-1} NAA level. Degree of PLBs formation decreased in all explants as the concentration of NAA increased and at 5 mgl^{-1} NAA level the degree of PLB formation was almost nil.

In case of *Cattleya* hybrid, the degree of PLBs formation was best in MSM. It was moderate in VW medium. In KC medium, the degree of PLB formation was found to be the least. The response was very good at 1 mgl^{-1} NAA level. All explants except root segments showed best response at this level of NAA. Moderate response was observed at 0.5 mgl^{-1} NAA level. Root explants showed the best response at 2 mgl^{-1} NAA level. Though the degree of response varied, callus segments showed PLBs formation at almost all levels (0.5-5 mgl^{-1}) of NAA, whereas other explants failed to show PLBs formation at higher levels of NAA.

In *Phalaenopsis* **Queen Emma,** the best degree of PLB formation was observed in MSM. The response was moderate in VW medium. In KC medium it was the least. *Phalaenopsis* hybrid showed a requirement for high NAA concentration. All explants except root segments showed good degree of PLB formation at 5 mgl^{-1} NAA level. The response was moderate at 4 mgl^{-1} NAA. The degree of PLB formation was least at lower NAA concentrations. Root segments however showed good response at 3 mgl^{-1} NAA level. In MS medium, root segments showed equal degree of PLB formation at 3 mgl^{-1}, 4 mgl^{-1} and 5 mgl^{-1} concentrations of NAA. White callus appeared from the explants at the point of contact with medium, which gradually turned green at some places and started producing PLBs.

To study the percentage of survival of explants, wet weight of the callus produced and percentage of PLB formation, shoot apices of *Dendrobium* and *Cattleya* hybrids and inflorescence apex of *Phalaenopsis* hybrid were inoculated on MS medium, VW medium or KC medium with 15 gl^{-1} sucrose, 8.5 gl^{-1} agar, 100 mgl^{-1} citric acid, 100 mgl^{-1} ascorbic acid, and varied concentrations of BAP and 2, 4-D.

Dendrobium **Queen Sonia** shoot apices were inoculated on VW medium. The percentage of survival was 95 at 0.1 mgl^{-1} 2, 4-D and 0.1 mgl^{-1} BAP level. The same result was obtained even at 1 mgl^{-1} 2, 4-D and 1 mgl^{-1} BAP. Percentage survival decreased with the increase in the concentration of both 2, 4-D and BAP. At higher levels of growth regulators the percent survival was low. On the contrary, the wet weight of the callus formed from each shoot apex explant increased with the increase in the concentration of both 2, 4-D and BAP. Wet weight was found to be maximum (482 mg) at 2 mgl^{-1} BAP and 5 mgl^{-1} 2, 4-D levels.

Percentage of PLB formation showed a decrease with the increase in the concentration of 2, 4-D and BAP. Maximum PLB formation (70%) was observed at 1 mgl^{-1} BAP and 1 mgl^{-1} 2, 4-D. No PLB formation was observed when 2, 4-D was used at higher levels than 4 mgl^{-1} concenration.

Dendrobium **Emma White** shoot apices were inoculated onto VW medium and they showed a decrease in the percentage of survival with an increase in the 2, 4-D level. At 0.1 mgl^{-1} 2, 4-D and 0.1 mgl^{-1} BAP level, 90% survival was recorded. At 1 mgl^{-1} BAP and 0.1 mgl^{-1} 2, 4-D, maximum survival (95%) was noticed. A minimum of 0.5% survival was recorded at 0.1 mgl^{-1} BAP and 5 mgl^{-1} 2, 4-D. 25% survival was recorded at 2 mgl^{-1} BAP and 0.1 mgl^{-1} 2, 4-D levels. Wet weight of the callus increased with the increase in the concentration of BAP and 2, 4-D. Minimum wet weight (292 mg) of callus was recorded at 0.1 mgl^{-1} BAP and 0.1 mgl^{-1} 2, 4-D. Maximum wet weight (542 mg) of callus was recorded at 2 mgl^{-1} BAP and 5 mgl^{-1} 2, 4-D.

Percentage of PLB formation decreased with the increase in the concentration of 2, 4-D and BAP. 65 percent PLBs formation was recorded at 0.1 mgl^{-1} BAP and 0.1 mgl^{-1} 2, 4-D level. At 1 mgl^{-1} BAP and 0.1 mgl^{-1} 2, 4-D, percent PLB formation was recorded 60. There was no PLBs formation at higher levels of 2, 4-D.

Cattleya **Naomi Kerns** shoot apices were cultured on MS medium. They showed 100 percent survival at 0.1 mgl^{-1} 2, 4-D and 0.1 mgl^{-1} BAP levels. But the percent survival decreased with the increase in the concentration of BAP and 2, 4-D. Percent survival was 90 at 1 mgl^{-1} BAP and 0.1 mgl^{-1} 2, 4-D. At 5 mgl^{-1} 2, 4-D and 0.1 mgl^{-1} BAP, the percent survival was as low as 5.

Wet weight of the callus formed increased with the increase in the concentration of BAP and 2, 4-D. Lowest wet weight (310 mg) was recorded at 1 mgl^{-1} BAP and 0.1 mgl^{-1} 2, 4-D levels. The highest wet weight (553 mg) was observed at 2 mgl^{-1} BAP and 5 mgl^{-1} 2, 4-D concentrations.

Percentage of PLB formation was maximum (90%) at 1 mgl^{-1} BAP and 0.1 mgl^{-1} 2, 4-D level; even at 2 mgl^{-1} BAP and 0.1 mgl^{-1} 2,4-D, 90% PLB formation was recorded. But at 2 mgl^{-1} BAP and 0.1 mgl^{-1} 2, 4-D, PLB formation was 45%. As there was an increase in the concentration of 2, 4-D, PLB formation decreased and no PLBs formation was recorded at 3, 4 and 5 mgl^{-1} levels of 2, 4-D.

Phalaenopsis **Queen Emma** inflorescence apices were cultured on MS medium. They showed 90% survival at 0.1 mgl^{-1} BAP and 1 mgl^{-1} 2, 4-D. At 0.1 mgl^{-1} BAP and 0.1 mgl^{-1} 2, 4-D, they showed 85% survival. But as the concentration of 2, 4-D was increased beyond 1 mgl^{-1} i.e., at 2, 3, 4 and 5 mgl^{-1} the percentage of survival was 75, 50, 35 and 15, respectively. At 2 mgl^{-1} BAP and 0.1 mgl^{-1} 2, 4-D; 80% survival was recorded. Percentage of survival gradually decreased as there was further increase in the concentration of 2, 4-D. Wet weight of the callus increased with the increase in concentration of 2, 4-D. At 0.1 mgl^{-1} 2, 4-D and 0.1 mgl^{-1} BAP level, 250 mg wet weight was recorded and at 0.1 mgl^{-1} BAP and 5 mgl^{-1} 2, 4-D, the wet weight was 306 mg. Wet weight was maximum at 2 mgl^{-1} BAP and 5 mgl^{-1}. At this level 458 mg callus was formed.

Percentage of PLB formation was maximum (80%) at 2 mgl^{-1} BAP and 0.1 mg 2, 4-D. It was 60% at 1 mgl^{-1} BAP and 2 mgl^{-1} 2, 4-D level and 45% at 0.1 mg BAP mgl^{-1} and 2 mgl^{-1} 2, 4-D. But at higher concentrations of 2, 4-D, percentage of PLBs formation was low and there was no PLB formation at 1 mgl^{-1} BAP and 5 mgl^{-1} 2, 4-D level.

To study the effects of NAA and BAP on percentage of survival, percentage of PLB formation, colour of PLBs the number of plantlets formed directly from callus segments of all the orchid hybrids were inoculated on different (MS, VW or KC) media with 8.5 gl^{-1} agar, 20 gl^{-1} sucrose and varied concentration of BAP and NAA.

Dendrobium **Queen Sonia** callus segments were inoculated on VWM. Percentage of survival was a maximum (100%) at 1 mgl^{-1} NAA and 1 mgl^{-1} BAP. At 1 mgl^{-1} NAA and 5 mgl^{-1} BAP concentration, 48% of survival was recorded. At 5 mgl^{-1} NAA and 1 mgl^{-1} BAP, 22% of survival was observed, and at 5 mgl^{-1} NAA and 15 mgl^{-1} BAP, percentage of survival was zero. Minimum percentage of survival (6) was recorded at 10 mgl^{-1} NAA and 5 mgl^{-1} BAP, and 15 mgl^{-1} NAA and 1 mgl^{-1} BAP. No survival of callus segments was recorded when the concentration of BAP and NAA was above 10 mgl^{-1}.

Percentage of PLB formation also decreased with the increase in the concentration of NAA and BAP. The 70% of PLBs formation was recorded at 1 mgl^{-1} NAA and 1 mgl^{-1} BAP. At 1 mgl^{-1} NAA and 5 mgl^{-1}, 20% of PLB formation was recorded. At 5 mgl^{-1} NAA and 1 mgl^{-1} BAP, 18% of PLB formation was observed. Minimum percentage of PLBs formation was observed at 10 mgl^{-1} NAA and 1 mgl^{-1} BAP, where 4% of PLB formation was recorded. There was no PLBs formation when the concentration of NAA was higher than 5 mgl^{-1}. Colour of the PLBs was found to be yellow or pale yellow or pale green or green. PLB formed with lower concentration of growth regulators gave rise to plants more often. If the concentrations were increased, PLBs were pale green and pale yellow. PLBs under the influence of high concentrations were yellow in colour. Pale yellow and yellow PLBs gradually turned green. After 12 weeks of culture, percentage of yellow PLBs turning green reduced with the increase in the concentration of growth regulators.

At 1 mgl^{-1} NAA and 1 mgl^{-1} BAP level, about five plantlets were formed directly from the callus segments. Rate of plantlet formation decreased with the increase in concentration of growth regulators. Under the influence of high concentrations, no plantlets were produced. When the PLBs produced were subcultured on VW (growth medium), green and pale green PLBs always formed plantlets. Pale yellow PLBs showed slow growth. Yellow PLBs either showed no growth or died.

Dendrobium **Emma White** showed similar results as that of the other *Dendrobium* hybrid, when its callus segments were inoculated onto VW medium. 98% survival was recorded at 1 mgl^{-1} NAA and 1 mgl^{-1} BAP level. At 1 mgl^{-1} NAA and 5 mgl^{-1} BAP level, 52% survival was recorded. At 5 mgl^{-1} NAA and 10 mgl^{-1} BAP, 10% survival was observed. At 1.5 mgl^{-1} NAA and 1 mgl^{-1} BAP, 2% survival was observed. Survival percentage decreased with the increase in the concentration of NAA and BAP. Percentage of PLB formation was 68% at 1 mgl^{-1} NAA and 1 mgl^{-1} BAP level. Percentage of PLB formation decreased with the increase in concentration of growth regulators. PLBs were not formed at higher levels of growth regulators. PLBs were green when the concentration of growth regulators was lower. As the concentration increased, PLBs turned pale green or yellow or pale yellow. When these PLBs were subcultured on VW medium, all green PLBs and many of the pale green PLBs formed plantlets. Yellow and pale yellow PLBs showed no further growth or died.

Cattleya **Naomi Kerns** callus segments were cultured on MS medium. At 1 mgl^{-1} NAA and 1 mgl^{-1} BAP, the percentage of survival was 100%, and it was 62% at 1 mgl^{-1} NAA and 5 mgl^{-1} BAP. Though the percentage of survival decreased with the increase in the concentration of BAP and NAA, the mortality rate was less as compared to *Dendrobium* hybrids. Percent survival was zero at 5 mgl^{-1} NAA and 15 mgl^{-1} BAP, and 15 mgl^{-1} NAA and

15 mgl^{-1} BAP levels. Percentage of PLB formation was 90% at 1 mgl^{-1} NAA and 1 mgl^{-1} BAP and it was 48% at 1 mgl^{-1} NAA and 5 mgl^{-1} BAP. At 5 mgl^{-1} NAA and 1 mgl^{-1} BAP, PLB formation was 32%. There was no PLBs formation at higher levels of growth regulators. PLBs produced under the influence of lower concentrations of growth regulators were green and pale green in colour. They were yellow and pale yellow when produced using higher concentration of growth regulators. Green and pale green PLBs produced plantlets when subcultured on MS medium. Yellow and pale yellow PLBs either showed slow growth or died.

Phalaenopsis **Queen Emma** callus segments were cultured on MS medium. At 5 mgl^{-1} NAA and 5 mgl^{-1} BAP level, a maximum of 80% survival was recorded. At 5 mgl^{-1} NAA and 1 mgl^{-1} BAP, 64% survival was recorded. At both these levels, green PLBs formed which eventually developed into plantlets on subculturing in MS medium. At 1 mgl^{-1} NAA and 1 mgl^{-1} BAP, 60% survival was recorded; where 20% PLB formation was seen; the PLBs were pale green and developed into plantlets directly. In this hybrid, green PLBs developed when the concentration of growth regulators was higher. All green and pale green PLBs produced plantlets directly, as well and on subculturing. At 10 mgl^{-1} NAA level, percentage of survival and percentage of PLBs formation ranged between 2 to 6%. At 15 mgl^{-1} NAA level, percent survival was zero.

Plantlet growth and formation of roots

A plantlet, for the purposes of this study, is defined as a differentiated PLB with at least one pair of leaves and several leaf primordial, with or without roots. Such plantlets measuring about 0.5 cm were inoculated on different media with varied concentrations of growth regulators (see Tables 39-59, Figs. 10, 11, 12, 14, 16, 18 and 36). See p-288.

Growth was non-parametrically studied with the intention to select suitable medium for growth. Studies were also made to examine the effect of modified rooting media containing banana pulp and growth regulators (see Table 51). Here also, non-parametric measurements were, used.

After identifying VW medium for *Dendrobiums* and MS medium for *Cattleya* and *Phalaenopsis* as the best media for plantlet growth, experiments were conducted to identify the most suitable growth regulator combination. Parameters like length of the plantlet, numbers of leaves and roots produced were recorded at the end of tenth week of culture.

Effects of NAA and BAP and nutrient media on the growth of the plantlets was studied by inoculating plantlets of all the orchid hybrids on VW, MS and KC basal media, with 8.5 gl^{-1} agar, 150 mgl^{-1} coconut water, and 20 gl^{-1} sucrose.

Dendrobium **Queen Sonia** plantlets showed best growth in VW medium, while moderate growth was observed in MS medium, and slow growth in KC medium. While growth increased with the increase in concentration of NAA, it reduced with increase in the concentration of BAP but show formation of side shoots. When the concentration of NAA was higher than 5 mgl^{-1} unusual growth was seen. At 10 mgl^{-1} NAA level plants were unusually tall with fewer narrow leaves and long internodes, and at 1 mgl^{-1} BAP and 5 mgl^{-1} NAA in VW medium healthy growth of plants with the formation of side shoots was observed. No growth was seen when the concentration of BAP and NAA was 10 mgl^{-1}.

Dendrobium **Emma White** plantlets showed very good growth in VW medium with moderate growth in MS medium and slower growth was recorded in KC medium. BAP was found to be useful for growth, only in lower concentrations. In the absence of BAP at 5 mgl^{-1} NAA, good growth of plantlets was recorded in both VW and MS media. But at 7 mgl^{-1} and 10 mgl^{-1} NAA concentrations, unusual growth of plantlets was observed. Increase in the BAP decreased the growth of plants associated with the formation of side shoots. No growth was observed when the concentration of BAP and NAA was 10 mgl^{-1} each. NAA and BAP were inhibitory at higher concentrations (see Figs. 45 and 46 for variations).

Cattleya **Naomi Kerns** plantlets showed very good growth in MS medium. In VW medium, the growth was good. Slower growth was observed in KC medium. In MS medium, at 3 and 5 mgl^{-1} NAA concentrations excellent growth was noticed. At 1 mgl^{-1}, 7 mgl^{-1} and 10 mgl^{-1} NAA levels, growth was moderate. At 1 mgl^{-1} BAP and 5 mgl^{-1} NAA, very good growth was recorded. At 3 mgl^{-1} BAP and 5 mgl^{-1} NAA, there was good growth and good formation of side shoots. There was no growth in the combination of high levels of NAA and BAP. Almost similar responses were elicited with both the growth regulators in VW and KC media (see Table 45-47).

Phalaenopsis **Queen Emma** plantlets showed the requirement of MS medium for good and healthy growth. Whereas in VW medium moderate growth was observed but KC medium induced slower growth. In MS medium, excellent growth was recorded at 1 mgl^{-1} BAP and 7 mgl^{-1} and 10 mgl^{-1} NAA. At 3 mgl^{-1} BAP and 7 mgl^{-1} NAA levels, there was good healthy growth besides side shoot formation. BAP alone in the absence of NAA did not induce growth, but at higher levels it induced side shoot formation (see Tables 48-50).

Effect of Banana Pulp (BP) was studied on the growth and rooting of the plantlets of all the orchid hybrids in different selected basal media with 8.5 gl^{-1} agar, 1mgl^{-1} BAP, 5 mgl^{-1} NAA, 500 mgl^{-1} activated charcoal, 20 gml^{-1} sucrose and 150 mlL^{-1} coconut water (see Table 51).

Dendrobium **Queen Sonia** showed good growth in MS medium with 150 mgl^{-1} BP; the same effect was seen even when 200 mgl^{-1} BP was used. But VW medium with 150 mgl^{-1} BP showed better result. However the growth was moderate in KC medium. But in all the media, while growth reached saturation at 150 mgl^{-1} BP, there was slow growth at lower levels (50 and 100 mgl^{-1} BP).

Dendrobium **Emma White** also responded in a similar way as the other *Dendrobium* hybrid. In case of both the *Dendrobium* hybrids good rooting was noticed. In all the media, rooting was good at 150 mgl^{-1} BP concentration. At 50 mgl^{-1} BP level, small and thin roots developed which however, were fewer in number. In MS medium with 200 mgl^{-1} BP, green PLBs formed from the base of the explants which later on formed plants.

Cattleya **Naomi Kerns** showed good growth and rooting in VW medium with 150 mgl^{-1} BP, but in MS medium with 150 mgl^{-1} BP the growth and rooting was excellent. When the concentration of BP was increased to 200 mgl^{-1} there was no further improvement in the growth and rooting of plantlets. This effect was seen in both VW and MS medium. But in KC medium moderate growth was observed.

In MS medium with 200 mgl^{-1} BP, green PLBs were produced from the basal regions of the explants, which eventually developed into plantlets.

Phalaenopsis **Queen Emma** showed excellent growth and good rooting, only in MS medium with 200 mgl^{-1} BP. The growth and rooting was poor in VW and KC medium. In MS medium with 50 mgl^{-1} BP, both the growth and the rooting were slow. With 100 mgl^{-1} BP and 150 mgl^{-1}, moderate and good growth, respectively.

Effects of cytokinins, kinetin and 2iP on the growth of planttets were studied in MS, VW and KC media, supplemented with 150 mlL^{-1} BP 500 mgl^{-1} activated charcoal, 8.5 gl^{-1} agar, and 20 gl^{-1} sucrose on the growth of plantlets were studied (see Tables 52-55). Increase in the length of the plant the number of leaves and roots produced, were observed at different levels of cytokinins. Plantlets measuring 1 cm long with 2 leaves and a few leaf primordia were inoculated on selected media.

Dendrobium **Queen Sonia** plantlets were cultured on VW medium. In the absence of kinetin, the length of the plant was found to be 2.1 cm with 4.1 leaves and 5.1 roots. At 0.5 mgl^{-1} kinetin level the plant length was found to be 2.2 cm, with 4 leaves and 6 roots with the increase in kinetin level, the growth, number of leaves and number of roots increased, but saturation point was reached at 2.5 mgl^{-1} concentration, where 2.9 cm length and 4.6 leaves and 8.2 roots were observed. At 3 mgl^{-1} kinetin level, the length was found to be 2 cm with 3 leaves and 8.2 roots. 2iP was found to be better than kinetin.

At 0.5 mgl^{-1} 2iP level, the length of plants was 2.5 cm with 4.2 leaves and 5.2 roots being recorded. As the concentration of 2iP was increased, the growth of the plantlets was improved. At 3 mgl^{-1} 2iP level, the length of the plants was found to be 3.2 cm with 9.0 roots and 5.0 leaves. But the leaves and roots were smaller, as compared to those produced in the presence of kinetin.

Dendrobium **Emma White** was cultured on VW medium. In the absence of kinetin, the length of the plant was 2 cm and 4 leaves and 5 roots were recorded. At 2 mgl^{-1} kinetin level, the plant length was 2.8 cm and 4.5 leaves and 7.1 roots were recorded. At 2.5 mgl^{-1} 2iP the length of the plant was 2.9 cm, and 4.6 leaves and 8.2 roots were recorded. At 3.0 mgl^{-1} 2iP, there was a decrease in the growth with 1.8 cm length, 3 leaves and 9.0 roots being recorded. Roots produced were smaller and thin. Kinetin gave slightly better results than 2iP. At 0.5 mgl^{-1} 2iP, 2.4 cm length, 4.2 leaves and 6.1 roots were recorded. At 2 mgl^{-1} 2iP level, 2.8 cm length, 4.2 leaves and 8.2 roots were recorded. At 2.5 mgl^{-1} 2iP level, 3.0 cm length 4.1 leaves and 10 roots were produced. At 3 mgl^{-1} 2iP, 3.1 cm length was recorded, 5 leaves and 10.6 roots were produced.

Cattleya **Naomi Kerns** plantlets were cultured on MS medium. In the absence of kinetin, 3 cm length, 5.1 leaves and 3.1 roots were observed. At 1 mgl^{-1} kinetin level, 3.2 cm length, 5.3 leaves and 3.4 roots were recorded. At 2.0 mgl^{-1} kinetin, 3.5 cm length, 5.5 leaves and 5.0 roots were recorded. With the further increase in the kinetin level there was decrease in the length and number of leaves but there was increase in the number of roots produced. At 3.0 mgl^{-1} kinetin level, 3.3 cm length, 5.1 leaves and 5.6 roots were observed. 2iP gave better results than kinetin.

Length of the plants, number of leaves and number of roots increased with the increase in the concentration of 2iP. At 0.5 mgl^{-1} 2iP level, 2.9 cm length was recorded with 5 leaves and 3.2 roots. At 1.0 mgl^{-1} 2iP level, 3 cm length of the plantlet, 5.2 leaves and 3.2 roots were recorded. At 2 mgl^{-1} 2iP level 3.8 cm length, 5.4 leaves and 4.0 roots were recorded. At 3.0 mgl^{-1} 2iP level, 3.9 cm length, 6.0 leaves and 5.3 roots were observed.

Phalaenopsis **Queen Emma** was cultured on MS medium. In the absence of kinetin 4.5 cm stem length, 2.0 leaves, and 2.2 roots were observed. As there was increase in the concentration of kinetin the growth of the plants also increased. At 1.0 mgl^{-1} kinetin level, 1.6 cm length, 2.1 leaves and 2.6 roots were recorded. At 2 mgl^{-1} kinetin level, 1.9 cm length was observed, and 2.2 leaves and 3.6 roots were produced. At 3.0 mgl^{-1} kinetin level, 3.0 cm length, 4.0 leaves and 3.8 roots were recorded. 2iP gave better results than kinetin. At 1 mgl^{-1} 2iP level, 2.1 cm length, 2.2 leaves and 2.2 roots were recorded. At 2 mgl^{-1} 2iP concentration, 2.8 cm length was observed and 3.2 leaves and 4.5 roots were observed. At 3 mgl^{-1} 2iP level, 3.4 cm length, 6.2 leaves and 5.4 roots were recorded.

Effects of different auxins, NAA, IBA and 2, 4-D, on the growth of plantlets in different basal media with 1 mgl^{-1} BAP, 150 mlL^{-1} BP, 250 mgl^{-1} activated charcoal, 8.5 gl^{-1} agar and 20 gl^{-1} sucrose, were studied. Length of the plantlets, and number of leaves and roots produced were taken as parameters for this study (see Tables 56-59).

Dendrobium **Queen Sonia** was cultured on VW medium. At 2 mgl^{-1} NAA level, 2.5 cm length, 6.1 leaves, and 4.2 roots were observed. At 6 mgl^{-1} NAA level, 4.0 cm length, 8.0 leaves, and 6.1 roots were observed.

At 8 mgl^{-1} NAA level, 5.5 cm length, 12.1 leaves, and 8.3 roots were recorded. In this case, the roots produced were unusual, thin and small.

Multiplication of Plantlets

The rate of multiplication of plantlets used in different selected media was studied. There was a range of responses from normal growth of the plantlets to different degrees of multiplication in terms of the number of plantlets produced at the end of 12 weeks of culture as compared to the number of plantlets originally used in the experiments (see Table 60-63).

Effects of auxins and cytokinins on the rate of multiplication of plantlets were studied by culturing the different plants on different selected media with 8.5 agar, 150 mgl^{-1} coconut water, 20 mgl^{-1} sucrose, and varied concentrations of NAA and BAP. Observations were made after 14 weeks of culture and the multiplication ratio was calculated. Multiplication ratio is the ratio of the total number of plants produced and the number of plants inoculated (see Figs. 7, 10, 14, 15, 19, 21 and 37).

It was found that both NAA and BAP were necessary for multiplication. In all cases, the rate of multiplication increased with the increase in the BAP concentration upto certain levels. But the multiplication ratio decreased as there was further increase in the concentration of the growth regulators. This clearly indicates that each hybrid requires a specific combination of growth regulators for proper growth. In the absence of BAP and the presence of only NAA there was little multiplication. Multiplication always occurred with the formation of side shoots from the base of the plants in the medium.

Dendrobium **Queen Sonia** was cultured on VW medium. In the presence of 10 mgl^{-1} NAA and no BAP, 1.8 and multiplication Ratio (MR) was recorded. In the presence of 1 mgl^{-1} BAP and 6 mgl^{-1} NAA, 2.5 MR was observed. At 2 mgl^{-1} BAP and 4 mgl^{-1} NAA level, MR was found to be 7. At 4 mgl^{-1} BAP and 1 mgl^{-1}, maximum MR (12.5) was recorded and this was found to be the best ratio of concentrations of BAP and NAA for good multiplication with proper growth. At 6 mgl^{-1} BAP and no NAA, 6.5 MR was recorded and at 8 mgl^{-1} BAP and without no NAA the MR was 6. But when the concentration of both NAA and BAP exceeded 8 mgl^{-1}, plants were found to be abnormal, showing small and narrow leaves, and thin roots.

Dendrobium **Emma White** was cultured on VW medium. In the absence of BAP and with the increase in the NAA concentration from 1 to 10 mgl^{-1}, there was slow increase in the MR from 1 to 1.9. MR was 1.2 at 2 mgl^{-1} NAA and no BAP. But with 1 mgl^{-1} BAP and 2 mgl^{-1} NAA MR was 2.5, and it was 3 at the same level of BAP (1 mgl^{-1}) and 4 mgl^{-1} NAA. At 2 mgl^{-1} BAP and 4 mgl^{-1} NAA concentration, MR was 6. Maximum MR for this hybrid was 11.5, recorded at 4 mgl^{-1} BAP and 1 mgl^{-1} NAA concentration. MR was 9 at 4 mgl^{-1} BAP and 2 mgl^{-1} NAA concentration. MR was 7 in the absence of NAA and in the presence of 6 mgl^{-1} BAP. At 8 mgl^{-1} BAP and 1 mgl^{-1} NAA, MR was 5. MR was 1.5 at 10 mgl^{-1} BAP and 10 mgl^{-1} NAA. Results indicate that both NAA and BAP are necessary for multiplication.

Cattleya **Naomi Kerns** was cultured on MS medium. In the absence of BAP but presence of only NAA, the MR was very low. MR increased from 1.4

to 1.9 with the increase in the concentration of NAA from 0 to 10 mgl^{-1}. MR was 2.5 at 1 mgl^{-1} each of NAA and BAP. Maximum MR of 9.5 was observed at a combination, of 2 mgl^{-1} BAP and 1 mgl^{-1} NAA. It was 9 at 4 mgl^{-1} BAP and 1 mgl^{-1} NAA concentration. At 8 mgl^{-1} BAP and 1 mgl^{-1} NAA concentration, 3.5 MR was observed. At 10 mgl^{-1} BAP and 2 mgl^{-1} NAA, 2.5 MR was recorded; and it was 1.8 at 10 mgl^{-1} each of NAA and BAP. There was increase in the MR with an increase in BAP upto 4 mgl^{-1} but MR decreased with further increase in the BAP concentration. A specific combination of concentrations of NAA and BAP was necessary for proper multiplication of plants.

Phalaenopsis **Queen Emma** was cultured on MS medium. In the absence of BAP but presence of only NAA, ranging between 1 mgl^{-1} to 6 mgl^{-1}, there was no multiplication. MR was 1.2 and 1.9 at 8 mgl^{-1} and 10 mgl^{-1} NAA concentrations, respectively. MR ranged between 1 to 1.9 when 1 mgl^{-1} BAP and 1 - 10 mgl^{-1} NAA were used in the medium. At 2 mgl^{-1} BAP and 4 mgl^{-1} NAA, the MR was 2, and it was 2.5 when 4 mgl^{-1} BAP and 2 mgl^{-1} NAA were used in the medium. 4.5 MR was observed at 6 mgl^{-1} BAP and 4 mgl^{-1} NAA concentrations. MR for this hybrid was found to be a maximum of 9.5 at 8 mgl^{-1} BAP and 4 mgl^{-1} NAA concentrations. At 10 mgl^{-1} BAP and 1 mgl^{-1} NAA, MR was 9, and it was 1.2 at 10 mgl^{-1} each of NAA and BAP. *Phalaenopsis* hybrid showed a requirement for higher concentrations of NAA and BAP for multiplication, as compared to other orchid hybrids studied. Though both these growth regulators were necessary the requirement of BAP was more for the multiplication of plantlets.

At 10 mgl^{-1} NAA level plants also showed unusual features. Stem length was 9.0 cm with 9.4 narrow leaves, long internodes and pale green leaves. Though 8.5 roots were produced, they were thin and small.

Growth was moderate in the presence of IBA. At 2 mgl^{-1} IBA, 2.4 cm length, 4 leaves and 3.1 roots were observed. At 6 mgl^{-1} IBA level, 3.3 cm length, 5.1 leaves and 4.0 roots were recorded. At 8 mgl^{-1} IBA level, 5.0 cm growth, 6.5 leaves and 5.0 roots were recorded. At 10 mgl^{-1} IBA, 7.3 cm length, 6.8 leaves and 5.2 roots were observed. The growth increased with the increase in the IBA level; the number of leaves and roots also increased, but in the presence of 8 and 10 mgl^{-1} IBA, unusual plants were produced.

Growth was moderate in the presence, of 2, 4-D also. But 6 mgl^{-1}, 8 mgl^{-1}, and 10 mgl^{-1} levels induced the formation of unusual plants. At 2 mgl^{-1} 2, 4-D, 2.5 cm length, 4.1 leaves and 2.0 roots were observed.

At 6 mgl^{-1} 2, 4-D level, 6.2 leaves and 3.0 roots were recorded. At, 10 mgl^{-1} 2, 4-D, 7.3 cm length, 6.6 leaves and 1.1 roots were recorded. Though the stem length and number of leaves increased with the increase in 2, 4-D concentration, the number of roots decreased.

Of all the auxins tried at different levels, *Dendrobium* **Queen Sonia** showed good, healthy and normal growth, without any abnormalities at 6 mgl^{-1} NAA level.

Dendrobium **Emma White** was cultured on VW medium. At 2 mgl^{-1} NAA level, 3 cm length, 6.0 leaves and 4.0 roots were recorded. At 4.0 mgl^{-1} NAA level, 4.1 cm length, 8.2 leaves and 6.1 roots were observed. At 6 mgl^{-1} NAA, healthy and normal plants were formed. At this level 4.5 cm length, 9.1 leaves and 8.2 roots were observed. At 8 mgl^{-1} NAA, 6.0 cm length, 12.3 leaves and 8.3 roots were recorded, but both leaves and roots were unusual. At 10 mgl^{-1} NAA level, 8.5 cm length, 10.1 leaves and 5.4 roots were observed. The growth was moderate in the presence of IBA. At 2.0 mgl^{-1} IBA level, 2.5 cm length, 4.1 leaves and 2.1 roots were observed. At 6 mgl^{-1} NAA level, 3.5 cm length was recorded with 5.4 leaves and 4.2 roots. At 10 mgl^{-1} IBA, 6.5 cm length was recorded with 6.2 leaves and 4.4 roots, but both leaves and roots were found to be abnormal. 2, 4-D induced growth only at 2 mgl^{-1} and 4 mgl^{-1} level. At 6 mgl^{-1}, 8 mgl^{-1} and 10 mgl^{-1} 2, 4-D, abnormal plants were formed. At 2 mgl^{-1} 2, 4-D, 2.5 cm length, 4.2 leaves and 2.2 roots were noticed. At 4 mgl^{-1} 2, 4-D, 3.5 cm length was recorded with 6.3 leaves and 2.2 roots. At 10 mgl^{-1} 2, 4-D level, 7.5 cm length was recorded with 6.8 leaves and 1.0 roots. Increase in 2, 4-D level increased the growth and number of leaves but the number of roots reduced. 2, 4-D was not found to be good growth regulator for the growth of plants. NAA was found to be the growth regulator best followed by IBA.

Cattleya **Naomi Kerns** plantlets were cultured on MS medium. NAA was found to be a better auxin than IBA and 2, 4-D. While IBA gave moderate results, 2, 4-D gave better results than IBA. However, the plants obtained by using higher concentrations of 2, 4-D were abnormal. At 2 mgl^{-1} NAA level, 3.1 cm length was observed and 4.0 leaves and 2.2 roots were produced. At 4 mgl^{-1} NAA level, the length of the plant was 4 cm with 6.1 leaves and 3.2 roots. Growth was better with the increase in the NAA level. At 10 mgl^{-1} NAA level, the length of the plants was 7.0 cm with 6.2 leaves and 5.1 roots. But the plants were abnormal with long internodes. At 2 mgl^{-1} IBA level, the length of plant was 2.8 cm with 4.2 leaves and

2.0 roots. Growth improved with the increase in IBA level. At 8 mgl^{-1} IBA concentration, the length of plant was 4.2 cm with 4.3 leaves and 4.2 roots. At 10 mgl^{-1} IBA, the length was 5.1 cm, and 4.4 leaves and 4.2 roots were produced. In this case, though there was increase in the stem length by 1 cm, the increase in the number of leaves was very less; and it was only 0.1. This lead to an increase in the internodal length. The other observation was that the plants produced under the influence of higher levels of IBA had smaller and narrower leaves, as compared to those produced in the presence of NAA. At 2 mgl^{-1} 2, 4-D the length of plant was 2.2 cm with 3.1 leaves and 2.1 roots. At 6 mgl^{-1} 2, 4-D concentration, the length was 3.4 cm with 4.2 leaves and 4.0 roots. At 10 mgl^{-1} 2, 4-D, the length was 5.6 cm, and 5.7 leaves and 4.3 roots were produced. Growth was better in terms of length of the stem and the number of leaves and roots in the presence of 2, 4-D than in the presence of IBA but the size of leaves and roots was smaller with unusual features.

Phalaenopsis **Queen Emma** was cultured in MS medium. NAA was a better auxin. Plants showed excellent growth in the presence of 10 mgl^{-1} NAA. It showed higher requirement of NAA than the orchids studied. IBA induced moderate growth even at higher (10 mgl^{-1}) level 2, 4-D induced slow growth at all levels studied. At 2 mgl^{-1} NAA concentration, the plant length was 2.1 cm with 2.1 leaves and 2.0 roots. At 6 mgl^{-1} NAA concentration, the length was 3.4 cm with 2.2 leaves and 5.2 roots. At 10 mgl^{-1} NAA, 3.7 cm length was recorded with 4.1 leaves and 8.1 roots. At 2 mgl^{-1} IBA, the plant length was 1.5 cm, and 2 leaves and 2 roots were observed. Hence there was no increase in the number of leaves but 2 roots were formed and stem length increased. At 6 mgl^{-1} IBA concentration, the length was 1.8 cm, and 2.2 leaves and 2.1 roots were observed. At 10 mgl^{-1} IBA, 2.2 cm length was recorded, and 4.2 leaves and 3.3 roots were observed. At 2 mgl^{-1} 2, 4-D concentration, the plant length increased to 1.5 cm and 1.9 roots were formed but there was no increase in the number of leaves. At 6 mgl^{-1} 2, 4-D concentration the plant length was 2 cm. The number of leaves marginally increased to 2.1 and 2.1 roots were observed. At 10 mgl^{-1} 2, 4-D the plant length was 1.5 cm with 2.2 leaves and 1.9 roots. There was a slight increase in the plant length with the increase in 2, 4-D concentration but above 8 mgl^{-1} level there was further decline in the growth of the plants in terms of length. Increase in the number of leaves was very little and root formation was also poor in the presence of 2,4-D.

Growth and maintenance of regenerated plants

Bottles containing plants to be hardened were kept in the green house for 3-4 days. Plants which had grown sufficiently in the media were taken out and grown in the greenhouse. In case of *Dendrobium* and *Cattleya,* 3-4 cm high plants were used for hardening. Whereas in *Phalaenopsis* 2-3 cm high plants were hardened.

All orchids were initially grown in community pots for 3-4 weeks, to give them tune to acclimatise to greenhouse conditions. Different media were used for community potting. Potting medium with charcoal and very small clay pieces was better. Coconut husk pieces and moss were used together in equal proportions, to induce good rooting and growth of the plants, but the mortality rate was high as moss retained too much water and caused the decay of plants (see Figs. 21 and 26).

Plants were taken out from the community pots after 3-4 weeks and grown in individual pots, with only one plant in each pot. The following types of potting media were used: (a) Charcoal pieces, (b) Broken burnt clay pieces, (c) Brick pieces, (d) Charcoal and clay pieces, (e) Charcoal, clay pieces and moss, (f) Charcoal, clay pieces and coconut husk pieces, (g) Coconut husk pieces, (h) Rock wool and clay pieces, (i) Clay balls and coconut husk pieces, (j) Fern blocks, (k) Wooden blocks and moss, and (l) Split coconut pericarp and coir.

Charcoal pieces of about 2 cm in diameter were proved to be best for rooting and growth. Brick pieces used with charcoal helped in retaining moisture. 5:1 ratio of charcoal and brick pieces was ideal for *Cattleya* hybrid. *Dendrobium* hybrid showed good growth in the medium containing clay balls and coconut husk pieces. Coconut husk was used after removing tannin contents by soaking in water for 10-15 days. *Phalaenopsis* hybrid showed good growth in the potting medium containing tree fern blocks.

All potted plants were kept in a greenhouse, maintaining high humidity. They were provided with nutrients in the form of foliar spray. A mixture of NPK (Nitrogen 30 gl^{-1}, phosphorous 20 gl^{-1}, potassium 20 gl^{-1}) and micronutrients of MS medium were dissolved in one litre of water. 0.1% of this mixture was prepared by dilution and was then sprayed on the plants, once a week. Watering was also done once a week.

INDEX TO TABLES

Sl. No.	Stages of Study	Observations and Results	S. No. of Tables
I	Selection of explants	Procedures, sizes and regions of explants used	1
II	Induction of Callus and PLBs	a) No. of days taken for callusing	2 - 5
		b) Percent and degree of callusing	6 - 16
III	Multiplication of callus and PLBs	Degree of multiplication	18
IV	PLB and plantlet induction from callus	a) Degree of PLB formation in all explants on all media	19 - 30
		b) Rate of PLB formation and plantlet induction from selected explants on selected media	32 - 34
		c) Rate of plantlet and PLB formation from allus on selected media	35 - 38
V	Growth and rooting of plantlets	a) On all media	39 - 50
		b) On all media with banana pulp	51
		c) On selected media for selection of hormones	52 - 55
		d) On selected media for selection of hormones	56 - 59
VI	Multiplication of plantlets	On selected media	60 - 63

Table 1

Details of Explants Used in the Experiments

Orchids	Position & size	Explants					
		Meristem	**Shoot apex**	**Nodal buds**	**Leaf**	**Internodal**	**Root**
A & B	1	GHO++ TCP +	GHO ++ TCP +	GHP ++ TCP	TCP ++ GHP +	GHP ++ TCP +	TCP ++ GHP +
	2	Terminal	Terminal	Axillary	Terminal	Intercalary	Terminal
	3	Entire	Entire	Entire	Segments	Segments	Segments
	4	1 x 0.5mm	2-3 cm	0.5 x 1 cm	1 x 1.5 cm	1-3 cm	0.5-1 cm
C	1	GHO ++ TCP +	GHO ++ TCP +	GHP ++ TCP	TCP ++ GHP +	GHP ++ TCP +	TCP ++ GHP +
	2	Terminal	Terminal	Axillary	Terminal	Intercalary	Terminal
	3	Entire	Entire	Entire	Segments	Segments	Segments
	4	1 x 0.5mm	1-3 cm	0.5 x 0.5 cm	0.5 x 1cm	1-2 cm	0.5 - 1 cm
D	1	TCP ++ GHP +	GHP ++	GHP ++	GHP ++ TCP +	GHP ++ TCP +	GHP ++ TCP +
	2	Terminal	Terminal*	Axillary*	Terminal	Intercalary*	Terminal
	3	Entire	Entire	Entire	Segments	Entire	Segments
	4	0.5 x 0.5mm	1-2 cm	0.5-1 cm	2 x 2 cm	0.5 cm	0.5 cm

GHO = Green House Grown Plant
TCP = Tissue Culture Plant
++ used after
+ used rarely
* = Inflorescence stalk

Orchids : A = *Dendrobium* **Queen Sonia**
B = *Dendrobium* **Emma White**
C = *Cattleya* **Naomi Kerns**
D = *Phalaenopsis* **Queen Emma**

Source, position & size of explants:

1. Source from which the explant was obtained.
2. Position from which explant was excised
3. Part / Parts of explant used
4. Size of part of explant inoculated

Table 2

No. of days taken by the shoot apices of *Dendrobium* **Queen Sonia** for the initiation of callus in different medium with 2 mgl^{-1} 2, 4-D, varied concentration of BAP, 100 mgl^{-1} citric acid, 8 gl^{-1} agar, 15gl^{-1} sucrose and 150 mlL^{-1} coconut water.

BAP in mgl^{-1}	NUTRIENT MEDIA		
	MS	VW	KC
1	41 ± 1.4	42 ± 2.8	62 ± 56
2	28 ± 2.8	36 ± 1.4	45 ± 2.8
3	25 ± 5.6	21 ± 4.2	39 ± 5.6
4	29 ± 2.8	22 ± 2.8	46 ± 4.2
5	3.5 ± 7.77	24 ± 9.19	48 ± 6.36

Table 3

No. of days taken by the shoot apices of *Dendrobium* **Emma White** for the initiation of callus in different medium with 2 mgl^{-1} 2, 4-D, varied concentration of BAP, 100 mgl^{-1} citric acid, 8 gl^{-1} agar, 15gl^{-1} sucrose and 150 mlL^{-1} coconut water.

BAP in mgl^{-1}	NUTRIENT MEDIA		
	MS	VW	KC
1	40 ± 2.8	41	65 ± 9.19
2	29 ± 1.4	35 ± 2.8	44 ± 2.8
3	26 ± 5.6	20 ± 4.2	38 ± 4.2
4	28 ± 4.2	21 ± 5.6	37 ± 5.6
5	34 ± 7.07	23 ± 8	47 ± 4.2

Note: *Each value is the average of two independent experiments and represents means ± SD in table 2-5.

Table 4

No. of days taken by the shoot apices of *Cattleya* **Naomi Kerns** for the initiation of callus in different medium with 2 mgl^{-1} 2, 4-D, varied concentration of BAP, 100 mgl^{-1} citric acid, 8 gl^{-1} agar, 15gl^{-1} sucrose and 150 mlL^{-1} coconut water.

BAP in mgl^{-1}	NUTRIENT MEDIA		
	MS	VW	KC
1	36 $\pm$ 2.8	46 $\pm$ 4.2	48 $\pm$ 5.6
2	30 $\pm$ 1.4	44 $\pm$ 4.2	36 $\pm$ 2.8
3	28 $\pm$ 4.2	33 $\pm$ 5.6	33 $\pm$ 1.4
4	24 $\pm$ 1.4	36 $\pm$ 1.4	36 $\pm$ 4.9
5	26 $\pm$ 7.07	39 $\pm$ 8	39 $\pm$ 7.77

Table 5

No. of days taken by the shoot apices of *Phalaenopsis Queen Emma* for the initiation of callus in different medium with 2 mgl^{-1} 2, 4-D, varied concentration of BAP, 100 mgl^{-1} citric acid, 8 gl^{-1} agar, 15gl^{-1} sucrose and 150 mlL^{-1} coconut water.

BAP in mgl^{-1}	NUTRIENT MEDIA		
	MS	VW	KC
1	33 $\pm$ 1.4	46 $\pm$ 2.8	68 $\pm$ 7.77
2	29 $\pm$ 4.2	43 $\pm$ 1.4	44 $\pm$ 1.4
3	26 $\pm$ 2.8	40 $\pm$ 4.2	42 $\pm$ 4.2
4	24 $\pm$ 5.6	34 $\pm$ 4.2	47 $\pm$ 4.9
5	18 $\pm$ 7.77	34 $\pm$ 9.19	48 $\pm$ 5.6

Table 6

Effects of different concentration of NAA on different explants of *Dendrobium* **Queen Sonia** in MS medium with 2 mgl^{-1} BAP, varied concentration of NAA, 8 gl^{-1} agar, 15 gl^{-1} sucrose, 100 mgl^{-1} citric acid and 150 mlL^{-1} coconut water. Observations were made in 10 weeks old cultures.

Explants	Effects	NAA in mgl^{-1}					
		0.5	**1**	**2**	**3**	**4**	**5**
Meristem	PC	61 ± 10.60	70 ± 14.4	82 ± 19.79	72 ± 15.52	68 ± 14.4	63 ± 16.9
	DC	2.1 ± 1.13	2.8 ± 1.55	3.6 ± 1.27	4.0 ± 1.9	4.2 ± 2.12	5.0 ± 2.5
Shoot apex	PC	55 ± 12.7	54 ± 11.3	77 ± 12.72	57 ± 16.9	53 ± 21.2	48 ± 22.6
	DC	1.8 ± 1.4	2.1 ± 1.69	4.6 ± 1.55	4.8 ± 1.69	4.9 ± 2.4	5.0 ± 2.8
Nodal bud	PC	55 ± 18	53 ± 16.9	78 ± 15.52	48 ± 19.79	46 24	40 25.45
	DC	2.0 ± 0.98	2.5 ± 1.9	3.0 ± 2.12	3.6 ± 2.4	4.2 ± 2.26	4.6 ± 2.5
Leaf Segment	PC	-	15 ± 8	20 ± 9.19	10 ± 12.72	10 ± 7.07	5 ± 4.2
	DC	-	1.2 ± 0.56	1.5 ± 0.707	1.0 ± 0.35	0.5 ± 0.35	0.5 ± 0.35
Internodal Segment	PC	23 ± 9.8	28 ± 11.3	29 ± 12.7	32 ± 15.52	18 ± 11.3	15 ± 8
	DC	1 ± 0.56	1.2 ± 0.98	2.0 ± 0.98	2.1 ± 1.69	2.2 ± 1.55	2.4 ± 2.12
Root Segment	PC	60 ± 14.84	68 ± 16.9	78 ± 15.52	68 ± 21.2	55 ± 18	53 ± 21.92
	DC	2.0 ± 0.98	2.4 ± 1.8	3.5 ± 1.9	4.0 ± 2.12	4.2 ± 2.26	4.5 ± 2.96

*Each value is the average of two independent experiments and represents means ± SD

PC = Percent Callusing DC = Degree of Callusing

Table 7

Effects of different concentration of NAA on different explants of *Dendrobium* **Queen Sonia** in VW medium with 2 mgl^{-1} BAP, varied concentration of NAA, 8 gl^{-1} agar, 15 gl^{-1} sucrose, 100 mgl^{-1} citric acid and 150 mlL^{-1} coconut water. Observations were made in 10 weeks old cultures.

Explants	Effects	NAA in mgl^{-1}					
		0.5	**1**	**2**	**3**	**4**	**5**
Meristem	PC	65 ± 5.6	89 ± 8	79 ± 9.8	68 ± 14.4	68 ± 14.4	68 ± 12.72
	DC	2.2 ± 0.56	2.9 ± 0.8	3.1 ± 1.13	3.8 ± 1.27	3.9 ± 1.4	4.8 ± 1.55
Shoot apex	PC	63 ± 8	80 ± 12.7	83 ± 9.8	58 ± 14.84	56 ± 11.3	50 ± 14.4
	DC	2.5 ± 1.13	3.0 ± 0.98	3.2 ± 1.27	3.8 ± 1.4	3.9 ± 1.69	4.6 ± 1.55
Nodal bud	PC	66 ± 5.6	75 ± 7.07	76 ± 15.52	49 ± 14.4	46 ± 11.3	43 ± 16.9
	DC	2.0 ± 0.56	2.8 ± 0.8	3.1 ± 0.98	3.3 ± 1.27	3.4 ± 1.13	4.0 ± 1.9
Leaf Segment	PC	-	29 ± 4.2	39 ± 8	75 ± 21.92	40 ± 15.2	10 ± 5.6
	DC	-	1.2 ± 0.56	2.0 ± 0.8	1.5 ± 0.8	1.0 ± 0.35	0.5 ± 0.4
Internodal Segment	PC	28 ± 9.8	30 ± 11.3	39 ± 15.52	25 ± 14.4	20 ± 11.3	10 ± 5.6
	DC	1 ± 0.4	1.1 ± 0.42	1.6 ± 0.56	1.8 ± 0.42	1.9 ± 0.8	2.0 ± 0.98
Root Segment	PC	60 ± 5.6	79 ± 15.5	80 ± 21.2	69 ± 18	60 ± 16.9	50 ± 14.4
	DC	1.2 0.98	2.5 ± 1.13	3.0 ± 1.13	4.0 ± 0.98	4.5 ± 1.4	4.6 ± 1.55

*Each value is the average of two independent experiments and represents means ± SD

Table 8

Effects of different concentration of NAA on different explants of *Dendrobium* **Queen Sonia** in KC medium with 2 mgl^{-1} BAP, varied concentration of NAA, 8 gl^{-1} agar, 15 gl^{-1} sucrose, 100 mgl^{-1} citric acid and 150 mlL^{-1} coconut water. Observations were made in 10 weeks old cultures.

Explants	Effects	NAA in mgl^{-1}					
		0.5	**1**	**2**	**3**	**4**	**5**
Meristem	PC	59 ± 19.79	68 ± 22.6	70 ± 24	66 ± 26.8	65 ± 21.92	59 ± 25.45
	DC	2.0 ± 18	2.5 ± 2.12	3.2 ± 2.9	3.9 ± 3.39	4.0 ± 3.5	4.9 ± 4.2
Shoot apex	PC	50 ± 18	53 ± 16.9	69 ± 22.6	56 ± 22.6	49 ± 26.8	48 ± 26.8
	DC	2.0 ± 1.55	2.2 ± 2.12	3.5 ± 2.26	4.0 ± 2.9	4.2 ± 3.1	4.4 ± 4.1
Nodal bud	PC	54 ± 18	50 ± 16.9	67 ± 23.3	48 ± 22.6	44 ± 24	38 ± 16.9
	DC	1.9 ± 1.4	2.0 ± 1.55	2.8 ± 1.69	2.9 ± 2.12	3.8 ± 2.9	4.0 ± 3.95
Leaf Segment	PC	-	10 ± 8	25 ± 19.79	30 ± 25.45	25 ± 21.92	-
	DC	-	10 ± 98	1.5 ± 1.27	1.0 ± 0.8	0.5 ± 0.4	-
Internodal Segment	PC	20 ± 15.52	25 ± 14.4	30 ± 21.2	26 ± 22.6	20 ± 15.52	10 ± 8
	DC	1.0 ± 0.98	1.2 ± 1.13	1.5 ± 1.27	1.6 ± 1.35	1.8 ± 1.8	1.9 ± 1.8
Root Segment	PC	49 ± 26.8	55 ± 36	59 ± 41	65 ± 38	50 ± 42	45 ± 41
	DC	1.5 ± 1.4	2.0 ± 1.9	3.0 ± 2.96	3.5 ± 3.5	3.8 ± 3.8	4.0 ± 3.95

*Each value is the average of two independent experiments and represents means ± SD

Table 9

Effects of different concentration of NAA on different explants of *Dendrobium* **Emma White** in MS medium with 2 mgl^{-1} BAP, varied concentration of NAA, 8 gl^{-1} agar, 15 gl^{-1} sucrose, 100 mgl^{-1} citric acid and 150 mlL^{-1} coconut water. Observations were made in 10 weeks old cultures.

Explants	Effects	NAA in mgl^{-1}					
		0.5	**1**	**2**	**3**	**4**	**5**
Meristem	PC	60 ± 16.9	90 ± 26.8	88 ± 25.42	73 ± 26.8	67 ± 24	58 ± 23.3
	DC	2.1 ± 1.4	2.5 ± 1.4	3.0 ± 1.55	3.8 ± 1.55	4.0 ± 2.12	4.9 ± 2.5
Shoot apex	PC	58 ± 15.52	86 ± 24	79 ± 21.2	70 ± 22.6	60 ± 19.79	50 ± 25.45
	DC	2.2 ± 1.4	2.2 ± 1.8	2.7 ± 1.27	3.6 ± 1.9	4.2 ± 2.12	5.0 ± 4.2
Nodal bud	PC	49 ± 24	72 ± 26.16	67 ± 26.8	62 ± 29.6	60 ± 28.28	45 ± 26.8
	DC	1.9 ± 0.8	2.0 ± 1.69	2.3 ± 1.4	2.8 ± 1.9	3.0 ± 2.96	3.8 ± 4.2
Leaf Segment	PC	-	29 ± 15.52	48 ± 22.6	29 ± 22.6	19 ± 5.52	10 ± 7.07
	DC	-	1.7 ± 98	3.1 ± 1.4	1.8 ± 0.98	1.0 ± 0.56	0.5 ± 0.35
Internodal Segment	PC	36 ± 14.4	58 ± 16.9	68 ± 16.9	56 ± 21.2	51 ± 19.79	44 ± 18
	DC	1.0 0.56	1.3 ± 0.707	1.8 ± 0.98	2.1 ± 1.2	2.6 ± 1.4	2.8 ± 1.4
Root Segment	PC	50 ± 11.3	55 ± 13.43	65 ± 9.8	70 ± 16.9	52 ± 14.4	45 ± 16.26
	DC	1.2 ± 9.8	2.2 ± 1.27	2.5 ± 1.55	3.0 ± 2.6	3.5 ± 2.96	4.5 ± 3.25

*Each value is the average of two independent experiments and represents means ± SD

Table 14

Effects of different concentration of NAA on different explants of *Cattleya* **Naomi Kerns** in KC medium with 2 mgl^{-1} BAP, varied concentration of NAA, 8 gl^{-1} agar, 15 gl^{-1} sucrose, 100 mgl^{-1} citric acid and 150 mlL^{-1} coconut water. Observations were made in 10 weeks old cultures.

Explants	Effects	NAA in mgl^{-1}					
		0.5	**1**	**2**	**3**	**4**	**5**
Meristem	PC	45 ± 10.6	47 ± 11.3	73 ± 15.52	50 ± 14.4	46 ± 11.3	40 ± 12
	DC	1.8 ± 0.56	1.8 ± 0.8	3.9 ± 0.98	4.0 ± 1.13	4.2 ± 2.5	4.5 ± 2.8
Shoot apex	PC	40 ± 5.6	45 ± 8	69 ± 11.3	48 ± 12.7	46 ± 11.3	35 ± 14.4
	DC	1.0 ± 0.42	1.2 ± 0.98	2.1 ± 1.13	2.6 ± 1.13	3.9 ± 2.12	4.0 ± 3.1
Nodal bud	PC	35 ± 9.8	44 ± 11.3	68 ± 12.72	49 ± 11.3	45 ± 14.4	33 ± 12.72
	DC	1.0 ± 0.42	1.2 ± 0.98	1.6 ± 1.2	2.0 ± 1.55	2.3 ± 1.9	2.6 ± 2.26
Leaf Segment	PC	-	25 ± 11.3	15 ± 9.8	10 ± 7.07	10 ± 8	-
	DC	-	1.0 ± 0.98	0.5 ± 0.4	0.5 ± 0.4	0.5 ± 0.42	-
Internodal Segment	PC	28 ± 12.7	40 ± 14.4	44 ± 16.9	36 ± 15.52	30 ± 16.9	25 ± 18
	DC	1.0 ± 0.98	1.5 ± 1.27	1.8 ± 1.69	1.9 ± 1.8	2.0 ± 1.9	2.2 ± 2.12
Root Segment	PC	38 ± 8	42 ± 14.4	60 ± 16.9	45 ± 15.52	40 ± 14.4	30 ± 12
	DC	1.6 ± 1.2	2.2 ± 1.9	3.0 ± 2.6	3.5 ± 2.96	3.6 ± 3.1	3.8 ± 3.39

*Each value is the average of two independent experiments and represents means ± SD

Table 15

Effects of different concentration of NAA on different explants of *Phalaenopsis* **Queen Emma** in MS medium with 2 mgl^{-1} BAP, varied concentration of NAA, 8 gl^{-1} agar, 15 gl^{-1} sucrose, 100 mgl^{-1} citric acid and 150 mlL^{-1} coconut water. Observations were made in 10 weeks old cultures.

Explants	Effects	NAA in mgl^{-1}					
		0.5	**1**	**2**	**3**	**4**	**5**
Meristem	PC	70 ± 10.60	80 ± 15.52	90 ± 16.9	65 ± 14.4	55 ± 9.8	45 ± 12
	DC	1.2 ± 0.98	2.0 ± 1.9	3.0 ± 2.96	3.8 ± 2.6	3.9 ± 3.5	4.9 ± 3.67
Shoot apex	PC	80 ± 8	85 ± 11.3	90 ± 21.2	70 ± 19.79	65 ± 22.6	35 ± 18
	DC	1.9 ± 1.4	2.6 ± 1.55	3.1 ± 1.9	4.3 ± 1.8	4.8 ± 2.5	5.0 ± 3.95
Nodal bud	PC	76 ± 11.3	77 ± 16.9	78 ± 25.45	69 ± 24	48 ± 19.79	40 ± 22.6
	DC	2.0 ± 1.27	2.0 ± 1.55	2.8 ± 1.69	3.1 ± 1.13	3.8 ± 2.26	4.8 ± 3.5
Leaf Segment	PC	40 ± 20.5	45 ± 22.6	50 ± 32.5	42 ± 31	40 ± 29.6	20 ± 15.52
	DC	1.0 ± 0.98	1.2 ± 1.13	1.8 ± 1.69	1.9 ± 1.55	2.8 ± 2.8	2.8 2.6
Internodal Segment	PC	55 ± 26.8	66 ± 28.28	79 ± 38	70 ± 35.3	50 ± 31	42 ± 29.6
	DC	1.2 ± 1.13	2.0 ± 1.69	2.7 ± 2.12	2.9 ± 2.26	3.0 ± 2.5	3.2 ± 2.6
Root Segment	PC	61 ± 38	76 ± 39.5	78 ± 41	53 ± 36	51 ± 35.5	41 ± 33.9
	DC	1.1 ± 0.98	2.0 ± 1.69	2.9 ± 2.12	3.6 ± 2.4	3.8 ± 2.12	4.2 ± 2.26

*Each value is the average of two independent experiments and represents means ± SD

Table 16

Effects of different concentration of NAA on different explants of *Phalaenopsis* **Queen Emma** in KC medium with 2 mgl^{-1} BAP, varied concentration of NAA, 8 gl^{-1} agar, 15 gl^{-1} sucrose, 100 mgl^{-1} citric acid and 150 mlL^{-1} coconut water. Observations were made in 10 weeks old cultures.

Explants	Effects	NAA in mgl^{-1}					
		0.5	**1**	**2**	**3**	**4**	**5**
Meristem	PC	55 ± 14.84	70 ± 24.45	75 ± 24	70 ± 71.92	65 ± 18	65 ± 17.67
	DC	2.2 ± 1.55	2.2 ± 1.4	2.9 ± 1.69	3.2 ± 2.96	3.8 ± 3.1	4.8 ± 3.39
Shoot apex	PC	50 ± 15.52	60 ± 22.6	70 ± 29.6	65 ± 21.2	60 ± 19.79	50 ± 16.9
	DC	1.8 ± 0.98	2.1 ± 1.13	2.8 ± 2.8	3.3 ± 2.96	4.1 ± 3.25	4.9 ± 3.67
Nodal bud	PC	50 ± 12.7	60 ± 16.9	72 ± 28.28	65 ± 23.45	60 ± 21.2	48 ± 18
	DC	1.8 ± 1.55	2.0 ± 1.9	3.0 ± 2.26	3.2 ± 2.8	4.2 ± 3.1	4.8 ± 2.96
Leaf Segment	PC	-	-	40 ± 38	29 ± 26.8	10 ± 9.8	5 ± 4.2
	DC	-	-	2.0 ± 1.9	1.5 ± 1.4	1.0 ± 0.8	1.0 ± 0.7
Internodal Segment	PC	-	35 ± 31	51 ± 41	46 ± 42	40 ± 38	15 ± 12.72
	DC	-	1.2 ± 0.7	1.8 ± 1.4	1.9 ± 1.13	2.1 ± 1.9	2.2 ± 1.8
Root Segment	PC	-	55 ± 41	68 ± 42	63 ± 49.49	54 ± 38	50 ± 36
	DC	-	1.9 ± 1.9	2.2 ± 2.12	3.1 ± 3.1	3.6 ± 2.96	3.9 ± 3.1

*Each value is the average of two independent experiments and represents means ± SD

Table 17

Effects of different concentration of NAA on different explants of *Phalaenopsis* **Queen Emma** in VW medium with 2 mgl^{-1} BAP, varied concentration of NAA, 8 gl^{-1} agar, 15 gl^{-1} sucrose, 100 mgl^{-1} citric acid and 150 mlL^{-1} coconut water. Observations were made in 10 weeks old cultures.

Explants	Effects	NAA in mgl^{-1}					
		0.5	**1**	**2**	**3**	**4**	**5**
Meristem	PC	80 ± 5.6	90 ± 7.07	75 ± 5.6	70 ± 12.72	55 ± 14.4	45 ± 12
	DC	1.2 ± 0.4	2.5 ± 0.56	3.3 ± 0.707	3.8 ± 0.8	4.9 ± 1.69	5.0 ± 2.26
Shoot apex	PC	75 ± 8	85 ± 10.6	90 ± 11.3	75 ± 12.7	70 ± 14.4	50 ± 9.8
	DC	1.1 ± 0.42	2.1 ± 0.56	2.3 ± 0.98	3.5 ± 1.55	4.5 ± 1.69	5.0 ± 1.8
Nodal bud	PC	77 ± 7.07	79 ± 9.8	82 ± 11.3	60 ± 14.4	43 ± 15.52	34 ± 14.4
	DC	1.0 ± 0.56	1.2 ± 0.8	2.8 ± 1.13	3.2 ± 1.55	3.9 ± 1.27	4.5 ± 0.98
Leaf Segment	PC	40 ± 21.2	46 ± 19.79	50 ± 14.42	42 ± 18	34 ± 19.79	20 ± 11.3
	DC	1.0 ± 0.8	1.8 ± 1.27	2.0 ± 1.13	2.2 ± 1.4	2.4 ± 0.98	2.5 ± 1.4
Internodal Segment	PC	-	48 ± 22.6	55 ± 21.2	60 ± 19.79	50 ± 22.6	42 ± 17.67
	DC	-	2.1 ± 0.98	2.6 ± 1.13	2.8 ± 1.69	2.9 ± 1.55	2.9 ± 1.8
Root Segment	PC	-	59 ± 29.6	69 ± 31	67 ± 24	51 ± 16.9	39 ± 19.79
	DC	-	2.4 ± 2.12	3.0 ± 2.8	3.2 ± 1.13	3.9 ± 1.55	4.1 ± 1.4

*Each value is the average of two independent experiments and represents means ± SD

Table 18

Effects of NAA and BAP in different media on the degree of multiplication of callus segments of different orchids. Observation made after 6 week of culture.

		ORCHIDS											
NAA in mgl^{-1}	BAP in mgl^{-1}	*D.* **Queen Sonia** Media			*D.* **Emma White** Media			*C.* **Naomi Kerns** Media			*P.* **Queen Emma** Media		
		MS	VW	KC	MS	VW	KC	MS	VW	KC	MS	VW	KC
0	0	1.1	1.0	1.0	1.2	1.0	1.0	1.2	1.0	1.0	1.1	1.0	1.0
	1	1.1	1.0	1.0	1.1	1.0	1.0	1.2	1.0	1.0	1.1	1.0	1.0
	2	1.2	1.1	1.0	1.3	1.0	1.0	1.5	1.4	1.1	1.1	1.0	1.0
	3	1.8	1.3	1.2	1.7	1.3	1.1	2.0	1.5	1.2	1.2	1.0	1.0
1	0	2.0	1.5	1.0	2.1	1.6	1.1	1.8	1.5	1.2	1.8	1.5	1.1
	1	2.2	2.0	1.5	2.3	1.8	1.5	2.5	1.8	1.3	1.9	1.6	1.5
	2	2.4	2.1	1.3	2.5	2.2	1.6	3.0	2.1	1.8	3.4	3.0	1.9
	3	3.4	2.0	1.8	2.9	2.2	1.6	4.0	2.9	1.9	4.1	2.9	2.0
2	0	2.0	2.0	1.5	2.1	1.9	1.6	1.1	1.0	1.0	1.0	1.0	1.0
	1	5	4.5	3.8	5	4.8	4.0	1.8	1.2	1.0	2.1	1.3	1.2
	3	3.9	3.2	2.2	3.8	2.9	2.4	5	4.2	3.8	2.2	2.0	1.8
3	0	2.2	1.6	1.1	2.1	1.5	1.2	2.0	1.5	1.2	2.2	1.5	1.0
	1	2.3	1.7	1.1	2.4	1.6	1.2	1.8	1.4	1.2	3.5	3.1	2.8
	2	4.9	3.8	2.6	4.9	3.8	2.5	2.5	1.8	1.2	5	4.0	3.3
	3	3.6	3.0	2.3	3.5	3.1	2.0	1.8	1.6	1.5	4.4	3.1	2.5

Table 19

Degree of PLB formation from different explants and callus segments of *Dendrobium* **Queen Sonia** in MS medium with 1 mgl^{-1} BAP and varied concentrations of NAA. Observations made after 10 weeks of inoculation.

NAA in mgl^{-1}	Explants and Callus Segments						
	Shoot apex	Inflores-cence stalk (apex)	Leaf segment	Internodal segment	Nodal bud	Root segment	Callus segment
0.5	+++++	++++	+++	+++	++++	++	++++
1	++++	++++	+++	+++	++++	++	+++++
2	+++	+++	++	++	+++	++++	+++
3	++	++	++	+	++	++	++
4	++	++	+	+	+	+	++
5	+	+	+	+	+	+	++

Table 20

Degree of PLB formation from different explants and callus segments of *Dendrobium* **Queen Sonia** in VW medium with 1 mgl^{-1} BAP and varied concentrations of NAA. Observations made after 10 weeks of inoculation.

NAA in mgl^{-1}	Explants and Callus Segments						
	Shoot apex	Inflores-cence stalk (apex)	Leaf segment	Internodal segment	Nodal bud	Root segment	Callus segment
0.5	+++++	+++++	++++	++++	+++++	+++	+++++
1	++++	++++	++	++	+++	+++	+++++
2	+++	+++	++	++	+++	+++++	+++
3	++	++	++	++	++	+++	+++
4	++	++	++	++	++	++	+++
5	++	++	+	+	+	+	+++

Table 21

Degree of PLB formation from different explants and callus segments of *Dendrobium* **Queen Sonia** in KC medium with 1 mgl^{-1} BAP and varied concentrations of NAA. Observations made after 10 weeks of inoculation.

NAA in mgl^{-1}	Explants and Callus Segments						
	Shoot apex	Inflores-cence stalk (apex)	Leaf segment	Internodal segment	Nodal bud	Root segment	Callus segment
0.5	++++	++++	++	++	++++	+	++
1	+++	+++	++	++	++	++	+++
2	++	++	+++	+++	++	++++	++++
3	++	++	++	++	++	++	++++
4	++	++	+	+	++	++	++
5	+	+	+	+	+	+	++

Table 22

Degree of PLB formation from different explants and callus segments of *Dendrobium* **Emma White** in MS medium with 1 mgl^{-1} BAP and varied concentrations of NAA. Observations made after 10 weeks of inoculation.

NAA in mgl^{-1}	Explants and Callus Segments						
	Shoot apex	Inflores-cence stalk (apex)	Leaf segment	Internodal segment	Nodal bud	Root segment	Callus segment
0.5	+++++	+++++	+++	+++	++++	+	++++
1	++++	++++	+++	+++	+++	+++	++++
2	++	+++	++	++	++	++++	+++
3	++	++	+	+	++	++	++
4	+	+	+	+	++	+	++
5	+	+	+	+	+	+	++

Table 23

Degree of PLB formation from different explants and callus segments of *Dendrobium* **Emma White** in VW medium with 1 mgl^{-1} BAP and varied concentrations of NAA. Observations made after 10 weeks of inoculation.

NAA in mgl^{-1}	Explants and Callus Segments						
	Shoot apex	Inflores-cence stalk (apex)	Leaf segment	Internodal segment	Nodal bud	Root segment	Callus segment
0.5	+++++	+++++	++++	++++	++++	+++	+++++
1	++++	++++	++++	+++	+++	+++	+++
2	+++	+++	+++	++	++	+++++	++
3	++	++	++	++	++	+++	++
4	++	++	++	++	++	++	++
5	++	++	+	+	++	+	+++

Table 24

Degree of PLB formation from different explants and callus segments of *Dendrobium* **Emma White** in KC medium with 1 mgl^{-1} BAP and varied concentrations of NAA. Observations made after 10 weeks of inoculation.

NAA in mgl^{-1}	Explants and Callus Segments						
	Shoot apex	Inflores-cence stalk (apex)	Leaf segment	Internodal segment	Nodal bud	Root segment	Callus segment
0.5	++++	++++	+++	+++	++++	++	+++
1	+++	+++	++++	+++	+++	++++	++++
2	++	++	++	++	++	+++	+++
3	++	++	+	+	+	+	++
4	+	+	+	+	+	+	++
5	+	+	+	+	+	+	+

Table 25

Degree of PLB formation from different explants and callus segments of *Cattleya* **Naomi Kerns** in VW medium with 1 mgl^{-1} BAP and varied concentrations of NAA. Observations made after 10 weeks of inoculation.

NAA in mgl^{-1}	Explants and Callus Segments						
	Shoot apex	Inflores-cence stalk (apex)	Leaf segment	Internodal segment	Nodal bud	Root segment	Callus segment
0.5	+++	+++	++	++	+++	++	++
1	+++++	++++	+++	+++	+++	++	++++
2	++	++	++	+	++	++++	++
3	++	++	+	+	++	++	++
4	+	+	+	+	+	+	++
5	+	+	+	+	+	+	++

Table 26

Degree of PLB formation from different explants and callus segments of *Cattleya* **Naomi Kerns** in MS medium with 1 mgl^{-1} BAP and varied concentrations of NAA. Observations made after 10 weeks of inoculation.

NAA in mgl^{-1}	Explants and Callus Segments						
	Shoot apex	Inflores-cence stalk (apex)	Leaf segment	Internodal segment	Nodal bud	Root segment	Callus segment
0.5	++++	++++	++	++	++++	++	+++
1	+++++	+++++	++++	+++	+++++	++	+++++
2	+++	++	++	++	++	++++	++++
3	++	++	+	+	+	++	+++
4	++	+	+	+	+	+	+++
5	+	+	+	+	+	+	++

Table 27

Degree of PLB formation from different explants and callus segments of *Cattleya* **Naomi Kerns** in KC medium with 1 mgl^{-1} BAP and varied concentrations of NAA. Observations made after 10 weeks of inoculation.

NAA in mgl^{-1}	Explants and Callus Segments						
	Shoot apex	Inflores-cence stalk (apex)	Leaf segment	Internodal segment	Nodal bud	Root segment	Callus segment
0.5	++	++	+	+	++	++	++
1	+++	+++	++	++	+++	++	+++
2	++	++	++	++	++	+++	++
3	++	++	+	+	+	++	++
4	+	+	+	+	+	+	++
5	+	+	+	+	+	+	+

Table 28

Degree of PLB formation from different explants and callus segments of *Phalaenopsis* **Queen Emma** in MS medium with 1 mgl^{-1} BAP and varied concentrations of NAA. Observations made after 10 weeks of inoculation.

NAA in mgl^{-1}	Explants and Callus Segments					
	Inflorescence stalk (apex)	Leaf segment	Internodal segment	Nodal bud	Root segment	Callus segment
0.5	+	+	+	+	+	++
1	++	+	+	++	+++	++
2	+++	+	++	++	++	+++
3	+++	++	++	+++	+++++	+++++
4	++++	+++	+++	++++	+++++	++++
5	+++++	++++	+++	+++++	+++++	++++

Table 29

Degree of PLB formation from different explants and callus segments of *Phalaenopsis* **Queen Emma** in VW medium with 1 mgl^{-1} BAP and varied concentrations of NAA. Observations made after 10 weeks of inoculation.

NAA in mgl^{-1}	Explants and Callus Segments					
	Inflorescence stalk (apex)	Leaf segment	Internodal segment	Nodal bud	Root segment	Callus segment
0.5	+	+	+	+	+	+
1	+	+	+	+	++	++
2	+++	++	++	+++	+++	++
3	+++	++	++	+++	+++++	+++++
4	+++	++	++	++	+++	++++
5	+++++	+++	+++	+++++	+++	++++

Table 30

Degree of PLB formation from different explants and callus segments of *Phalaenopsis* **Queen Emma** in KC medium with 1 mgl^{-1} BAP and varied concentrations of NAA. Observations made after 10 weeks of inoculation.

NAA in mgl^{-1}	Explants and Callus Segments					
	Inflorescence stalk (apex)	Leaf segment	Internodal segment	Nodal bud	Root segment	Callus segment
0.5	+	+	+	+	+	+
1	+	+	+	+	++	+
2	+	+	+	+	++	++
3	++	+	+	+	++++	++
4	+++	+	+	++	++	++
5	+++	++	++	++	+++	++++

Table 31

Effects of growth regulators on the shoot apex explants of *Dendrobium* **Queen Sonia**. Observation made after 10 weeks of inoculation.

BAP in mgl^{-1}	Observation	Conc. of 2,4-D mgl^{-1}					
		0.1	1	2	3	4	5
0.1	% survival	95 ±1.4	90 ±11.3	65 ±12.7	45 ±11.3	15 ±9.8	05 ±4.2
	Wet weight of the callus produced (mg)	286 ±8	299 ±15.52	301 ±19.79	305 ±22.6	325 ±16.9	326 ±15.52
	% of PLB formation	60 ±5.6	45 ±7.09	20 ±11.3	10 ±5.6	0	0
1	% survival	70 ±5.6	95 ±4.2	90 ±5.6	40 ±14.4	20 ±19.79	05 ±1.4
	Wet weight of the callus produced (mg)	291 ±26.8	302 ±25.45	314 ±16.9	325 ±15.52	412 ±14.4	472 ±11.3
	% of PLB formation	65 ±5.6	70 ±4.2	40 ±9.8	20 ±8	0	0
2	% survival	45 ±18	65 ±16.9	60 ±11.3	35 ±7.07	10 ±4.2	05 ±1.4
	Wet weight of the callus produced (mg)	316 ±22.6	325 ±16.9	356 ±15.52	372 ±16.9	392 ±11.3	482 ±16.9
	% of PLB formation	15 ±5.6	10 ±7.07	10 ±8	5 ±1.4	0	0

Table 32

Effects of growth regulators on the shoot apex explants of *Dendrobium* **Emma White**. Observation made after 10 weeks of inoculation.

BAP in mgl^{-1}	Observation	Conc. of 2,4-D mgl^{-1}					
		0.1	1	2	3	4	5
	% survival	90 ±5.6	85 ±8	70 ±4.2	50 ±4.2	20 ±18	05 ±4.2
0.1	Wet weight of the callus produced (mg)	292 ±12.72	312 ±14.4	352 ±16.9	412 ±25.45	410 ±19.79	415 ±26.8
	% of PLB formation	65 ±9.8	45 ±11.3	25 ±12.7	5 ±4.2	0	0
	% survival	95 ±1.4	60 ±5.6	65 ±4.2	45 ±14.4	20 ±28.28	10 ±5.6
1	Wet weight of the callus produced (mg)	399 ±21.2	402 ±18	413 ±16.9	429 ±19.79	436 ±15.52	472 ±22.6
	% of PLB formation	60 ±14.4	40 ±11.3	15 ±12.7	0	0	0
	% survival	25 ±12.7	75 ±14.4	75 ±15.52	40 ±11.3	25 ±24	10 ±5.6
2	Wet weight of the callus produced (mg)	431 ±12.7	450 ±9.8	471 ±22.6	511 ±11.3	532 ±24	542 ±25.45
	% of PLB formation	15 ±11.3	10 ±8	0	0	0	0

Table 33

Effects of growth regulators on the shoot apex explants of *Cattleya* **Naomi Kerns** in MS medium. Observation made after 10 weeks of inoculation.

BAP in mgl^{-1}	Observation	Conc. of 2,4-D mgl^{-1}					
		0.1	1	2	3	4	5
	% survival	100 ±5.6	90 ±7.07	70 ±12.72	55 ±5.6	35 ±18	05 ±4.2
0.1	Wet weight of the callus produced (mg)	313 ±9.8	328 ±12.7	369 ±15.52	373 ±14.4	390 ±19.79	395 ±18
	% of PLB formation	80 ±16.9	65 ±11.3	25 ±12.72	0	0	0
	% survival	90 ±12.72	75 ±11.3	70 ±18	35 ±21.2	25 ±19.79	10 ±5.6
1	Wet weight of the callus produced (mg)	310 ±21.2	342 ±16.9	356 ±22.6	374 ±22.6	392 ±16.9	421 ±21.2
	% of PLB formation	85 ±14.4	60 ±15.52	20 ±11.3	0	0	0
	% survival	90 ±12.6	95 ±24	65 ±19.79	45 ±15.52	15 ±4.2	10 ±4.2
2	Wet weight of the callus produced (mg)	401 ±21.2	452 ±24	449 ±22.6	479 ±25.45	512 ±28.28	553 ±36.76
	% of PLB formation	45 ±9.8	25 ±16.9	15 ±12.72	0	0	0

Table 34

Effects of growth regulators on the shoot apex explants of *Phalaenopsis* **Queen Emma** in MS medium. Observation made after 10 weeks of inoculation.

BAP in mgl^{-1}	Observation	Conc. of 2,4-D mgl^{-1}					
		0.1	1	2	3	4	5
	% survival	85 ±15.52	90 ±16.9	75 ±21.2	50 ±11.3	35 ±9.8	15 ±12.7
0.1	Wet weight of the callus produced (mg)	250 ±14.4	262 ±11.3	271 ±22.6	284 ±16.9	292 ±25.5	306 ±14.4
	% of PLB formation	05 ±1.4	25 ±11.3	45 ±14.4	15 ±8	10 ±7.07	5 ±4.2
	% survival	80 ±25.45	75 ±24	65 ±21.2	35 ±15.52	15 ±12.72	05 ±1.4
1	Wet weight of the callus produced (mg)	272 ±15.52	290 ±14.4	307 ±12.7	342 ±11.3	348 ±19.79	370 ±21.2
	% of PLB formation	15 ±11.3	20 ±14.4	60 ±21.2	50 ±9.8	05 ±4.2	0
	% survival	80 ±26.8	85 ±49.5	35 ±25.45	25 ±21.2	10 ±9.8	05 ±1.4
2	Wet weight of the callus produced (mg)	309 ±36.76	380 ±48.08	392 ±49.5	399 ±42.42	441 ±43.84	458 ±41
	% of PLB formation	80 ±11.3	45 ±14.4	25 ±15.52	25 ±11.3	5 ±1.4	0

Table 35

Effects of NAA and BAP on survival of callus segments and PLB formation from callus segments of *Dendrobium* **Queen Sonia** in VW medium. Observation made after 12 weeks of culture.

NAA in mgl^{-1}	BAP in mgl^{-1}	No. of callus segments	Survival rate (%)	PLB formation (%)	Colour of PLBs	No. of plantlets formed from each callus segment directly	PLB formation on subculturing
1	1	50	100	70	G	5	Plantlets formed
	5	50	48	20	PG	3	Plantlets formed
	10	50	36	12	PY	2	Slow growth
	15	50	06	02	Y	0	Died
5	1	50	22	18	PG	2	Plantlets formed
	5	50	08	06	PY	1	No growth
	10	50	10	04	PY	-	Died
	15	50	0	0	-	-	-
10	1	50	10	04	PG	1	Plantlets formed
	5	50	06	0	-	-	-
	10	50	0	0	-	-	-
	15	50	0	0	-	-	-
15	1	50	06	0	-	-	-
	5	50	0	0	-	-	-
	10	50	0	0	-	-	-
	15	50	0	0	-	-	-

G - Green, PG - Pale Green, PY - Pale Yellow, Y - Yellow

Table 36

Effects of NAA and BAP on the survival of callus segments and PLB formation from callus segments of *Dendrobium* **Emma White** in VW medium. Observation made after 12 weeks of culture.

NAA in mgl^{-1}	BAP in mgl^{-1}	No. of callus segments	Survival rate (%)	PLB formation (%)	Colour of PLBs	No. of plantlets formed from each callus segment directly	PLB formation on subculturing
1	1	50	98	68	G	4	Plantlets formed
	5	50	52	30	PG	2	Plantlets formed
	10	50	34	14	PY	1	No growth
	15	50	10	04	PY	1	Died
5	1	50	24	14	PG	2	Plantlets formed
	5	50	10	04	PY	1	No growth
	10	50	10	02	PY	-	Died
	15	50	0	0	-	-	-
10	1	50	06	02	PG	1	Plantlets formed
	5	50	04	0	-	-	-
	10	50	0	0	-	-	-
	15	50	0	0	-	-	-
15	1	50	02	02	Y	-	Died
	5	50	0	0	-	-	-
	10	50	0	0	-	-	-
	15	50	0	0	-	-	-

G - Green, PG - Pale Green, PY - Pale Yellow, Y - Yellow

Table 37

Effects of NAA and BAP on the survival of callus segments and PLB formation from callus segments of *Cattleya* **Naomi Kerns** in MS medium. Observation made after 12 weeks of culture.

NAA in mgl^{-1}	BAP in mgl^{-1}	No. of callus segments	Survival rate (%)	PLB formation (%)	Colour of PLBs	No. of plantlets formed from each callus segment directly	PLB formation on subculturing
1	1	50	100	90	G	6	Plantlets formed
	5	50	62	48	G	5	Plantlets formed
	10	50	40	22	PG	2	Plantlets formed
	15	50	0	0	-	1	-
5	1	50	48	32	G	4	Plantlets formed
	5	50	22	16	PG	2	No growth
	10	50	04	0	-	-	-
	15	50	0	0	-	-	-
10	1	50	16	12	Y	2	No growth
	5	50	04	0	-	-	-
	10	50	02	0	-	-	-
	15	50	02	0	PY	-	Died
15	1	50	04	02	PY	-	Died
	5	50	02	0	-	-	-
	10	50	02	0	-	-	-
	15	50	0	0	-	-	-

G - Green, PG - Pale Green, PY - Pale Yellow, Y - Yellow

Table 38

Effects of NAA and BAP on the survival of callus segments and PLB formation from callus segments of *Phalaenopsis* **Queen Emma** in MS medium. Observation made after 12 weeks of culture.

NAA in mgl^{-1}	BAP in mgl^{-1}	No. of callus segments	Survival rate (%)	PLB formation (%)	Colour of PLBs	No. of plantlets formed from each callus segment directly	PLB formation on subculturing
1	1	50	60	20	PG	1	Plantlets formed
	5	50	40	35	PG	1	Plantlets formed
	10	50	68	48	G	2	Plantlets formed
	15	50	44	40	G	4	Plantlets formed
5	1	50	64	12	G	1	Plantlets formed
	5	50	80	40	G	2	Plantlets formed
	10	50	62	24	G	4	Plantlets formed
	15	50	12	10	G	-	Plantlets formed
10	1	50	06	0	-	-	-
	5	50	04	0	-	-	-
	10	50	04	0	-	-	-
	15	50	02	02	G	1	Plantlets formed
15	1	50	0	0	-	-	-
	5	50	0	0	-	-	-
	10	50	0	0	-	-	-
	15	50	0	0	-	-	-

G - Green, PG - Pale Green, PY - Pale Yellow, Y - Yellow

Table 39

Effect of NAA and BAP on the plantlet growth of D. **Queen Sonia** in MS medium with 150 mlL^{-1} CW. Observation made in 12 weeks cultures.

BAP in mgl^{-1}	NAA in mgl^{-1}					
	0	**1**	**3**	**5**	**7**	**10**
0	+	+++	+++	+++++	+++++	+++++
1	+	+++	+++	++++	+++++	+++++
3	+	+++	+++	+++	++++	++++
5	+	++	++	++	++	++
7	+	++	++	++	++	++
10	+	++	++	++	+	+

Table 40

Effect of NAA and BAP on the growth of plantlets of D. **Queen sonia** in VW medium with 150 mlL^{-1} CW. Observation made in 12 weeks cultures.

BAP in mgl^{-1}	NAA in mgl^{-1}					
	0	**1**	**3**	**5**	**7**	**10**
0	+++	+++	++++	+++++	+++++	+++++
1	++	+++	++++	+++++	++++	++++
3	++	+++	++++	++++	+++	+++
5	+	+++	++	++	++	++
7	+	+++	++	++	++	++
10	+	++	++	++	+	+

Table 41

Effect of NAA and BAP on the growth of plantlet of D. **Queen Sonia** in KC medium with 150 mlL^{-1} CW. Observation made in 12 weeks cultures.

BAP in mgl^{-1}	NAA in mgl^{-1}					
	0	**1**	**3**	**5**	**7**	**10**
0	+	++	+++	+++	+++	++++
1	+	++	++	++++	+++	+++
3	+	++	+++	++++	+++	+++
5	+	++	++	+++	+++	+++
7	+	+	++	++	++	+
10	+	++	++	++	++	++

Table 42

Effect of NAA and BAP on the growth of plantlets of D. **Emma White** in MS medium with 150 mlL^{-1} CW. Observation made in 12 weeks cultures.

BAP in mgl^{-1}	NAA in mgl^{-1}					
	0	**1**	**3**	**5**	**7**	**10**
0	+	+++	+++	+++++	+++++	+++++
1	+	++	+++	++++	+++++	+++++
3	+	++	+++	++++	+++	+++
5	+	++	++	++	++	+
7	+	++	++	++	+	+
10	+	++	+	+	+	+

Table 43

Effect of NAA and BAP on the growth of plantlet of D. **Emma White** in VW medium with 150 mlL^{-1} CW. Observations made in 12 weeks cultures.

BAP in mgl^{-1}	NAA in mgl^{-1}					
	0	1	3	5	7	10
0	+++	+++	++++	+++++	+++++	+++++
1	+++	+++	++++	++++	++++	+++++
3	++	+++	++++	+++++	++++	+++
5	++	+++	++	++	++	++
7	+	++	+++	+++	++	++
10	+	++	++	++	+	+

Table 44

Effect of NAA and BAP on the growth of plantlets ofD. **Emma White** in KC medium with 150 mlL^{-1} CW. Observations made in 12 weeks cultures

BAP in mgl^{-1}	NAA in mgl^{-1}					
	0	1	3	5	7	10
0	+	++	+++	+++	+++	++++
1	+	++	++	++++	+++	+++
3	+	++	+++	++++	+++	+++
5	+	+	++	+++	+++	+++
7	+	+	++	++	++	+
10	+	++	++	++	++	++

Table 45

Effect of NAA and BAP on the growth of plantlet of Cattleya **Naomi Kerns** in MS medium with 150 mlL^{-1} CW. Observations made in 12 weeks cultures.

BAP in mgl^{-1}	NAA in mgl^{-1}					
	0	**1**	**3**	**5**	**7**	**10**
0	++	++++	+++++	+++++	++++	++++
1	++	++++	++++	+++++	+++	+++
3	++	+++	+++	+++++	+++	++
5	+	++	+++	++++	++	++
7	+	++	+++	+++	++	++
10	++	++	+	+	+	+

Table 46

Effect of NAA and BAP on the growth of plantlets of Cattleya **Naomi kerns** in VW medium with 150 mlL^{-1} CW. Observations made in 12 weeks cultures.

BAP in mgl^{-1}	NAA in mgl^{-1}					
	0	**1**	**3**	**5**	**7**	**10**
0	+	+++	+++++	+++++	+++	+++
1	++	+++	+++	+++++	+++	+++
3	++	++	+++	++++	++	++
5	+	++	++	+++	++	+
7	+	+	+++	+++	++	+
10	+	++	+	+	+	+

Table 47

Effect of NAA and BAP on the growth of plantlet of Cattleya **Naomi Kerns** in KC medium with 150 mlL^{-1} CW. Observations made in 12 weeks cultures.

BAP in mgl^{-1}	NAA in mgl^{-1}					
	0	1	3	5	7	10
0	+	++	+++	++++	++++	+++
1	+	++	++	++++	+++	+++
3	+	++	++	+++	++	++
5	+	++	++	+++	++	++
7	+	+	++	++	++	+
10	+	++	+	+	+	+

Table 48

Effect of NAA and BAP on the growth of plantlets of Phalaenopsis **Queen Emma** in MS medium with 150 mlL^{-1} CW. Observations made in 12 weeks cultures.

BAP in mgl^{-1}	NAA in mgl^{-1}					
	0	1	3	5	7	10
0	+	++	++	+++	++++	+++++
1	+	++	++	+++	+++++	+++++
3	+	++	++	++	+++++	++++
5	+	++	+++	+++	+++	++++
7	+	++	+++	++	+++	+++
10	+	+	+	++	++	++

Table 49

Effect of NAA and BAP on the growth of plantlet of Phalaenopsis **Queen Emma** in VW medium with 150 mlL^{-1} CW. Observations made in 12 weeks cultures.

BAP in mgl^{-1}	NAA in mgl^{-1}					
	0	**1**	**3**	**5**	**7**	**10**
0	+	++	++	+++	++++	++++
1	+	+	++	+++	+++++	+++++
3	+	+	++	++	+++	+++
5	+	++	++	++	+++	+++
7	+	+	++	++	++	++
10	+	+	++	++	+	+

Table 50

Effect of NAA and BAP on the growth of plantlets of *Phalaenopsis* **Queen Emma** in KC medium with 150 mlL^{-1} CW. Observations made in 12 weeks cultures.

BAP in mgl^{-1}	NAA in mgl^{-1}					
	0	**1**	**3**	**5**	**7**	**10**
0	+	++	++	+++	+++	++++
1	+	+	++	+++	++++	++++
3	+	+	++	++	++	++
5	+	+	+	++	++	++
7	+	+	+	++	++	++
10	+	+	+	++	+	+

Table 51

Effects of banana pulp (BP) on the growth and rooting of the plantlets of different orchids in different media and growth regulators. Observations made in 12 weeks cultures.

Different media with 1 mgl^{-1} BAP and 5mgl^{-1} NAA	BP in mlL^{-1}	Orchids			
		D. Queen Sonia	*D.* Emma White	*C.* Naomi Kerns	*P.* Queen Emma
MS	50	++	++	+++	++
	100	+++	+++	++++	+++
	150	+++++	+++++	+++++	+++++
	200	+++++	+++++	+++++	+++++
VW	50	+++	+++	++	++
	100	++++	++++	++++	++
	150	+++++	+++++	+++++	+++
	200	+++++	+++++	+++++	++
KC	50	++	++	++	++
	100	+++	+++	+++	+++
	150	++++	++++	+++	+++
	200	++++	++++	+++	++

Table 52

Effects of different concentrations of Cytokinins on the growth of the plantlets of Dendrobium **Queen Sonia** in VW medium with 5 mgl^{-1} NAA. Observations made in 10 weeks culture.

Cytokinins	in mgl^{-1}	Length of plantlet cm ± SE	No. of leaves ± SE	No. of roots ± SE
Kinetin	0	21 ± 0.28	4.1 ± 1.09	5.1 ± 1.19
	0.5	2.2 ± 0.29	4.0 ± 1.34	6.0 ± 0.98
	1	2.5 ± 0.078	4.2 ± 1.34	5.4 ± 1.09
	1.5	2.5 ± 0.39	4.4 ± 1.49	6.0 ± 2.09
	2.0	2.7 ± 0.79	4.5 ± 1.76	7.0 ± 2.39
	2.5	2.9 ± 0.56	4.6 ± 1.8	8.2 ± 2.47
	3.0	2 ± 0.69	3.0 ± 97	8.2 ± 2.47
2iP	0	2.2 ± 0.078	4.0 ± 0.89	5.0 ± 2.09
	0.5	2.5 ± 0.28	4.2 ± 1.34	5.2 ± 1.97
	1	2.6 ± 0.79	4.1 ± 1.27	6.0 ± 1.67
	1.5	2.6 ± 0.49	4.4 ± 0.98	6.3 ± 1.8
	2.0	2.9 ± 0.89	4.6 ± 2.09	7.1 ± 2.29
	2.5	3.1 ± 0.89	4.0 ± 1.76	8.4 ± 2.39
	3.0	3.2 ± 0.98	5.0 ± 1.49	9.0 ± 2.5

*Each value is the average of two independent experiments and represents means ± SE.

Table 53

Effects of different concentrations of Cytokinins on the growth of the plantlets of *Dendrobium* **Emma White** in VW medium with 5 mgl^{-1} NAA. Observations made in 10 weeks culture.

Cytokinins	in mgl^{-1}	Length of plantlet cm ± SE	No. of leaves ± SE	No. of roots ± SE
Kinetin	0	2 ± 0.078	4.0 ± 0.98	5.0 ± 1.59
	0.5	2.2 ± 0.29	4.1 ± 0.98	5.2 ± 1.8
	1	2.4 ± 0.28	4.3 ± 1.19	6.0 ± 1.76
	1.5	2.6 ± 0.49	4.3 ± 1.09	6.6 ± 1.97
	2.0	2.8 ± 0.69	4.5 ± 0.98	7.1 ± 2.09
	2.5	2.9 ± 0.98	4.6 ± 1.49	8.2 ± 2.39
	3.0	1.8 ± 0.56	3 ± 1.76	9.0 ± 2.47
2iP	0	2.4 ± 0.28	4.1 ± 1.09	6.0 ± 1.34
	0.5	2.4 ± 0.39	4.2 ± 1.09	6.1 ± 1.97
	1	2.5 ± 1.34	4.3 ± 1.59	6.2 ± 1.76
	1.5	2.6 ± 1.27	4.4 ± 1.69	7.1 ± 2.47
	2.0	2.8 ± 1.19	4.2 ± 1.34	8.2 ± 2.68
	2.5	3.0 ± 1.09	4.1 ± 1.49	10.0 ± 2.8
	3.0	3.1 ± 1.27	5.0 ± 1.76	10.6 ± 2.79

*Each value is the average of two independent experiments and represents means ± SE.

Table 54

Effects of different concentrations of Cytokinins on the growth of the plantlets of *Cattleya* **Naomi Kerns** in MS medium with 5 mgl^{-1} NAA. Observations made in 10 weeks culture.

Cytokinins	in mgl^{-1}	Length of plantlet cm ± SE	No. of leaves ± SE	No. of roots ± SE
Kinetin	0	3 ± 0.56	5.1 ± 0.98	3.1 ± 0.89
	0.5	3.1 ± 0.69	5.2 ± 0.98	3.3 ± 1.19
	1	3.2 ± 0.98	5.3 ± 1.34	3.4 ± 1.27
	1.5	3.4 ± 0.89	5.4 ± 1.79	4.2 ± 1.34
	2.0	3.5 ± 1.19	5.5 ± 1.89	5.0 ± 1.49
	2.5	3.4 ± 1.09	5.2 ± 1.69	5.4 ± 1.59
	3.0	3.3 ± 1.27	5.1 ± 2.47	5.6 ± 1.49
2iP	0	2.9 ± 0.28	5.0 ± 1.69	3.0 ± 1.19
	0.5	2.9 ± 2.29	5.0 ± 1.69	3.2 ± 1.97
	1	3 ± 0.49	5.2 ± 1.8	3.2 ± 1.97
	1.5	3.2 ± 0.56	5.5 ± 1.97	3.3 ± 1.49
	2.0	3.8 ± 0.69	5.4 ± 2.29	4.0 ± 2.09
	2.5	3.8 ± 0.98	5.2 ± 2.09	5.2 ± 1.97
	3.0	3.9 ± 1.56	6.0 ± 2.29	5.3 ± 2.29

*Each value is the average of two independent experiments and represents means ± SE.

Table 55

Effects of different concentrations of Cytokinins on the growth of the plantlets of Phalaenopsis **Queen Emma** in MS medium with 5 mgl^{-1} NAA. Observation made in 10 weeks culture.

Cytokinins	in mgl^{-1}	Length of plantlet cm ± SE	No. of leaves ± SE	No. of roots ± SE
Kinetin	0	1.5 ± 0.49	2.0 ± 0.39	2.2 ± 0.49
	0.5	1.5 ± 0.98	2.0 ± 0.49	2.4 ± 0.38
	1	1.6 ± 0.56	2.1 ± 0.49	2.6 ± 0.56
	1.5	1.8 ± 0.28	2.2 ± 0.98	3.2 ± 0.69
	2.0	1.9 ± 0.56	2.2 ± 0.98	3.6 ± 0.98
	2.5	2.2 ± 0.98	2.4 ± 1.09	3.6 ± 1.09
	3.0	3.0 ± 1.09	4.0 ± 2.09	3.8 ± 1.19
2iP	0	1.8 ± 0.79	2.0 ± 0.39	2.0 ± 0.39
	0.5	1.9 ± 0.89	2.2 ± 0.49	2.2 ± 0.49
	1	2.1 ± 1.34	2.2 ± 0.56	3.2 ± 0.98
	1.5	2.5 ± 1.59	2.4 ± 0.98	4.6 ± 0.98
	2.0	2.8 ± 1.69	3.2 ± 1.09	4.5 ± 1.09
	2.5	3.0 ± 1.59	4.4 ± 1.97	5.0 ± 1.19
	3.0	3.4 ± 1.76	6.2 ± 2.09	5.4 ± 2.09

*Each value is the average of two independent experiments and represents means ± SE.

Table 56

Effects of different concentrations of Auxins on the growth of the plantlets of *Dendrobium* **Queen Sonia** in VW medium with 1 mgl^{-1} BAP. Observations made in 12 weeks cultures.

Auxins	in mgl^{-1}	Length of plantlet cm ± SE	No. of leaves ± SE	No. of roots ± SE
NAA	2	2.5 ± 0.49	6.1 ± 0.98	4.2 ± 0.39
	4	3.5 ± 0.89	8.2 ± 1.49	5.0 ± 0.69
	6	4.0 ± 1.09	8.0 ± 1.34	6.1 ± 0.79
	8	5.5 ± 0.98	12.1 ± 2.47	8.3 ± 1.49
	10	9.0 ± 1.76	9.4 ± 2.47	8.5 ± 1.34
IBA	2	2.4 ± 0.56	4.0 ± 0.69	3.1 ± 0.49
	4	3 ± 0.98	4.1 ± 1.09	3.2 ± 0.98
	6	3.3 ± 0.98	5.1 ± 1.27	4.0 ± 1.19
	8	5.0 ± 1.34	6.5 ± 1.76	5.0 ± 1.76
	10	7.3 ± 1.97	6.8 ± 2.9	5.2 ± 1.8
2,4-D	2	2.5 ± 0.79	4.1 ± 1.19	2.0 ± 1.49
	4	3.2 ± 1.49	5.0 ± 1.59	2.2 ± 1.59
	6	4.2 ± 1.8	6.2 ± 1.97	8.0 ± 2.47
	8	5.1 ± 2.09	6.2 ± 2.39	3.0 ± 2.5
	10	7.3 ± 2.47	6.6 ± 2.9	1.1 ± 1.09

*Each value is the average of two independent experiments and represents means ± SE.

Table 57

Effects of different concentrations of Auxins on the growth of the plantlets of *Dendrobium* **Emma White** in VW medium with 1 mgl^{-1} BAP. Observations made in 12 weeks cultures.

Auxins	in mgl^{-1}	Length of plantlet cm ± SE	No. of leaves ± SE	No. of roots ± SE
NAA	2	3.0 ± 0.39	6.0 ± 0.49	4.0 ± 0.28
	4	4.1 ± 0.49	8.2 ± 1.34	6.1 ± 0.56
	6	4.5 ± 0.39	9.1 ± 1.49	8.2 ± 0.49
	8	6.0 ± 0.98	12.3 ± 2.09	8.3 ± 1.19
	10	8.5 ± 1.19	10.1 ± 2.47	5.4 ± 1.09
IBA	2	2.5 ± 0.49	4.1 ± 0.69	2.1 ± 0.39
	4	3.2 ± 0.56	4.2 ± 1.97	3.4 ± 0.89
	6	3.5 ± 0.98	5.4 ± 1.76	4.2 ± 1.49
	8	4.0 ± 1.27	6.1 ± 2.29	5.2 ± 1.34
	10	6.5 ± 2.09	6.2 ± 2.29	4.4 ± 1.69
2,4-D	2	2.5 ± 1.49	4.2 ± 1.27	2.2 ± 1.76
	4	3.5 ± 1.8	6.3 ± 2.09	2.2 ± 1.97
	6	4.5 ± 2.8	6.6 ± 2.29	3.1 ± 2.47
	8	5.2 ± 2.9	6.6 ± 2.8	2.2 ± 2.09
	10	7.5 ± 2.9	6.8 ± 2.9	01 ± 0.98

*Each value is the average of two independent experiments and represents means ± SE.

Table 58

Effects of different concentrations of Auxins on the growth of the plantlets of *Cattleya* **Naomi Kerns** in MS medium with 1 mgl^{-1} BAP. Observations made in 12 weeks cultures.

Auxins	in mgl^{-1}	Length of plantlet cm ± SE	No. of leaves ± SE	No. of roots ± SE
NAA	2	3.1 ± 0.79	4.0 ± 0.49	2.2 ± 1.19
	4	4.6 ± 1.19	6.1 ± 0.79	3.2 ± 1.09
	6	3.3 ± 1.09	4.2 ± 1.19	4.1 ± 1.34
	8	3.9 ± 1.27	4.3 ± 1.09	5.0 ± 2.39
	10	7.0 ± 2.29	6.2 ± 1.97	5.1 ± 2.29
IBA	2	2.8 ± 0.56	4.2 ± 0.39	2.0 ± 0.98
	4	3.3 ± 0.89	6.4 ± 0.69	2.2 ± 0.98
	6	3.1 ± 1.19	4.3 ± 1.69	3.1 ± 1.27
	8	4.2 ± 1.59	4.3 ± 1.59	4.2 ± 1.49
	10	5.1 ± 1.8	4.4 ± 1.8	4.3 ± 1.69
2,4-D	2	2.2 ± 0.89	3.1 ± 1.19	2.1 ± 0.98
	4	3.1 ± 1.49	4.0 ± 1.97	2.2 ± 1.69
	6	3.4 ± 2.29	4.2 ± 2.09	4.0 ± 2.8
	8	4.3 ± 2.5	4.3 ± 2.79	4.2 ± 2.79
	10	5.6 ± 2.68	5.7 ± 2.8	4.3 ± 2.9

*Each value is the average of two independent experiments and represents means ± SE.

Table 59

Effects of different concentrations of Auxins on the growth of the plantlets of *Phalaenopsis* **Queen Emma** in MS medium with 1 mgl^{-1} BAP. Observations made in 12 weeks cultures.

Auxins	in mgl^{-1}	Length of plantlet cm ± SE	No. of leaves ± SE	No. of roots ± SE
NAA	2	2.1 ± 0.49	2.1 ± 0	2.0 ± 0.56
	4	3.0 ± 0.39	2.2 ± 0.07	4.2 ± 1.27
	6	3.4 ± 0.56	2.2 ± 0.28	5.2 ± 1.34
	8	3.6 ± 0.79	3.0 ± 1.27	6.0 ± 1.97
	10	3.7 ± 0.89	4.1 ± 1.19	8.1 ± 2.39
IBA	2	1.5 ± 0.56	2.0 ± 0	2.0 ± 0.49
	4	1.6 ± 0.79	2.1 ± 0	2.0 ± 0.49
	6	1.8 ± 0.49	2.2 ± 1.19	2.1 ± 0.39
	8	2.0 ± 0.79	3.1 ± 1.09	3.2 ± 2.09
	10	2.2 ± 0.98	4.2 ± 0.98	3.3 ± 1.8
2,4-D	2	1.5 ± 0.49	2.0 ± 0	1.9 ± 0.49
	4	1.5 ± 0.56	2.0 ± 0	2.0 ± 0.98
	6	2.0 ± 0.89	2.1 ± 0.078	2.1 ± 1.49
	8	2.0 ± 0.98	2.2 ± 0.28	1.9 ± 1.76
	10	1.5 ± 0.79	2.2 ± 0.28	1.9 ± 1.8

*Each value is the average of two independent experiments and represents means ± SE.

Table 60

Effects of auxins and cytokinins on the multiplication ratio (MR) of *Dendrobium* **Queen Sonia** plantlets in VW medium with 150 mlL^{-1} coconut water. Observations made in 12 weeks culture.

BAP in mgl^{-1}	NAA in mgl^{-1}						
	0	1	2	3	4	6	10
0	1	1	1	1.2	1.5	1.8	1.8
1	1.5	1.5	1.5	1.5	2.5	1.5	1.6
2	1.5	2	2	7	4	2	1.4
4	5	12.5	8.5	8	4	2	2
6	6.5	2	4.5	5	3	2	1.4
8	6	5	3.5	4.5	2	2	1.5
10	4	3	2	2	1.5	1.6	1.8

Table 61

Effects of auxins and cytokinins on the multiplication ratio (MR) of Dendrobium **Emma White** plantlets in VW medium with 150 mlL^{-1} coconut water. Observations made in 12 weeks culture.

BAP in mgl^{-1}	NAA in mgl^{-1}						
	0	1	2	3	4	6	10
0	1	1	1.2	1.4	1.6	1.8	1.9
1	1.5	2	2.5	3	1.5	1.6	1.8
2	2	2	3	6	4	2	1.5
4	4.5	11.5	9	9.5	4	2	2
6	7	2	5	5	3.5	1.5	1.6
8	5.5	5	3.5	4	2	1.4	1.6
10	4	4	3	2	1.8	1.6	1.5

Table 62

Effects of auxins and cytokinins on the multiplication ratio (MR) of *Cattleya* **Naomi Kerns** plantlets in MS medium with 150 mlL^{-1} coconut water. Observations made in 12 weeks culture.

BAP in mgl^{-1}	NAA in mgl^{-1}						
	0	**1**	**2**	**3**	**4**	**6**	**7**
0	1.4	1.5	1.5	1.6	1.8	1.8	1.9
1	2	2.5	2.5	2.5	1.5	1.5	1
2	5	9.5	7.5	5	4.5	2.5	2
4	6	9	6.5	4	4	2	1.5
6	4.5	4	3	2.5	2	1.8	1.8
8	4	3.5	3	3	2.5	2	1.5
10	4	3	2.5	2	2	1.5	1.8

Table 63

Effects of auxins and cytokinins on the multiplication ratio (MR) of *Phalaenopsis* **Queen Emma** plantlets in MS medium with 150 mlL^{-1} coconut water. Observations made in 12 weeks culture.

BAP in mgl^{-1}	NAA in mgl^{-1}						
	0	**1**	**2**	**3**	**4**	**6**	**7**
0	1	1	1	1	1	1.2	1.5
1	1.5	1.5	2	2	1.6	1.8	1.9
2	1.5	1.5	2	2	1.8	1.6	1.4
4	2	2	2.5	2.5	2	1.8	1.6
6	3	3.5	4	4.5	4	3	2
8	7	7.5	7.5	9.5	3.5	2	1
10	7.5	9	6.5	5	3	1.6	1.2

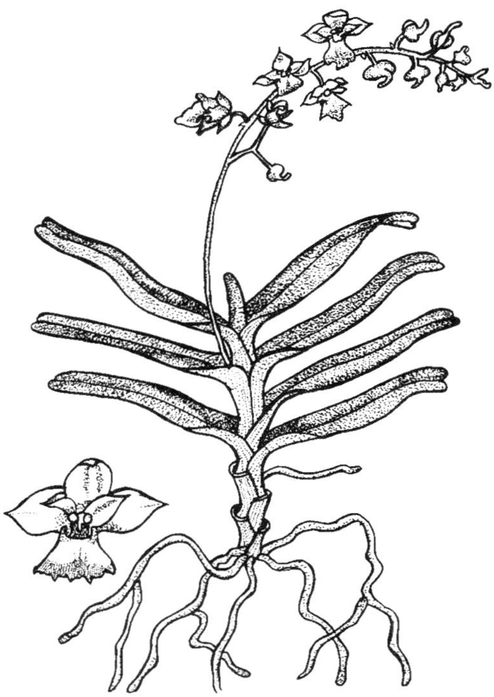

Fig. a. *Aerides crispa*

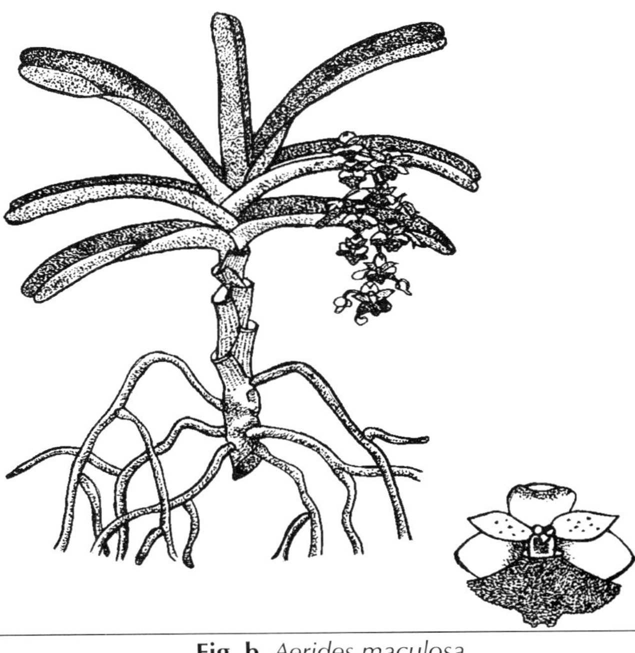

Fig. b. *Aerides maculosa*

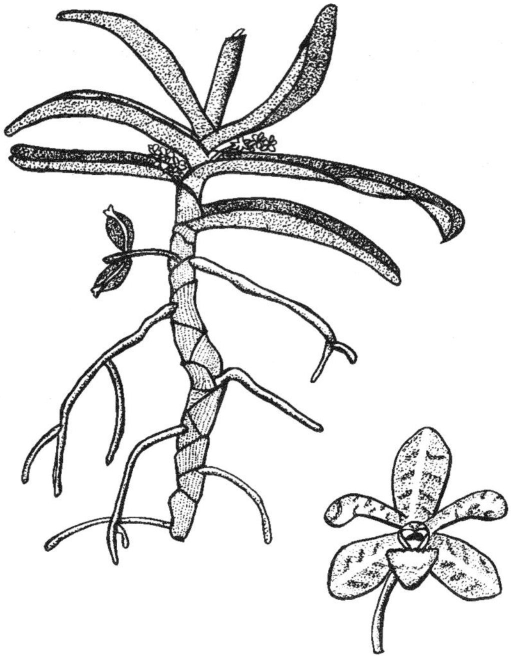

Fig. c. *Acampe praemorsa*

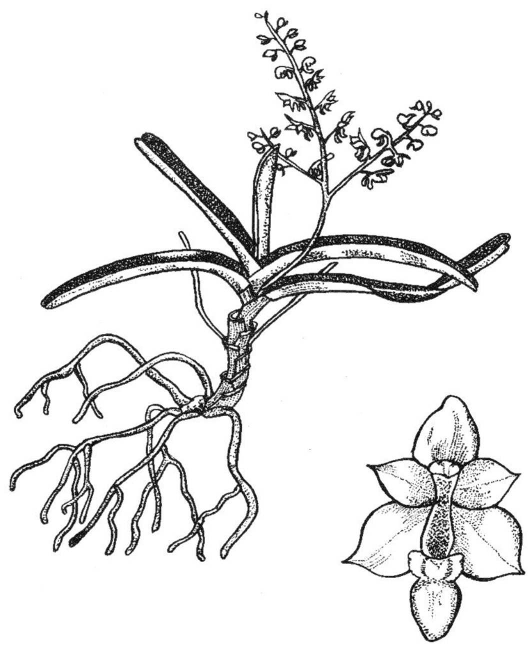

Fig. d. *Aerides ringens*

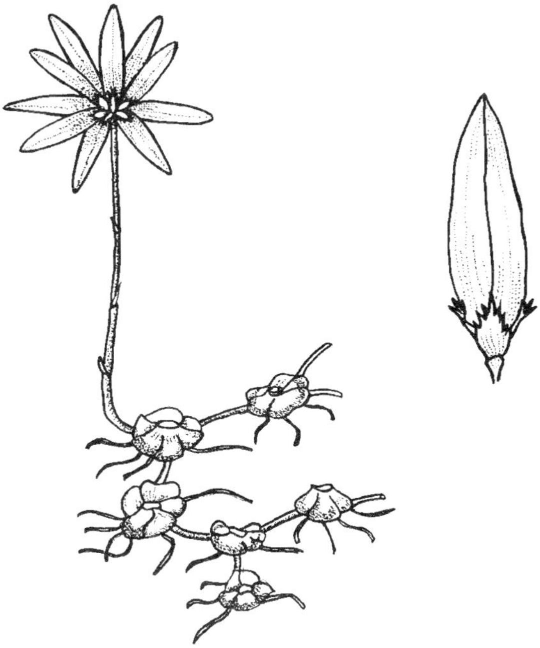

Fig. e. *Bulbophyllum fimbriatum*

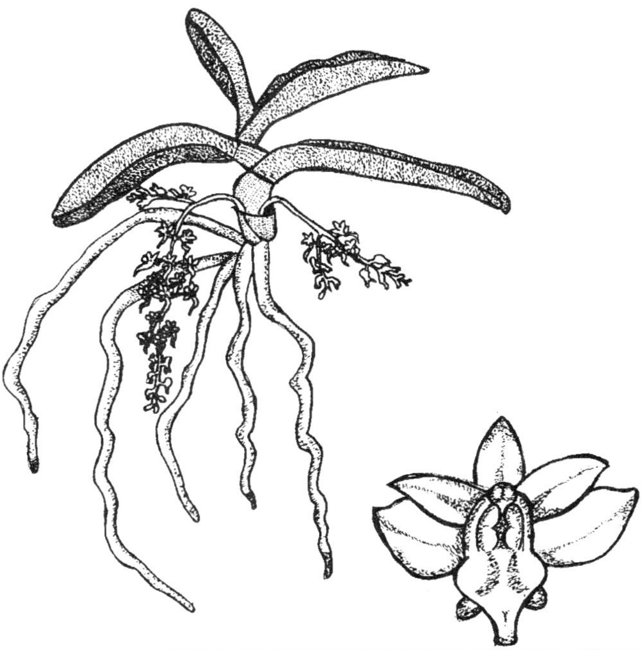

Fig. f. *Diplocentrum congestum*

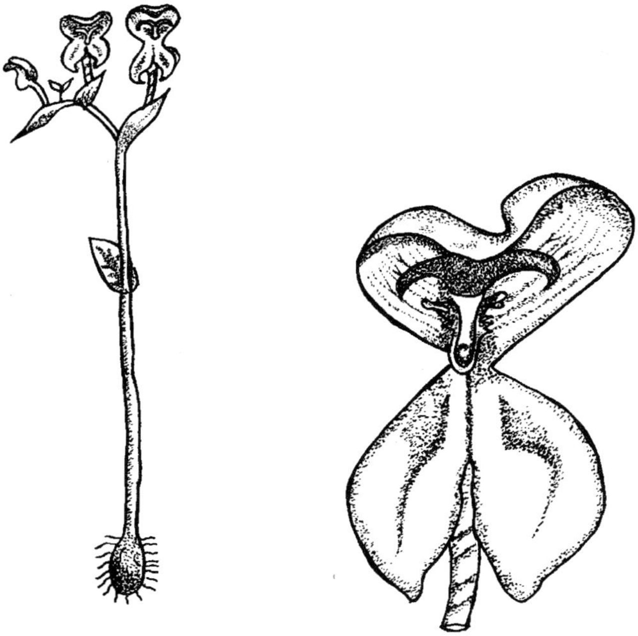

Fig. g. *Disperis zeylanica*

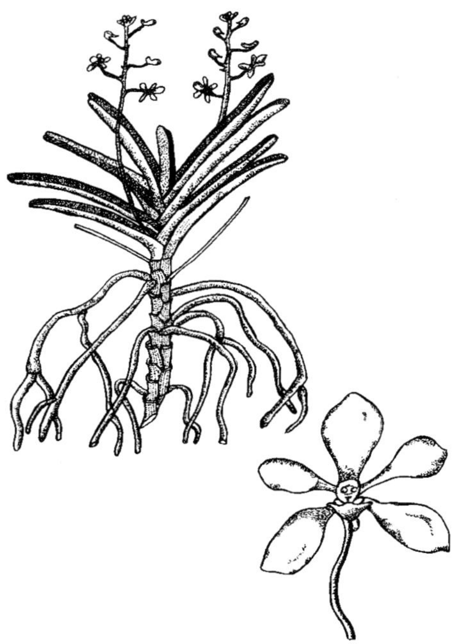

Fig. h. *Vanda testacea*

REFERENCES

Abraham A (1991). Orchids in India, their cultivation propagation and commercialization. **pp**. 1-2. Vij, S.P. and Kohli, R.K (ed) *Proceedings of National Seminar on Biology, Improvement, Propagation and Commercialization of Indian orchids,* IIHR, Bangalore, India.

Adelberg J, Pollock R, Rajapakse N and Young R (1998). Micropropagation decontamination, transcontinental shipping and hydroponic growth of *Cattleya* while sealed in semipermeable membrane vessels. *Scientia Horticulture.* **73**(**1**): 23-35.

Alberts AA (1953). Use offish emulsion for the germination of orchid seed. *Orchid J.* **2**: 464-466.

Allen MF, Allen BE and Friese CF (1989). Responses of the non-mycotrophic plant *Salsola kali* to invasion by vesicular-arbuscular mycorrhizal fungi. *New Phytologist.* **111**: 45-49.

Allenberg H (1976). Notizen Zur Keimung, Meristemkultur and Regeneration von Erdochideen. *Die Orchidee.* **27**: 28-31.

Alloufa MMAI, Arauzo KCL and Dutra MDFB (1998). Clonal Propagation of *Oncidium cebolata* using tissue culture. **pp**. 63. *Proceedings of IX International Congress on Plant Tissue and Cell Culture* Jeruselum, Israel.

Amore TD and Kamemoto A (1992). Yield and morphology of diploid *Dendrobium* hybrids and their corresponding amphidiploids. *Lindleyana.* **7**(**3**): 162-167.

Anonymous (1887). Techniques: Blue *Vanda*-Conservation of a rare orchid through micropropagation. *Bot. Gard. Conservation News* 1: 47-49 (reprinted from a publication of the Tropical Botanic Garden and Research Institute, Plant Biotechnology Division, Palode, Trivandrum, India.

Apte SS and Bhargava SC (I998). Indo-Dutch cooperation in floriculture. *Indian Floriculture.* **43**(**3**): 7-9.

Ardelean M, Pamfil D, Zavoi A, Uriziceanu I and Cosa A (1995). Influence of human bioenergy on micropropagation of *Cymbidium* species-I. Effect of different types of human bioenergy applied. *Seria Agriculture Sci. Horticultura* (Romania) **48**(**1**): 83-89.

Arditti J, Ball EA and Reisinger DM (1977b). A method for vegetative propagation of *Phalaenopsis. Amer. Orchid Soc. Bull.* **46**(**3**): 236-237.

Arditti (1984). A history of orchid hybridization, seed germination and tissue culture. *Bot. J. Linn. Soc.* **89**: 359-381.

Arditti J (ed) (1982). Orchid biology: Reviews and perspectives, Vol. 2. Cornell Univ. Press, Ithaca, N.Y.

Arditti J (1966a). Orchids. *Sci. Amer.* **214**: 70-78.

Arditti J (1966b). The use of tomato juice media for orchid seed germination Mal. *Orchid Rev.* **8**: 67-68.

Arditti J (1967). Factors affecting the germination of orchid seeds. *Bot. Rev.* 33: 1-97.

Arditti J (1992). Fundamentals of Orchid Biolgy, Joho Wiley, New York.

Arditti J and Ernst R (1993). Micropropagation of orchids. John Wiley and Sons, Inc. New York, Chichester, Brisbane, Toranto, Singapore.

Arditti J and Arditti M (1985). Mericloning: *The 25th Aniversary. Orchid Rev.* **93**: 412-413.

Arditti J, Molisch SK and Ball EA (1977a). Culture: A new means of clonal propagation. *Aust. Orchid Rev.* **38**: 175-179.

Armaleo D, Ye G-N, Klein TM, Shark KB, Sanford JC and Johnston SA (1990). Biolistic nuclear transformation of *Saccharomyces cerevisiae* and other fungi, *Curr. Genet.*, **17**: 97-103.

Baghavan V (1966). Nutrition, growth and morphogenesis of plant embryos *Biol. Rev.* **41**: 1-58.

Bajaj YPS (1990). Somaclonal variation - Orgin, induction, cryopreservation, and implication in plant breeding. Bajaj, Y.P.S (ed) Biotechnology in agriculture and forestry, Vol-2. Somaclonal variation in crop improvement-I. Springer-Verlag, Berlin, Heidelberg, New York.

Baker KM, Mathes MC and Wallance BJ (1987). Germination of *Pontiera* and *Cattleya* seeds and development *of Phalaenopsis* Protocorms. *Lindleyana,* **2**(**2**): 77-83.

Baldarelli RM and Lengyel JA (1990). Transient expression of DNA after ballistic introduction into *Drosophila*embryos, *Nucl. Acids Res.*, **18**: 5903-5904.

Ball EA, Reisinger DM and Arditti J (1974-1975). Clonal propagation of *Phalaenopsis. Malay. Orchid Rev.* **12**: 6-9.

Ball EA, Arditti J and Churchill M-E (1971). Clonal propagation of orchids from leaf tips. *Orchid Rev.* **79**: 281-288.

Banik MS, Hadiuzzaman H, Rahman MM, Haque and Islam AS (1986). Clonal propagation of Arundina bambusifolia Lindl. — a terrestrial orchid *Malay. Orchid Rev.* (Singapore) **20**: 34-37

Bapat VA and Narayanaswami S (1977). Rhizogenesis in a tissue culture of the orchid *Spathoglottis plicata. Bull. Torrey Bot. Club.* **104**:2-4.

Baryley I (1956). Use of tree forn in potting orchids. *Amer. Orchid Soc. Bull.* **25**: 9-14.

Beechey C (1970). Propagation of orchids from aerial *roots Ann. Orchid Soc. Bull.* **41**: 1085-1088.

Begum AA, Aamaki M and Kako S (1994). Formation of PLBs and shoot development through *in vitro* culture of outer tissue of *Cymbidium* PLBs. *J. of the Japanese Soc. of Hort. Sci.* **63**(**3**): 663-673.

Bending GD and Read DJ (1997). Lignin and soluble phenolic degradation by ectomycorrhizal and ericoid mycorrhizal fungi. *Mycological Research.* **101**: 1348-1354.

Benthem G (1881). Notes on Orchidaceae *J. Linn. Soc. Bot.,* 18: 281-360. Bentham G and Hooker, JD, 1883. *Genera Plantaraurn.* **3**: 513-515.

Bergman FJ (1972). Shoot tip multiplication of orchid clones. The status of "meristem" propagation through 1971. *Am. Hortic.* 51, (**20-23**), 41-44.

Bernard (1909). L'evolution dans la symbiose, les orchidees et leur champignons commesaux. *Ann. Sci. Nat. Bot. Ser.* (**9**): 1-196.

Bernard N (1899). Sur lagermination de *Ncottia nidus-apis. C.R. Acad. Sci. Paris* **128**: 1253-1255.

Bertsch W (1966). A new frontier: Orchid propagation by meristem tissue culture. **pp**. 225-229 L.R. de Garmo (ed) *Proc. 5th World Orchid Conf.* Long Beach, California. (1965).

Bezalel E (1995). Floriculture Industry in Israel. **pp**. 19-20. Prakash, J. and Bhandary, KR (ed.). *Floriculture Technology Trades and Trends.* Oxford and IBH Publishing Co. Pvt. Ltd., New Delhi, Bombay, Calcutta.

Bhojwani SS and Razdan MK (1996). *Plant Tissue Culture. Theory and Practice.* Elsevier science BV, Amsterdam.

Biermann B and Linderman RG (1983). Use of vesicular-arbuscular mycorrhizal roots, intraradical vesicles and extraradical vesicles as inoculum. *New Phytologist.* **95**: 97-105.

Black PM (1977). *Beautiful orchids,* Hamlyn, London, New York, Sydney, Totonto.

Blowers AD, Bogorad L, Shark KB and Sanford JC (1989). Studies on *Chlamydomonas* chloroplast transformation: foreign DNA can be stably maintained in the chromosome, *Plant Cell.* **1**: 123-132.

Bonga JM (1982). Tissue culture techniques. In Bonga J.M. and Durzan D.J. (ed). *Tissue culture in forestry.* Martinus Nizhoff/Junk. The Hague.

Bopaiah AK (1986). Cytological, Cytochemical and *in vitro* seed culture studies in some membres of the subtribe *Cytropedinae* of the tribe *Cymbidieae* (Orchidaceae). Ph.D thesis submitted to Karnatak University, Dharwad.

Boullard B (1997). One century of research on orchid mycorrhizas. *Academie d' Agriculture de France.* **83**(**2**): 41-46.

Bower B and Birch RG (1992). Transgenic sugarcane plants via microprojectile bombardment, *Plant Journal.* **2:**, 409-416.

Boynton JE, Gillham NW, Harris EH, Hosler JP, Johnson AM, Jones AR, Randolph-Anderson BL, Robertson D, Klein TM, Shark KB and Sanford J C (1988). Chloroplast transformation in *Chlamydomonas* with high velocity microprojectiles. *Science.* **240**: 1534-1538.

Boynton JE, Gillham NW, Harris EH, Hosler JP, Johnson AM, Randolph-Anderson BL, Robertson D, Klein TM and Sanford JC (1987). Transformation of a chloroplast ATPase mutant of *Chlamydomonas* using high velocity microprojectiles to deliver a wild type chloroplast DNA fragment. *Molecular Biology of Chloroplasts and Mitochondria*, Cold Spring Harbor, New York.

Bruce WB, Christensen AH, Klein T, Fromm M and Quail PH (1989). Photoregulation of a phytochrome gene promoter from oat transferred into rice by particle bombardment. *Proc. Natl. Acad. Sci. USA*. **86:** 9692-9696.

Brundrett MC (1991). Mycorrhizas in natural ecosystems. In: Macfayden A, Begon M. and Fitter A.H. (eds.) Advances in Ecological Research, Vol. 21. Academic Press, London. pp. 171-313.

Bubeck SK (1973). A study of *Paphiopedilum* meristem culture. Ph.D. thesis., Rutgers University. University Microfilms International, Ann Arbor, Mich.

Burkill HM (1964). The role of the Singapore Botanic Gardens in the development of orchid hybrids. In: Proc. 4th World Orchid Conf. Singapore, **pp**. 1-15

Cao J, Wang Y-C, Klein TM, Sanford J and Wu R (1990). Transformation of rice and maize using the biolistic process, from *Symposium on Plant Gene Transfer, UCLA Symposium*, 21-33, Alan R. Liss, Inc.

Capesius L and Meyer Y, 1977. Isolation of nuclei from protoplasts of orchids. *Cytobiologia*. **15**: 485-490.

Champagnat M and Morel G, 1969. Multiplication vegetative *de Cattleya* a partir de bourgeons cultives *in vitro* des tissue de tuberculesd *'Ophrys. C.R. Acad. Sci. Paris*. **274**: 3379-3380.

Champagnat M and Morel G (1972). La culture *in vitro* des tissue de tubercules d'Ophrys. C.R.- Acad. Sci. Paries. **275**: 3379-3380.

Champagnat M, Morel G, Chabut P and Cognet AM (1966). Recherches morphologiques et histologiques sur la multiplication vegetative de quickques orchidees du genre *cymbidium Rev. Gen. Bot*. **73**: 706-746.

Champagnat MG, Morel G and Mounetou B (1970). La multiplication vegetative des Cattleya a partir de jeunes feuiles cultiviees ascptiquement *in vitro. Ann. Sci. Nat. Bat. Biol. Veg. Ser.* **12**(**11**): 97-114.

Chan YL, Lin KH, Sanjaya, Liao LJ, Chen WH and Chan MT (2005). Gene stacking in Phalaenopsis orchid enhances dual tolerance to pathogen attack.*Transgenic Res*. **14**(**3**): 279-88.

Chang C and Chang WC (1998). Plant regeneration from callus *of Cymbidium ensifolium* var. *misericors. Plant Cell Reports* **17**: 251-255.

Charest PJ, Calero N, Lachance D, Datla RSS, Duchesne LC and Tsang EWT (1993). Microprojectile-DNA delivery in conifer species factors affecting assessment of transient gene expression using the ?-glucuronidasereporter gene. *Plant Cell Rep*. **12**: 189-193.

Chathurvedi HC, Prasad RN Sharma AK, Misra P and Sharma M, 1995. Tissue culture propagation of ornamental plants. **pp. In** 367-377, Prakash, J and Bhandary, K.R (Ed) *Floriculture; Technology, Trades and Trends*. Oxford and IBH Publishing Co. Pvt. Ltd., New Delhi, Bombay, Calcutta.

Chaturvedi HC and Sharma AK, 1986. Mericloning of orchids through culture of tips of leaves and roots, 469-472 S.P. Vij *(ed.), Biology, conservation, and culture of orchids.* Affiliated East-West Press, New Delhi.

Cheah KT and Sagawa Y, 1978. *In vitro* propagation of *Aranda* Wendy Scott and *Aranthera* James Storei. *Hort Science.* **13**: 661-662.

Chen WH, Hsich RM, Tsai WT and Wu CC, 1990. Application of biotechnology in the improvement of *Phalaenopsis.* **In** *Poceedings on Nagoya Intl. Orchid Show,* Nagoya.

Chen WH, Chen TM, Fuy YM, Hsich RM and Chen WS, 1998. Studies on Somaclonal Variation in *Phalaenopsis. Plant Cell Reports.* **18**: 7-13.

Chen WH, Chen TW and Fu YM, 1998. RAPD analysis of somaclonal variation in *Phalaenopsis* orchid. **pp**. 134. *Proceedings of IX International Congress on Plant Tissue and Cell Culture* Jeruselum, Israel.

Chen WH, Fu YM, Hsieh RM, Wu CC, Chyou MS and Tsai WT, 1995. Modern breeding in *Phalaenopsis* orchid. *Taiwan-Sugar.* **42**(**3**): 17-22.

Chen WH, Tsai WT, Wu CC and Hsieh RM, 1991. Electrofusion and cell division of *Phalaenopsis* protoplasts. *Taiwan-Sugar.* **38**(**2**): 14-18.

Chen YQ and Piluek C, 1995. Effects of thidiazuron and N-benzylaminopurine on shoot 'regeneration of *Phalaenopsis. Plant Growth Regulation.* **16**(**1**): 99-101.

Chennaveeraiah MS and Patil SJ, 1975. Morphogenesis in seed cultures of *Spathoglottis. Current Science.* **44**(**2**): 68.

Chia TF, Hew CS, Loh CS and Lee YK, 1988. Carbon/nitrogen ratio and greening and protocorn formation in orchid callus tissue. *Hort. Science.* **23**: 559-601.

Chibbar RN, Kartha KK, Leung N, Qureshi J and Caswell K, 1991. Transient expression of marker genes in immature zygotic embryos of spring wheat (*Triticum aestivum*) through microprojectile bombardment, *Genome.* **34**, 453-460.

Chopra VL, 1995. Indian floriculture: An overview 8. Prakash J and Bhandary, K.R., (ed) *Floriculture; Technology, Trades and Trends.* Oxford and IBH Publishing Co. Pvt. Ltd., New Delhi, Bombay, Calcutta.

Chow HT, 1986. *In vitro* propagation of*Nervilia aragooma Gaud. J. of the college of sci. and Engineering. National Chung Hsing University. Thaiwan.* **23**: 47-58.

Chow HT, WC, Hsich and Chang CS,1982. *In vitro* propagation of *Anocctochilus formosanus. J. Sci. Eng.* **19**: 155-166.

Christensen M, 1989. A view of fungal ecology. *Mycologia* **81**: 1-19.

Christou P, Ford TL and Kofron M, 1992. The development of a variety-independent gene-transfer method for rice, *TIBTECH,* **10**: 240-246.

Christou P, McCabe DE and Swain WF, 1988. Stable transformation of soybean callus by DNA-coated gold particles, *Plant Physiol.* **87**: 671-674.

Christou P, McCabe DE, Martinell BJ and Swain WF, 1990. Soybean genetic engineering – commercial production of transgenic plants, *TIBTECH.* **8**: 145-151.

Christou P, 1990. Soybean transformation by electric discharge particle acceleration, *Physiol. Plant.* **79**: 210-212.

Christou P, Swain WF, Yang N-S and McCabe DE, 1989. Inheritance and expression of foreign genes in transgenic soybean plants, *Proc. Natl. Acad. Sci. USA.* **86**: 7 500-7504.

Chung JD, Lee JH, Jec SO and Kim CK (1998). Effect of medium composition on multiple shooting and subsequent growth of mericlone from rhizome of shoot tip cultures of temperate *Cymbidium* species. *J. of the Korean Soc. for Hortl. Sci.* **39**(**3**): 343-349.

Churchill ME (1971). Production of orchids from seeding leaf tips. *Orchid Dig.* **34**: 71-273.

Churchill ME, 1972. Tissue culture of orchids. II Methods for root tips. *Amer. Orchid Soc. Ball.* **41**: 726-730.

Churchill ME (1973). Tissue culture of orchids. I. Methods for leaf tips. *New Phytol.* **72**: 161-166.

Churchill M-E, Bal EA and Arditti J (1970). Production of orchid plants from seeding leaftips. *Orchid Dig.* **34**: 271-273.

Churchill ME, Arditti J and Ball EA (1971). Clonal propagation of orchids from leaf tips. *Amer. Orchid Soc. Bull.* **42**: 109-113.

Clement E (1973a). Agermination of *Odontoglossum* and other seed without fungal aid. Orchid Rev. **81**: 123-129.

Clement E (1973b). The non-symbiotic and symbiotic germination of orchid seeds Orchid Rev. **81**: 152-154.

Cocozza TM, De-Lucia B and Serra G (1994). Influence of "Oasis" substratum on *Phalaenopsis* soilless culture. *Acta-Horticulturae.* **361**: 464-469.

Cordier C, Pozo MJ, Barea JM, Gianinazzi S and Gianinazzi-Pearson V (1998). Cell defence responses associated with localized and systematic resistance to *Phytophthora parasitica* induced by an arbuscular mycorrhizal fungus. Molecular Plant-Microbe Interactions **11**: 1017-1028.

Creissen G, Smith C, Francis R, Reynolds H and Mullineaux P (1990). *Agrobacterium*- and microprojectilemediated viral DNA delivery into barley microspore-derived cultures, *Plant Cell Rep.*, **8**: 680-683.

Cribb PJ, Greatwood J and Hunt PP (1985). *Hand book on orchid Nomenclature and Registration (3rd Edition), International Orchid Commission,* Florida, USA.

Cullen J (1992). *The Orchid Book.* Cambridge University Press, London.

Cummings DJ, Domenico JM and Sanford JC (1990). Mitochondrial DNA from *Podospora anserina*: transformation to senescence via projectile injection of plasmids, *Molec. Biol. Aging.* **123**: 91-101.

Czerevezenko TM and Kushnir GP (1986). *Orchidei v kulture.* Haukova Dumka, Kiev, Ukraine.

Dall Rosa M and Laneri U (1977). Modification of nutrient solutions for germination *in vitro* of some cultivated orchids and for the vegetative propagation of *Cymbidium* cultivars. *Am. Orchid Soc. Bull.* **46**: 813-820.

Daniell H, Vivekananda J, Nielsen BL, Ye GN, Tewari KK and Sanford JC (1990). Transient foreign gene expression in chloroplasts of cultured tobacco cells after biolistic delivery of chloroplast vectors, *Proc. Natl. Acad. Sci. USA.* **87**: 88-92.

Daniels Hetrick BA, Leslie JF, Thompson Wilson G and Gerschefske Kitt GD (1988). Physical and topological assessment of effects of a vesicular-arbuscular mycorrhizal fungus on root architecture of big bluestem. *New Phytologist* **110**: 85-96.

De Fossord RV (1976). Tissue culture for plant propagators. University of New England, Armidale (Australia).

Debergh PC and Maene (1981). A scheme for commercial propagation of ornamental plants by tissue culture, Scientia Hot. **14**: 335-345.

Debergh PC and Read PE (1993). Micropropgation. **pp** 1-13. Debergh, P.E. and Zimmerman, R.H. (ed). *Micropropagation: Technology and Application,* Kluwer Academic Pbls, London, Dordrecht, Boston.

Debuchy R, Purton S and Rochaix J-D (1989). The argininosuccinate lyase gene of *Chlamydomonas reinhardtii*: an important tool for nuclear transformation and for correlating the genetic and molecular maps of the ARG7 locus, *EMBO J.* **8**: 2803-2809.

Dekhuijzen HM (1971). Sterilization of cytokinins. **pp**. 129-132 Van Bragt, J, Mosselm, DAA, Pierik, R.L.M and Veldstra, H (ed). Effects of sterilization on components in nutrient media. *H. Veerman and Zonen.* Wageningen.

Desphande V (1992). Flowers and business, both blossom, *Times of India* (daily newspaper) June 27th.

Devi J, Borthakur B and Deka PC (1997). Clonal propagation of *Dendrobium moschatum* and *Cymbidium aloifolium* through shoot tip culture *J. Orchid Soc. India* **11**(**1-2**): 19-21.

Devi J, Nath M, Devi M, Deka PC (1990). Effect of different media on germination and growth of some north-east Indian species of *Dendrobium J. Orchid Soc. India* **4**(**1-2**): 45-59.

Devi T and Deka PC (2001). Orchid. **In pp**. 29-62 Parthasarathy V.A, Bose, T.K and Das P(eds). *Biotechnology of Horticultural Crops* (Vol-3).

Diwakar KM (1987). Cytoembryological investigations on certain orchids of tribe *Epidendreae* and *Vandae*. Ph.D thesis submitted to Karnatak University, Dharwad.

Dodds JH and Roberts LW (1995). *Experiments in plant tissue culture* III Edition. Cambridge University Press.

Dressier RL (1981). *The Orchids: Natural history and classification,* Harward University Press, London.

Duan JX and Yagawa (1995). Floral induction and development in *Phalaenopsis In vitro. Plant Cell, Tissue and Organ Culture.* **43**: 71-74.

Duan JX and Yagawa S (1994). *In vitro* flowering of *Doriella, Phalaenopsis and Dendroboium*. **pp**. 87-96 in Proceedings of Nagoya International Orchid Conference, Nagoya, Japan.

Duchesne LC and Charset PJ (1991). Transient expression of the β-glucuronidase gene in embryogenic callus of *Picea mariana* following microprojection, *Plant Cell Rep.* **10**: 191-194.

Duchesne LC, Lelu MA, Von Aderkas P and Charest PJ (1993). Microprojectile-mediated DNA delivery in haploid and diploid embryogenic cells of *Larix* spp., *Can. J. For. Res.* **23**: 312-316.

Eason WR, Newman EI and Chuba PN (1991). Specificity of interplant cycling of phosphorus: the role of mycorrhizas. *Plant and Soil.* **137**: 267-274.

Eichholtz DA, Hasegawa PM and Robitaille HA (1982). Effects of gentamicin on growth of shoot initiation from cultured tobacco callus and *Salpiglossis* leaf discs. *In vitro* **18**: 12-14.

Ellis DD, McCabe D, McInnis S, Ramachandran R, Russell DR, Wallace KM, Martinell BJ, Roberts DR, Raffa KF and McCown BH (1993). Stable transformation of *Picea glauca* by particle acceleration, *Bio/Technol.* **11**: 84-89.

Ellis DD, McCabe D, Russell D, Martinell B and McCown BH (1991). Expression of inducible angiosperm promotors in a gymnosperm, *Picea glauca* (white spruce), *Plant Molec. Biol.* **17**: 19-27.

Engelke AL, Hamzi HQ and Skoog F (1973). Cytokinin-Giberellin regulation of shoot development and leaf form in tobacco plantlets. *Amer. J. Bot.* **60(6)**: 491-495.

Ernst R (1967). Effect of select organic nutrient additives on growth *in vitro* of *Phalaenopsis* seedlings. *Am. Orchid Soc. Bull.* **36**: 694-704.

Ernst R (1974). The use of activated charcoal in asymbiotic seeding culture of Paphiopedilum. *Am Orchid Soc. Ball.* **43**: 35-38.

Ernst R (1975). Studies in asymbiotic culture of orchids. *Am. Orchid Soc. Bull.* **44**: 12-18.

Ernst R (1976). Charcoal or glasswool in asymbiotic culture. **pp** 379-383 *Proc. 8th World Orchid Conf. Frankfiirt.* (1974).

Ernst R (1994). Effects of thidiazuron on *in vitro* propagation of *Phalaenopsis* and *Doritaenopsis* (Orchidaceae). *Plant cell, Tissue and organ culture.* **39(3)**: 273-275.

Ernst R, Arditti J and Healey PL (1970). The nutrition of orchid seedlings. Am. Orchid Soc. Bull. **39**: 599-605, 691-700.

Falkiner FR (1990). The criteria for choosing an antibiotic for control of bacteria in plant tissue culture. *Newsl. IAPTC.* **60**:13-23.

Fast G (1973a). Uber einige Methoden zur Vermehrung von Orchideen. *Jahresbericht der Fachhochschule Weibenstephan,* pp 1-5.

Fast G (1973b). Die Vermehrung von *Oncidium papilio* durch Treibspitzen-Kultur and Besprechung einiger Nahrmedian. *Die Orchidee.* **24**: 240-246.

Ferchandez GCJ (1992). Residual Analysis and data transformations: Important tools in statistical analysis. *Hort Sci.* **27(4)**: 277-300.

Fernando KKS (1979). Clonal propagation of *Dendrobiwn* Caesar Red Lip by meristem culture. *Trop. Agric. (Columbo, Sri Lanka).* **135**: 111-113.

Finney DJ (1989). Was this your statistical text book? V. Transformation of data, *Exptl. Agri.* **25**: 165-175.

Fischer CEC (1928). *In Gamble's Flora of Presidency of Madras* (3rd Reprint 1957). BSI, Calcutta.

Fischer P (1986). Cut-flower production from *Phalaenopsis* grown in expanded clay. Fine-grain expanded clay gave the best results. *Zierpflazenbau.* **26(24)**: 964.

Fiske MD (1982). The Eric E. Young somatic hybrid orchid prize. *Amer. Orchid Soc. Bull.* **51**: 43-46.

Fitch MM, Manshardt RM, Gonsolves D, Slightom JL, Quemada H and Sanford JC (1990). Stable transformation of papaya via microprojectile bombardment, *Plant Cell Rep.* **9**: 189-194.

Fonnesbech M (1972b). Organic nutrients in the media for propagation *of Cymbidium in vitro*. *Physiol Plant*. **27**: 360-364.

Fonnesbech M (1972a). Growth hormones and propagation of *Cymbidium in vitro*. *Physiol Plant*. **27**: 310-312.

Fox TD, Sanford JC and McMullin T W, 1988. Plasmids can stably transform yeast mitochondria lacking endogenous mtDNA, *Proc. Natl. Acad. Sci. USA*. **85**: 7288-7292.

Fridborg G, Pe'terson M, Landstorm KE and Erikson T (1978). The effect of activated charcoal on tissue cultures: Absorption of metabolites inhibiting morphogenesis, *Physiol. Plant*. **43**: 104-106.

Fromm ME, Morrish F, Armstrong C, Williams R, Thomas J and Klein TM (1990). Inheritance and expression of chimeric genes in the progeny of transgenic maize plants, *Bio/Technol*. **8**: 833-839.

Fu FML (1978). Clonal propagation of Aranda, Assocenda, and Cattleya by leaf tissue culture. *Card. Bull*. (Singapore). **31**: 132-138.

Fu FML (1979a). Studies on the tissue culture of orchids. Clonal propagation of *Phalaenopsis* by lateral buds from flower stems. *Orchid Rev*. **87**: 308-310.

Fu FML (1979b). Studies on the tissue culture of orchids. Clonal propagation of *Aranda Ascocenda,* and *Cattleya* by leaf tissue culture. *Orchid Rev*. **87**: 343-346.

Fu XD, Qian XH, Mao BZ and Li DB (1997). Effects of some factors on the growth and differentiation of *Cymbidium aloifoliwn* protocorm. *J. of Zhegiang Agricultural University*. **23**(**5**): 547-550.

Fujii K, Kawano M and Kako S (1999). Effects of Benzglademine and α- Naphthalene-acetic acid on cell division and nuclear DNA contents in outer tissue of *Cymbidium* explants cultured *in vitro. J. Japan. Soc. Hort. Sci*. **68**(**1**): 41 -48.

Fujii K, Kawano M and Kako S (1999). Effects of benzyladenine and α- Naphthalene-acetic acid on the formation of protocorn like bodies (PLBs) from explants of outer tissue of *(Cymbidium)* PLBs cultured *in vitro. J. Japan. Soc. Hort. Sci*. **68**(**1**): 35-40.

Gallaud I (1905). Études sur les mycorrhizes endophytes. *Revue General de Botanique* **17**: 5-48, 66-83, 123-136, 223-239, 313-325,425-433, 479-500.

Gamborg OL, Murashige T, Thorpe TA, and Vasil IK (1976). Plant tissue culture media. *In vitro*. **12**: 473-478.

Gandawijaja D (1980). Effects of nitrate and ammonium on growth of tissue culture of *Dendrobium phalaenopsis* Fitzg. *Ann. Bagor*. **7**: 63-69.

Ganesh DS, Sreenath HL and Jayashree H (1996). Micropropagation of *Vanilla* through node culture. *J. Plantation Crops*. **24**(**1**): 16-22.

Gautheret RJ (1942), Hetero-auxines et cultures de tissues vegetaux. *Bull. Soc. Chim. Bial*. **24**: 13-41.

Gautheret RJ (1959). La culture des tissue vegetaux. Masson, Paris.

Gautheret RJ (1983). Plant tissue culture: A history. *Bot. Mag. Tokyo*. **96**: 398-410.

Gautheret RJ (1985). History of plant tissue and cell culture. A personal account. **pp**. 1-59 I.K. Vasil [ed.], *Cell culture and somatic cell genetics,* Vol 2. Academic Press, Orlando, Fla.

Gavinlertavatana P and Prutpongse P (1993). Commercial Micropropagation in Asia In. Debergh P.C. and Zimmerman, R.H (ed) *Micropropagation, Technology and Application.* Kluwer Academic Pbls. London. Dordrecht. Boston.

George EF and Sherrington PO (1984). *Plant propagation by Tissue Culture.* Exegetics Ltd. Eversley, Basingstoke, Hants.

George PS and Ravishankar GA (1997). *In vitro* multiplication of *Vanilla planifolia* using axillary bud explants. *Plant Cell Reports.* **16**: 490-494.

Ghosh SP (1998). Commercial floriculture. *Indian Horticulture.* **43(1)**: 7-11

Goff SA, Klein TM, Roth BA, Fromm ME, Cone KC, Radicella JP and Chandler VL (1990). Transactivation of anthocyanin biosynthetic genes following transfer of B regulatory genes into maize tissues, *EMBO J.* **9**: 2517-2522.

Goh CJ and Wong PF (1990). Micropropagation of the monopodial orchid hybrid *Aranda* 'Deborah' using inflorescence explants. *Scientia Horticulturae.* **44(3-4)**: 315-321.

Goh CJ (1970). Tissue culture of *Vanda* Miss Joaquim. *J. Singapore Nat. Acad. Sci.* **2**: 31-33.

Goh CJ (1971). The influence of pH on orchid culture. Mal. Orchid Rev. **10**: 32-35. Goh CJ, 1973. Meristem culture *of Aranda* Deborah Mal. Orchid Rev. **12**: 9-12.

Goh CJ (1973). Meristem culture of *Aranda* Deborah. *Malay. Orchid Rev.* (Singapore) **11**: 10-14.

Goh CJ (1981). Clonal propagation of orchids through tissue culture. *Malay. Orchid Rev. (Singapore).* **15**: 38-49.

Gopalkrishnan LP and Seeni S (1987). Isolation, culture and fusion of protoplasts in some selected orchids. **pp** 288 K. Saito and R. Tanaka [eds.], *12th World Orchid Conf.* Tokyo.

Gordon-Kamm W J, Spencer TM, Mangano ML, Adams TR, Daines RJ, Start WG, O'Brien JV, Chambers SA, Adams Jr, WR, Willetts NG, Rice TB, Mackey CJ, Krueger RW, Kausch AP and Lemaux PG (1990). Transformation of maize cells and regeneration of fertile transgenic plants, *Plant Cell.* **2**: 603-618.

Govil S and Gupta SC (1994). Micropropagation industry in industry in India. 'Retrospects and Prospects. J. Sci. and Ind. Res. **53**: 768-777.

Grandmaison J, Olah GM, van Calsteren MR and Furlan V (1993). Characterisation and localisation of plant phenolics likely involved in the pathogen resistance expressed by endomycorrhizal roots. *Mycorrhiza.* 3: 155-164.

Green T (I983). Mericlone mutant - *Brassolaeliocattleya* Greenwich 'Irish Flair' *Am OrchidSoc. Bull.* **52**: 1294.

Grewal HS, Arora JS and Singh K (1996). Propagation of Ornamental Plants through tissue culture. **pp.** 356-372 Islam, A (ed) *Plant Tissue Culture.* Oxford and IBM publishing Co. Pvt. Ltd., Calcutta, New Delhi.

Griesbach RJ (1983). The use of indoleacetylamino acids in the vitro propagation of *Phalaenopsis* orchids. *Scientia Hortic.* (Amsterdam). **19**: 363-366.

Griesbach RJ (1984). The *in vivo* propagation of *Phalaenopsis orchids. Aust. Orchid Rev.* **49**: 41-43.

Gu Z, Arditti J and Nyman LP (1987). *Vanilla planifolia:* Callus induction and plantlet production *in vitro. Lindleyana.* **2**: 48-52.

Guo LH, Stepien PP, Tso JY, Brousscau R, Narang S, Thomas DY and Wu R (1984). Synthesis of human insulin gene VIII of expression vector for fused proinsulin in *Escherichia coli. Gené.* **29(1-2)**: 251-254.

Haas NF (1977). *Beitrage zur asympiotischen samenkeimung and Meristemvemchrung terrestrischer Orchideen Europas, Amerikas* and *Salidafricas.* Ph.D. thesis., Justus Liebig Univeritat, Giessen, Germany.

Haas-von Schmude NF (1983). Klonale Messenvermehrung von *Phalaenopsis. Die Orchidee.* **34**: 242-248.

Haas-Von Schmude NF (1985). Tissue culturing *Phalaenopsis* using leaves and leaf segments. **pp.** 311 K. Tan [ed.], *Proc. 11th World Orchid Conf.* Miami. (1984).

Haas-Von-Schmude NF (1984). *Phalaenopsis* propagation by meristem culture. *Deutscher-Gartenbau.* **38(20)**: 878-879.

Haberlandt G (1902). Kulturversuche mit isolierten pflanzenzellen. *Sitzundber. Akad. Wiss. WienMath.Naturwiss. Jl. Abt.* 1, **111**: 69-92.

Hagio T, Blowers AD and Earle ED (1991). Stable transformation of sorghum cell cultures after bombardment with DNA-coated microprojectiles, *Plant Cell Rep.,* **10**: 260-264.

Haring MA, and De Block M (1990). New roads towards chloroplast transformation in higher plants, *Physiol.Plant.* **79**: 218-220.

Harley JL and Smith SE (1983). Mycorrhizal Symbiosis. Academic Press, London.

Hasegawa A (1987). Studies on the propagation of oriental *Cymbidium. Mem. Fac. Agric. Kagawa Univ.* **50**: 1-108.

Hasegawa and Goi M (1987). Rhizome formation in *Cymbidium goeringii* Reichenbach fil. And *Cymbidium kanran* Makino in shoot-tip-culture. *J. Jpn. Soc. Hortic. Sci.* **56**: 70-78.

Hasegawa Obashi H and Goi M (1985). Effects of BA, rhizome length, mechanical treatment and liquid shaking culture on the shoot formation from rhizome in *Cymbidium faberi* Rolfe. *Acta Hortic. (Wageningen).* **166**: 25-40.

Heller R (1949). Sur l'emploi du papier filtre sans cendres camme support pour less cultures de tissus vegetaux. Compt. Rend. Soc. Biol. **143**: 335-357.

Heller R (1953). *Recherches sur la nutrition minerals des tissue vegetaux cultives in vitro. These paris et Ann. Sci. Nat. Bot. Veg. Paries.* **14**: 1-223.

Henderson MR and Anderson H (1956). Malayan orchid hybrids. Singapore: Government Printing Office.

Hermann S and Sell Y (1991). Morphogenic potential of nodal cuttings of the flowering shoots of *Phalaenopsis* cultivated *in vitro. Beitrage zur Biologie der pflangen.* **66(2)**: 249-269.

Hew CS and Yip KC (1986). Isolation of orchid cells. *Malay. Orchid Bull.* **3**: 49-54.

Hew CS, Ting SK and Chia TF (1988). Substrate utilization by *Dendrobium* tissues. *Bot. Gaz.* **149(2)**: 153-157.

Hieber AD, Mudalige-Jayawickrama RG, Kuehnle AR (2006) "Color genes in the orchid Oncidium Gower Ramsey: identification, expression, and potential genetic instability in an interspecific cross". *Planta*. Feb. **223**(**3**): 521-31.

Hogberg P, Plamboeck AH, Taylor AFS and Fransson PMA, 1999. Natural C-13 abundance reveals trophic status of fungi and host-origin of carbon in mycorrhizal fungi in mixed forests. Proceedings of the National Academy of Sciences of the United States of America. **96**: 8534-8539.

Holdgate DP (1977). Propagation of ornamentals by tissue culture. Pages 18-42, 207-248 in J. Reinert and Y.P.S. Bajaj (eds). *Plant cell tissue and organ culture*. Springer Verlag, New York.

Holters J and Zimmer K (1990b). Shoot regeneration from root tips of orchids *in vitro* III. Model plant *Mormodes histrio Linden at Rchb.f. Garten-bauwissenschaft*. **55**(**5**): 201-205.

Holters J and Zimmer K (1990a). Shoot regeneration from root tips of orchids *in vitro*. IV. Organogenesis on *Mormodes histrio* roots. *Gartenbawissenschaft*. **55**(**6**): 264-267.

Homma Y and Asahira T (1985). New means *of Phalaenopsis* propagation with internodal sections of flower stalk. *;. Japanese Soc. for Hort. Sci.* **54**(**3**): 379-387.

Hooker JD (1888). Orchidaccae: Flora of British India Vols. V and VI. BSI, Howrah.

Huang L, 1984. Alternative media and method for *Cattleya* propagation by tissue culture. *Amer. Orchid Soc. Bull.* **53**(**2**): 167-170.

Huang L (1988). A procedure for sexual multiplication of *Paphiopedilums in vitro. Amer. Orchid Soc. Bill.* **57**(**3**): 274-278.

Hunt PF (1987). *The orchids,* Spring Books (Octopus Publishing Group PLC) London.

Hunt PP (1986). The nomenclature and registration of orchid hybrids and specific and generic levels. **pp.** 7-374. *Intraspecific classification of wild and cultivated plants*. Clarendon press. Oxford, UK.

Ichihashi S (1987). Systematic investigation of orchid culture media. **pp.** 60-64 *The Proceedings of the world orchid symposium,* Hiroshima. Japan

Ichihashi S and Islam MO (1999). Effects of complex organic additives on callus growth in three Orchid genera, *Phalaenopsis, Doritaenopsis,* and *Neqfmetia J. Japan. Soc. Hort. Sci.* **68**(**2**): 10-17.

Ichihashi S (1979). Clonal propagation of *Ascocenda* through shoot meristem culture. *Bull. Aichi Univ. Educ. Nat. Sci.* **28**: 143-156.

Ichihashi S and Kako S (1973). Studies on clonal propagation *of Cattleya* through tissue culture method. II. Factors affecting survival and growth of shoot meristem *of Cattleya in vitro. J. Jpn. Soc. Hortic. Sci.* **42**: 264-270.

Iida A, Seki M, Kamada M, Yamada Y and Morikawa H (1990). Gene delivery into cultured plant cells by DNA-coated gold particles accelerated by a pneumatic particle gun, *Theor. Appl. Genet.* **80**: 813-816.

Inamura H and Saiton, T (1987). Studies on the aseptic culture *mAeridesjaponicum. Bull of the Shimane Agricultural experiment station,* **22**: 48-65.

Intuwong O and Sagawa Y (1973). Clonal propagation of *Sarcanthinae* orchids by aseptic culture of inflorescences. *Am. Orchid Soc. Bull.* **42**: 209-215.

Intuwong O and Sagawa Y (1974). Clonal propagation by shoot tip culture. *Am. Orchid Soc. Bull.* **43**: 893-895.

Intuwong O, Kunisaki JT and Sagawa Y (1972a). Vegetative propagation of *Phalaenopsis* by flower stalk cuttings. *Okika O Hawaii* [Hawaii Orchid]. **1**: 13-18.

Intuwong O, Kunisaki JT and Sagawa Y (1972). Vegetative propagation of *Phalaenopsisby* flower stalk cuttings. Na Okika of Hawaii. **1**: 13-18.

Irawati Harjadi SS, Suseno H and Idris S (1977). Tissue culture *ofAranthera* James Storiei. *OrchidRev.* **85**: 138-142.

Ishii Y, Jakamura T, Goi M and Tanaka M (1998). Callus induction and somatic embryogenesis *of Phalaenopsis. Plant Cell Reports.* **17(6-7)**: 446-450.

Islam MO, Ichihasni S and Matsui S (1998). Control of growth and development of protorn like body derived from callus by carbon sources in *Phalaenopsis. Plant Biotechnology.* **15(4)**: 183-187.

Isley P (1965). Meristem tissue propagation techniques and protentialities. Orchid Rev. **73**: 371-376.

Ito I (1955). Germination of seeds from immature pod and subsequent growth of seedlings in *Dendrobium mobile* lindl. *Sci. Rep. Saikyo Univ.* **7**: 35-42.

Ito I (1960). Culture of orchid seedings by way of completing the growth of ovaries of cut flowers. *Japan. OrchidSoc. Bull.* **6**: 4-7.

Jaedong C, Hee LT, OK JS and Kil KC (1998). Effects of medium composition on multiple shooting and subsequent growth of mericlone from rhizome of shoot tip cultures of temperature *Cymbidium* species. J. of Korean Soc. For Hort. Sci. **39(3)**: 343-349.

Jain SK (1980). *Orchids and Mountain flora of India.* 67th session of Indian Science Congress Association, Calcutta.

Johnson ET, Yi H, Shin B, Oh BJ, Cheong H, Choi G (1999). Cymbidium hybrida dihydroflavonol 4-reductase does not efficiently reduce dihydrokaempferol to produce orange pelargonidin-type anthocyanins. Plant J. **19(1)**: 81-5.

Johnston SA, Anziano PQ, Shark K, Sanford JC, and Butow RA (1988). Mitochondrial transformation in yeast by bombardment with microprojectiles, *Science.* **240**: 1538-1541.

Joshi PC and Ball EA (1968b). Growth values and regenerative potentially in mesophyll cultures *ofArachis hypogeaZ. Pflanzenphysial.* **9**: 5-75.

Joshi PC, and Ball EA, 1968a. Growth of isolated palisade cells *ofArachis hypoge Dev.* **5/0/17**: 308-325.

Kanase A, Sugimoto Y, Hirai J, Oyamada T, Jakano T (1993). The process involved in the histogenesis and organogenesis of *Cymbidium* protocorm like bodies *in vitro.* **pp.** 64-70 The proceedings of Nagoya, Intl. Orchid Conference, Nagoya, Japan.

Karim BA, Ghani A, Haris HAB and Ugang NBH (1992). Micropropagation of *Mokara, Lindleyana.* **7(1)**: 7-10.

Kartha KK, Chibbar RN, Georges F, Leung N, Caswell K, Kendall E and Qureshi J (1989). Transient expression of chloramphenicol acetyltransferase (CAT) gene in barley cell cultures and immature embryosthrough microprojectile bombardment, *Plant Cell Rep.* **8**: 429-432.

Kaur S and Sharma CM (1997). Selection of best medium for *in vitro* propagation of *Dendrobium. Advances in plant Sciences.* **10**(**1**): 1-5.

Kendrick B (1992). The Fifth Kingdom. Mycologue Publications, Waterloo.

Kerbauy GB (1984a). Regeneration of protocorn-like bodies through *in vitro* culture of root tips of*Catasetum* (Orchidaccae). Z *Pflanzenphysical.* **113**: 287-291.

Kerbauy GB (1984b). Plant regeneration of*Oncidium Varicosum* (Orchidaceae) by means of root tip culture. *Plant Cell Reports.* **3**: 27 - 29.

Kerbauy GB (1991). *In vitro* conservation of *Cattleya* root tip cells into protocorm like bodies. *J. of plant physiology.* **138**(**2**): 248-251.

Keshavamurthy KR and Yoganarasimhan SN (1990). Flora of Coorg (Kodagu), Karnataka, India. Vimsat publishers, Bangalore.

Khaw CH, Ong HT and Nair H (1978a). Hormones in the nutrition of orchid tissues in mericloning. **pp.** 60-65 *Proc. Symp. Orchidology.* Orchid Society of Southeast Asia, Singapore.

Khoshoo TN (1995). Ornamental Horticulture in India: problems and prospects. **pp.** 10-18. Prakash, J and Bhandary, K.R. (ed). *Floriculture; Technology Trades and Trends.* Oxford and IBH Publishing Co. Pvt. Ltd., New Delhi, Bombay, Calcutta.

Kikkert JR (1993). The Biolistic PDS-1000/He device, *Plant Cell Tissue Organ Cult.* **33**: 221-226.

Kim (1984). Studies on clonal propagation on the *Cymbidium* through organ culture *in vitro J. Korean Soc. Hortic. Soc.* **25**: 65-71.

Kim KW and Kako S (1984). Studies on *Cymbidium* clonal propagation by floral organ culture *in vitro J. of the Korean soc. of Horticultural Science.* **25**(**1**): 65-71.

Kim KW and Kako S (1982). Effect of plant growth regulators on organ formation in the *Cymbidium* shoot apex culture *in vitro. J. Jpu. Soc. Hortic. Sci.* **51**: 106-114.

Kim YJ, Hong YP, Park YK, cheung SK and Kim EY (1998). *In vitro* propagation of *Cymbidium Kansam. Research Reports of the Rural Development Administration Hort.* **30**(**3**): 77-82.

Kindle KL, Schnell R A, Fernández E and Lefebvre PA (1989). Stable nuclear transformation of *Chlamydomonas* using the *Chlamydomonas* gene for nitrate reductase, *J. Cell Biol.* **109**: 2589-2601.

Kishi F and Takagi K (1997). Efficient method for the preservation and regeneration of orchid Plbs. *Scientia Horticulture.* **68**(**1/4**): 149-152.

Klein RM and Klein DT (1970). Research methods in plant science. Natural History Press. Garden City. N.Y.

Klein TM, Fromm M, Weissinger A, Tomes D, Schaaf S, Slette, M and Sanford JC, (1988a). Transferof foreign genes into intact maize cells with high-velocity microprojectiles, *Proc. Natl. Acad. Sci. USA.* **85**. 4305-4309.

Klein TM, Goff SA, Roth BA and Fromm ME (1990). Applications of the particle gun in plant biology, in Proceedings of the VIII International Congress on Plant Tissue and Cell Culture, June 24-29, pp. 56-66, Amsterdam.

Klein TM, Gradziel T, Fromm ME and Sanford JC (1989). Factors influencing gene delivery into *Zea mays* cells by high-velocity microprojectiles, *Bio/Technol.* **6**: 559-563.

Klein TM, Harper EC, Svab Z, Sanford JC, Fromm ME and Maliga P (1988c). Stable genetic transformation of intact *Nicotiana* cells by the particle bombardment process, *Proc. Natl. Acad. Sci. USA.* **85**: 8502-8505.

Klein TM, Kornstein L, Sanford JC and Fromm ME (1989a). Genetic transformation of maize cells by particle bombardment, *Plant Physiol.* **91**: 440-444.

Klein TM, Roth BA and Fromm ME (1989b). Advances in direct gene transfer into cereals, in "Genetic Engineering; Principles and Methods," Vol. 11, J. K. Setlow (ed.), Plenum Press, New York.

Klein TM, Roth BA and Fromm ME (1989c). Regulation of anthocyanin biosynthetic genes introduced into intact maize tissues by microprojectiles, *Proc. Natl. Acad. Sci. USA.* **86**: 6681-6685.

Klein TM, Wolf ED, Wu R and Sanford JC (1987). High-velocity microprojectiles for delivering nucleic acids into living cells, *Nature.* **327**: 70-73.

Knudson L (1922). Nonsymbiotic germination of orchid seeds. *Bot. Gaz.* **73**: 1-15.

Knudson L (1919). Validity of detached root cap cells *Am. J. Bot.* **6**: 309-310.

Knudson L (1921). La germination do simbiotica de las semillas de orginidias. *Bol. Rec'. Soc. Exp. Thist. Nat.* **21**: 250-260.

Knudson L (1946). A new nutrient soslution for the germination of the orchid seed, *Am. Orchid Soc. Bull.* **14**: 214-217.

Knudson L (1951). Nutrient soltuion for orchids, Botan, Gaz. **112**: 528-532.

Koch L (1974). Untersuchungen zur vegetativen Vermebrung bci *Phalaenopsis in vitro.* Diss, Technical Univ., Hannover, Germany.

Koh MC, Goh CJ and Loh CS (1988). Protoplast isolation and culture of *Aranda* hybrids. *Malay. Orchid Rev.* **22**: 70-78.

Kohlenbach HW and Wernicke W (1978). Investigations on inhibitory effect of agar and function of active carbon in anther culture Z *Pflanzen-Physiol.* **86**: 463-472.

Kononowicz H and Janick J (1984). *In vitro* propagation of *Vanilla planifolia. Hort. science.* **19**: 58-59.

Koziel MG, Beland GL, Bowman C, Carozzi NB, Crenshaw R, Crossland L, Dawson J, Desai N, Hill M, Kadwell S, Launis K, Lewis K, Maddox D, McPherson K, Meghji MR, Merlin E, Rhodes R, Warren GW, Wright and M, Evola S (1993). Field performance of elite transgenic maize plants expressing an insecticidal protein derived from *Bacillus thuringensis, Bio/Technol.* **11**: 194-200 (1993).

Krans JE and Monteriro WR (1989). Formation of Protocrm like bodies from root apices of *Catasetum pileatum. Ann. of Botany.* **64**(**5**): 491-498.

Kraus JE and Kerbauy GB (1987). Some requirements for regeneration of root tips of *Catasetum pileatum.* **pp.** 19, Zenda, A.A. (ed). Proceedings of International Congress of Plant tissue culture-Tropical species, Bogota.

Kraus JE and Monteiro WR (1989). Formation of PLBs from root apices *of Catasetum pilteatum* cultivated *in vitro-I.* Morphological aspects. *Annals of Botany.* **65**(**5**): 491-498.

Krikorian AD (1982), Cloning higher plants from aseptically cultured tissues and cells. *Biol. Rev.* **57**:151-218.

Krikorian AD and Berquam DL (1969). Plant cell and tissue cultures: The role of Haberlandt. *Bot. Rev.* **35**: 59-88.

Kubota S, Kato T, and Yoneda K (1993). The effects of the concentration of fertilizer application and the physico-chemical properties of sphagnum moss and clay pots on the growth of Phalaenopsis. *J. of the Japanese Society for Hort. Sci.* **62**(**3**): 601-609.

Kubota S and Yoneda K (1994). Effect of time of nitrogen application on growth and flowering in *Phalaenopsis* plant. *Japanese J. of Tropical Agriculture.* **38**(**1**): 73-77.

Kuehnle AR and Sugii N (1992). Transformation of *Dendrobium* orchid using particle bombardment of protocorms, *Plant Cell Rep.* **11**: 484-488.

Kukulczanka K, Czastka B and Marczonek A (1987). The effect of the substrates on the *in vitro* culture *ofCymbidium, Acta Agrobotanica.* **40**(**1-2**): 41-51.

Kukulczanka K, Kromer K and Roginska B (1989). Joint effects of growth regulators and peptone on regeneration of *Vuylstekeara* 'Cambria' orchid under tissue culture conditions. *Acta. Hortic.* **251**: 161-166.

Kukulczanka K and Wojciechowska V (1983). Propagation of two *Dendrobium* species by *in vitro culture. Acta Hortic.* (*Wageningen*). **131**: 105-1 10.

Kumar A and Kumar VA (1998). *Clonal Tissue Culture of Important Fruit Crops.* International Book Distributing Co., Charbagh, Lucknow.

Kumar PKS, Suresh Kumar PK (1992). Potting media and post-transplantation growth of *Dendrobium* hybrid seedlings. *J. of the orchid Society of India.* **6**(**1-2**): 131-133.

Kunisaki JT, Kim KK and Sagawa Y (1972). Shoot tip culture of *Vanda. Am. Orchid Soc. Bull.* **41**: 435-439.

Kushnir GP (1986). Microclonal propagation of *Dandrobium phalaenopsis* Fitzg. **pp.** 63-64 *Conservation and culture of orchids. Abst. 3rd All Union Orchid Conf.* Main Botanical Garden, Academy of Sciences, Moscow.

Kusumoto M (1979a). Effects of combinations of growth regulations, and of organic supplements on the growth of Cattleya, plantlets cultured *in vitro. J. Jpn. Soc. Hortic. Sci.* **47**: 492-501.

Kusumoto M (1979b). Effects of combination of growth regulators, and of organic supplements on the proliferation and organogenesis of *Cattleya* protocorm like bodies cultured *in vitro J. Jap. Soc. Hort. Sci.* **47**: 502-510.

Kusumoto M (1981). Studies on the tissue culture of orchids-II. The primary culture of *Epidendrium, Epiphronitis, Militonia* and proliferation (Sic) propagation and organogenesis of the formed protocorm like bodies. *Bull Coll. Agric. Ver. Med. Nibon Univ.* **8**: 118-124.

Kusumoto M and Furukawa J (1977). Effect of organic matter on the growth of *Cymbidium* protocorns cultured *in vitro. J. Jap. Soc. Hortic. Sci.* **45**: 421-426.

Kyte L and Kleyn J (1996). Plants from Test Tubes (III Edition). Timber Press, Oregon.

La Rue CD (1953). Studies on growth and regeneration of gametophytes and sporophytes of gymnosperms. Abnormal and pathological plant growth *Brookhaven Symp., Biol.* **6**: 187-208.

Lam TW, Erust R, Arditti J and Ichihashi S, 1991. The effects of complex additives and 6-(r,r-dimethylallylamino)-purine on the proliferation *of Phalaenopsis,* Protocorms. *Lindleyana.* **6**(**1**): 24-26.

Lam-Chan LT and Lee SM (1996). Inflorescence culture of some tropical orchid varieties. *Singapore J. of Primary Industries*. **24**: 5-29.

Larkin PJ, Brettell RL, Ryan SA, Davies PA, Pallotta MA, and Scowcroft WR (1985). Somaclonal variation: Impact on plant biology and breeding strategies. **pp.** 3-100 Zaitlin, M., Day, P and Hollaender, A (ed). *Biotechnology in plant science, Relevance to Agriculture in the eighties*. Academic Press.

Larkin PJ and Scowcraft WR (1981). Somaclona Variation a novel source of variability from cell culture for plant improvement. *Theor. Appl Genet*. **60**: 197-214.

Lavrentyeva AN (1980). Method of the clonal propagation *of Cymbidium*. **pp.** 121-124 K. Kaazik, M. Liyk, Y. Martin, T. Reive, and V. Roost [eds.], *Conservation and cultivation of orchids*. Academy of Sciences of the Estonian SSR, Tallinn Botanical Garden, Tallian.

Leduc N, Iglesias VA, Bilang R, Zhang S, Gisel A, Sautter C and Potrykus I (1993). Gene transfer to flower meristem cells by particle bombardment. Keystone Symposium On Evolution And Plant Development, Taos, New Mexico, USA, January.

Lewis JD and Koide RT (1990). Phosphorus supply, mycorrhizal infection and plant offspring vigour. Functional Ecology. **4**: 695-702.

Li L, Qu R, de Kochko A, Fauquet C, and Beachy RN (1993). An improved rice transformation system using the biolistic method, *Plant Cell Rep*. **12**: 250-255.

Liau CH, You SJ, Prasad V, Hsiao HH, Lu JC, Yang NS and Chan MT (2003). Agrobacterium tumefaciens-mediated transformation of an Oncidium orchid. Plant Cell Rep. **21(10)**: 993-8.

Lim HC, Park NB, Choi DC, Jin SG, Park KH and Choi BJ (1993). Studies on micropropagation *in vitro* and hardening culture in *Dendrobium moniliforme*. *RDA-Journal of Agricultural science, Biotechnology*. **35**(**2**): 221-225.

Lim-Ho CL (1981). Expreimental findings of the tissue culture of orchid hybrids at the Singapore Botanic Gardens. Gardners'Bull. **34**(**1**): 148-160.

Lim-Ho CL and Lee GC (1987). Clonal propagation *ofOncidium* from dormant buds on flower stalk. *Malay. Orchid Rev.* (Singapore). 48-52.

Lim-Ho CL, Teo-Lee GC and Phua LK (1984). Clonal propagation or orchids from flower buds. In pp. 98-101 A.N. Rao [ed.], Proc. Fifth Asean Orchid Cong. Singapore.

Lin CC (1986). *In vitro* culture of flower stalk internodes of *Phalaenopsis* and *Dor itaenopsis, Lindley ana*. **1**: 158-163.

Lin YS, Chen WH Hsieh RM Tsal WT Fu YM Wu CC and Chyou MS (1995). Kenyamyein sensitivity of *Phalaenopsis*. Report of the Jaiwan Sugar Research Institute. **147**: 11-19.

Lindley J (1830). The Genera and species of Orchidaceous plants. London. Lindley, J. 1857. Conribution to the Orchidology of India. *J. Linn. Soc.* **1**: 170-190.

Loh CS, Rao AN and Goh CJ (1975). Clonal propagation from leaves in the orchid *Aranda J. Singapore Natl. Acad. Sci.* **4**: 97-99.

Loh CS and Rao AN (1985). Isolation and culture of mesophyll protoplasts *of Aranda* Noorah Alsagoff. Malay. Orchid Rev. (Singapore). **19**: 34-37.

Loh CS, Rao AN and Goh CJ (1975). Clonal propagation from leaves in the orchid *Aranda. J. Singapore Natl. Acad. Sci.* **4**: 97-99.

Loo SW (1945). Cultivation of excised stem tips *of Asparagus in vitro. Am. J. Bot.* **32**: 13-17.

Ludwig SR, Bowen B, Beach L and Wessler SR (1990). A regulatory gene as a novel visible marker for maize transformation, *Science*. **247**: 449-450.

Ma ZJ, Wang LT, Lu YG, Sun P and Liu MQ (1998). Study on commercial production of *Phalaenopsis. F. of Zhegiang Forestry college.* **15**(**2**): 192-196.

Manorama P, Rao AN, Goh CJ and Loh CS (1986). Leaf callus development in *Aarnda* and *Dendrobium*. **pp.** 102-109 A.N. Rao (ed). Proc. Fifth Asian Orchid Congress (1984), Singapore.

Marchner H (1986). Mineral nutrition of higher plants. Academic Press. Orlando.

Marston ME (1966). Vegetative propagation of plants using tissue culture techniques. Not-tingham Univ. School of Agri. Report. 77-80.

Marston ME (1967). Clonal multiplication of orchids by shoot meristem culture. Hort. Sci. **19**: 80-86.

Marston ME (1969). Vegetative propagation *in vitro*. **pp.** 393-394 W.J. Whittington (ed), *Root growth*. Butterworth, London.

Marston ME and Vourairai P (1967). Multiplication of orchid clones by shoot meristem culture. A review of the literature. Univ. Nottingham, Dept, Hort. Misc. **17**:1-8.

Mathews VH and Rao CS (1985). *In vitro* culture of vanda hybrid *(Vanda* TMA X *Vanda* Miss Joaquim) II. Studies on seedling explants. *Proc. Indian Natl. Sci. Acad.* **4**: 496-504.

Mauro M, Sabapathi D, Smith RA (1994). Influence of benzylaminopurine and alpha-naphthaleneacetic acid on multiplication and biomass production of *Cattleya aurantiaca* shoot explants. *Lindleyana* **9**(**3**): 169-173.

Mayer L (1956). Wachstum and Organbildung an *in vitro* kultivierten Segmenten von *Pelargonium zonale* and *Cyclamen persicum*. *Planta* **47**: 401-446.

Mayfield SP and Kindle KL (1990). Stable nuclear transformation of *Chlamydomonas reinhardtii* by using a *C. reinhardtii* gene as the selectable marker, *Proc. Natl. Acad. Sci. USA*. **87**: 2087-2091.

McCabe DE and Martinell BJ (1993). Transformation of elite cotton cultivars via particle bombardment of meristems, *Bio/Technol.* **11**: 596-598.

McCabe DE, Swain WF and Martinell BJ (1988a). Pollen-mediated plant transformation (European Patent Publication Number 0 270 356).

McCabe DE, Swain WF, Martinell BJ and Christou P (1988b). Stable transformation of soybean (*Glycine max*) by particle acceleration, *Bio/Technol.* **6**: 923-926.

McCown BH, McCabe DE, Russell DR, Robison DJ, Barton KA and Raffa KF (1991). Stable transformation of *Populus* and incorporation of pest resistance by electric discharge particle acceleration, *Plant Cell Rep.* **9**: 590-594.

McElroy D, Zhang W, Cao J and Wu R (1990). Isolation of an efficient actin promoter for use in rice transformation, *Plant Cell*. **2**: 163-171.

Me Lellan RI (1958). Orchid potting media. *Amer. Orchid Soc. Bull.* **25**: 81-83.

Men S, Ming X, Wang Y, Liu R, Wei C and Li Y (2003). Genetic transformation of two species of orchid by biolistic bombardment. Plant Cell Rep. Feb. **21**(**6**): 592-8.

Mendel RR, Müller B, Schulze J, Kolesnikov V and Zelenin A (1989). Delivery of foreign genes to intact barley cells by high-velocity microprojectiles, *Theor. Appl. Genet.* **78**: 31-34.

Miller RM, Hetrick BAD and Wilson GWT (1997). Mycorrhizal fungi affect root stele tissue in grasses. Canadian Journal of Botany. **75**: 1778-1784.

Mishiba K, Chin DP and Mii M (2005). Agrobacterium-mediated transformation of Phalaenopsis by targeting protocorms at an early stage after germination. **24**(**5**): 297-303.

Mitra A, Dey S and Sawarkar SK (1998). Photoautotrophic *in vitro* multiplication of the orchid *Dendrobium* under CO_2 enrichment. *Biologia-Plantarum.* **41**(**1**): 145-148.

Mitra GC (1971). Studies on seeds, shoot tips and stem discs of an orchid grown in aseptic culture. *Indian J. Exp. Biol.* **9**: 79-85.

Moore D (1849). On growing orchids from seeds. *Card Chron.* **35**: 549.

Morel G (1960). Producing virus-free Cymbidiums. Am. Orchid Soc. Bull. **29**: 495-497.

Morel G (1963). La culture *in vitro* du meristeme apical de certaines orchidees. *C.R. Acad. Sci. Paris* **256**: 4955-4957.

Morel G (1964a). Tissue culture - a new means of clonal propagation of orchids. *Am. Orchid Soc. Bull.* **33:** 473-478.

Morel G (1964b). La culture *in vitro* du meristeme apical. *Rev. Cytol. Biol. Veg.* **27**: 304-314.

Morel G (1965). Clonal propagation of orchids by meristem culture. *Cymbidium Soc.* **20**: 3-11.

Morel G (1970). Neues ouf dem Gebiet der Meristem-Forschung. *Die Orchidee* **20**: 433-443

Morel G (1974). Clonal multiplication of orchids. **pp** 169-222 C.L. Withner (ed), *The orchids.: Scientific studies.* Wiley-Interscience, New York,

Morel G and Champagnat M, 1969. Different modes d 'evolution de l'apex au cours de la formation de protocormes. *Proc. 2nd European orchid Cong., Paris,* pp. 20-26.

Morel G and Martin C (1952). Guerison de Dahlias atteints d'une maladie a virus. *Compt. Rend.* **235**: 1324-1325.

Morel G and Martin C (1955). Guerison des plantes atteintes de maladies a virus; par culture de maladies a virus; par culture de meristemes apicaux. *Report of 14th Intern. Hort. Congr. Netherlands,* pp. 303-310.

Morikawa H, Iida A and Yamada Y (1989). Transient expression of foreign genes in plant cells and tissues obtained by a simple biolistic device (particle-gun), *Appl. Microbiol. Biotechnol.* **31**: 320-322.

Morin C, Samson J and Dessureault M (1999). Protection of black spruce seedlings against Cylindrocladium root rot with ectomycorrhizal fungi. Canadian Journal of Botany. **77**: 169-174.

Muir WHA, Hilderbrandt C and Ricker AJ (1954). Plant tissue cultures produced from single isolated pant cells. *Science.* **119**: 877-878.

Mujib A and Jana BK (1994). Clonal propagation of *Dendrobium* 'Madame pompaour' through apical meristem culture. *Advances in plant sciences.* **7**(**2**): 340-346.

Murashige T (1961). Suppression of shoot formation in cultured tobacco cells by Gibberellic acid. *Science*. **144**: 280.

Murashige T (1974). propagation through tissue culture. *Ann. Rev. Plant Physiol* **25**: 473-477.

Murashige T (1977). Manipulation on organ culture in plnat tissue culture. Bot. Bull. Acad. Sin. (Taipei). **18**: 1-24.

Murashige T and Skoog F (1962). A revised medium for rapid growth and bioassays with tobacco tissue cultures. *Physiol. Plant.* **15**: 473-497.

Murashige T and Skoog F (1962). A revised medium for rapid growth and bioassays with tobacco tissue cultures. *Physiol Plantarum*. **15**: 473-497.

Mythili PK, Subba Rao MV and Manga V (1995). Cytology of explants, Calli and regenerants in five inbred lines of pearl millet, *Pennisetum glaucwn Cytologia*. **60**: 23-29.

Nagaraju V and Panthasarathy VA (1994). Propagation of *Phains tankervilliae* using seed and protocorm explants. *J. OrchidSoc. India*. **8(1/2)**: 77-78.

Nagaraju V and Parthasarathy VA (1995). *In vitro* Propagation of Phaius and babmoo orchid by shoot tip culture. *Annals of Plant Physiology*. **9**(**2**): 102-104

Nagaraju V and Parthasarathy VA (1995). Micropropagation of *Phaius* by seed and protocorm. *Annals of Plant Physiology*. **9**(**2**): 105-108.

Narayana Swamy S (1994). Plant cell and Tissue culture. Tata Me Graw Hill Publishing Co. Ltd., New Delhi.

Nelson T (1990). New horses for monocot gene jockeys, *Plant Cell*. **2**: 589.

Newsham KK, Fitter AH and Watkinson AR (1995). Arbuscular mycorrhiza protect an annual grass from root pathogenic fungi in the field. *Journal of Ecology* **83**: 991-1000.

Niimi Y, Akaza C and Hayata Y (1995). Organogenesis and plant regeneration from callus cultures in *Neofinetia falcata*. **pp.** 81-88 The proceedings of NIOC, Nagoya, Japan.

Niimoto DH and Sagawa Y (1961). Ovule development in *Dendrobium. Am. Orchid Soc. Bull.* **49**: 372-373.

Nissen SJ and Sutler EG (1990). Stability of IAA and IBA in nutrient medium to several tissue culture procedures. *Hon. Science*. **25**: 800-802.

Northern RT (1962). Hoem orchid growing, New York: D. van Nostrand.

Nuraini I and Mohammed SJ (1992). Micropropagation of orchids using scape nodes as the explant material. Recent advances in horticultural science in the tropics, Universiti Pertanian Malaysia, Selangor, Malaysia, 7-9 Aug. 1990 (edited by Wan Mohamad, W.O., Rosli Mohamad, Siti Hajar Ahmad, Khoo Khay Chong). *Acta Horticulturae*. **292**: 169-172.

Oard JH, Paige DF, Simmonds JA and Gradziel TM (1990). Transient gene expression in maize, rice, and wheat cells using an airgun apparatus, *Plant Physiol.*, **92**, 334-339.

Ong PY (1969). Physiological studies on pod development and seed germination of *Vanda* Miss Joaquim. *Honors Res. Exercise,* Univ. Singapore.

Pack KY and Jun SI (1994). Effect of source gradient and sucrose metabolism on organogenesis through rhizome culture of temperature *Cymbidium*. **pp.** 56. *Proceedings of VIII International Congress of Plant Tissue and Cell Culture,* Firenze, Italy.

Pack KY, Shim GB, Kim JJ (1989). The effects of natural products and time of exposure to BAP on organogenesis of temperature cymbidium rhizomes. *Korean soc. for Horticultural sci.* **7**(**1**): 192-193.

Paek KY and Yeung BC (1991). The effects of 1-naphthaleneaectic acid and N^6-benzyladenine on the growth of *Cymbidium forrestii* rhizomes *in vitro*. *Plant, cell, tissue and organ culture*. **24**: 65-71.

Paek KY, GJ, Park MR and Hirajima S (1993). Effect of sucrose gradient and changes of free sugar content during organogensis through rhizome culture of temperate *Cymbidium*. **pp.** 55-63. The Proceedings of Nagoya International Orchid conference, Nagoya, Japan.

Paek KY, Park JD and Shim GB (1993). Growth response of several terrestrial orchids to Fe-EDTA level *in vitro*. *J. of the Korean Society for Hort. Sci.* **34**(**4**): 308-314.

Pais MS, Barroso J, Feverceiro J and Oliveira M (1983). Intergeneric fusion of terrestrial orchid protoplasts induced by different fusion promoting agents. *Experientia (suppl 6th Int. Protoplast Symp)*. **45**: 74-75.

Park YS, Kakuta S, Kano A and Kabe MO (1996). Efficient propagation of protocorm like bodies of *Phalaenopsis* in liquid medium. *Plant, Cell, Tissue and organ culture* **45**: 79-85.

Pascoe IG (1991). History of systematic mycology in Australia. In: Short P. (ed.) History of Systematic Botany in Australasia. Australian Systematic Botany Society Inc., Canberra. pp. 259-264.

Paul CA, Rajeevan PK, Anitha, Paul C (1992). Influence of media on growth parameters in *Dendrobium, J. OrchidSoc India*. **6**(**1-2**): 125-130.

Peakall R and James SH (1989). Chromosome numbers of some Australian Orchids. *Lindleyana*. **4**(**2**): 85-88.

Pemphishey KT (1997). Commercialization of orchids and exports. *J. of orchid Soc. of India*. **11**(**1-2**): 23-27.

Pessoa C and Pessoa A (1985). Culture of the Brazilian Cattleya labiata var. warneri. Ameri Orchid Soc. Bull., **54**(**5**): 553-557.

Phatak SD (1983). Cytotaxonomical and numerical chemotaxonomic studies in some members of the genus Habenaria. Ph.D thesis submitted to Karnatak University, Dharwad.

Philip VJ and Padikkala J (1989). The role of IAA in the conversion of root meristems to shoot meristems in *Vanilla Planifolia*. J. plant physiol. **135**: 233-236.

Philip VJ and Nainar SAZ (1986). Clonal propagation of *Vanilla planifolia* (Salisb.) using tissue culture. *J. Plant Physiol.* **122**: 211-215.

Pieper W and Zimmer K (1976). Clonal propagation of *Phalaenopsis in vitro. Acta Horticulturae* **64**: 21-23:

Pierik RLM (1971). Plant tissue culture as motivation for the symposium. **pp.** 3-13. Van Bragt, J, Mossel, DAA, Pierik, RLM and Veldstra, H. (ed.) Effects of sterilization on components in nutrient media. H. *Veenman and Zonen*. Wageningen.

Pierik RLM (1987). *In vitro Culture of Higher plants*. Martinus Nighoff Pbls. Dordrecht. Boston. Lancaster.

Poole HA and Sheehan TJ (1977). Effects of media and supplementary micro element fertilizers on growth and chemical composition of *Cattleya* Amer. Orchid Soc. Bull (February). 155-160.

Posthumus AC (1971). Auxins. **pp.** 125-128 *J. Van* Bragt D.A.A. Mossel R.L.M. Pierik and H. Veldtsra (ed.), *Effects of sterilization on components in nutrient media.* Misc. Paper 9 (1971), Landbouwgeschool, Wageningen, and H. Veenman and Zonen N.V., Wageningen, The Netherlands.

Potrykus I (1989). Gene transfer to cereals: an assessment, *TIBTECH*. **7**: 269-273.

Potrykus I (1990). Gene transfer to plants: assessment and perspectives, *Physiol. Plant.* **79**: 125-134.

Potrykus I (1991). Gene transfer to plants: assessment of published approaches and results, *Annu. Rev. Plant Physiol. Plant Mol. Biol.* **42**: 205-225.

Prakash L, Lee CL, Loh CS and Goh CJ (1996). *In vitro* propagation of commercial orchids. As assessment of current methodologies and development of a Novel approach - Through section culture. **Pp**. 42-49 Islam, A.S. (ed) *Plant Tissue Culture*. Oxford and IBH Publishing Co. Ltd., New Delhi, Calcutta.

Prakesh CS and Varadarajan U (1992). Genetic transformation of sweet potato *(Ipomoea batatas)* by particle bombardment, *Plant Cell Rep.* **11**: 53-57.

Prasad S and Kumar U (1998). *Commercial Floriculture*. Agrobotanica, Bikanir, India.

Price GC and Earlev ED (1984). Sources of orchid protoplasts for fusion experiments. *Am. OrchidSoc. Bull.* **53**: 1035-1043.

Ramsay RR, Dixon KW and Sivasitamparam K (1986). Patterns of infections and endophytes associated with western Australian orchids. *Lindleyana*. **1**: 203-214.

Randolph ABL, Boynton JE, Gillham NW, Harris EH, Johnson AM, Dorthu MP, Matagne and Further RF (1993). Characterization of the respiratory deficient *dum*1 mutation of *Chlamydomonas reinhardtii*and its use as a recipient for mitochondrial transformation, *Mol. Gen. Genet.* **238**: 235-244.

Rao AN (1977). Tissue Culture in Orchid Industry. **pp.** 18-42. Reinert, J and Bajaj, Y.P.S (ed). *Plant cell, tissue and organ culture.* Springer-verlag Berlin Heidelberg, New York.

Rao AN and Avadhani PN (1964). Some aspects of vitro culture of Vanda seeds. **pp.** 94-202 B.C. Yeoh (ed.). *Proc 4th World Orchid Conf.* (Singaore 1963), Straits Times Press (M) Ltd., Singapore.

Ravishankar GA and Vekataraman LV (1997). Biotechnological Applications of Plant Tissue and Cell culture. Oxford and IBH Publishing Co. Pvt. Ltd., Calcutta.

Reggiardo MI, Arana JL, Orsaria LM, Permingeat HR, Spitteler MA and Vallejos RH (1991). Transient transformation of maize tissues by microprojectile bombardment, *Plant Sci.* **75**: 237-243.

Reichembach HG (1883). On proliferous roots of orchids. The report on the orchid conference./ *Roy. Hortic. Soc.* **7**: 18.

Reisinger DM, Ball EA and Arditti J (1976). Clonal propagation of Phalaenopsis by means of flower-stalk node cultures. *Jpn. Orchid Soc. Bull.* **22**: 9-14 (in Japanese).

Remert RA and Mohr HC (1967). Propagation of *Cattleya* by tissue culture of lateral bud meristems. *Proc. Amer. Soc. Hort. Sci.* **91**: 664671.

Robbins WJ (1922). Cultivation of excised root tips and stem tips under sterile conditions. *BatGaz.* **73**: 376-390.

Robertson D, Weissinger AK, Ackley R, Glover S and Sederoff RR (1992). Genetic transformation of Norway spruce (*Picea abies* (L.) Karst) using somatic embryo explants by microprojectile bombardment, *Plant Molec. Biol.* **19**: 925-935.

Rotor Jr. G (1949). A method for vegetative propagation of *Phalaenopsis* species and hybrids. *Am. Orchid Soc. Bull.* **18**: 738-739.

Rubluo A, Chavez V, Martinez AP and Martinez-Vazquez O (1993). Strategies for the recovery of endangered orchids and cacti through *in vitro* culture. *Biological Conservation.* **63**(**2**): 163-169.

Russell DR, Wallace KM, Bathe JH, Martinell BJ, and McCabe DE (1993). Stable transformation of *Phaseolus vulgaris* via electric-discharge mediated particle acceleration, *Plant Cell Rep.* **12**: 165-169.

Russell JA, Roy MK and Sanford JC (1990). Cell injury as a limiting factor in stable biolistic plant transformation, *In Vitro Cell. Develop. Biol.* 26: 43A.

Russell JA, Roy MK and Sanford JC (1992a). Major improvements in biolistic transformation of suspension cultured tobacco cells, *In Vitro Cell. Develop. Biol.* **28P:** 97-105.

Russell JA, Roy MK and Sanford JC (1992b). Physical trauma and tungsten toxicity reduce the efficiency of biolistic transformation, *Plant Physiol.* **98**: 1050-1056.

Russon JB (1965). The meristem stakes and mistakes. *Orchid Rev.* **73**: 141-142.

Sagarik R (1994). Thai orchid species seed propagation results. **pp.** 40-46. *Proceedings ofNagoya International orchid congress, Nagoya, Japan.*

Sagawa Y and Sehgal IP (1967). Aseptic stem propagation of *Vanda Miss* Joaquim. *Bull. Pacific OrchidSoc. Hawaii.* **25**: 17-18.

Sagawa Y and Kunisaki JT (1982). Clonal propagation of orchids by tissue culture. **pp.** 683-684 *Proc. 5th Cong. Plant Tissue and Cell Culture.*

Sagawa Y and Shoji T (1967). Clonal propagation of *Dendrobium* through meristem culture. *Am. Orchid Soc. Bull.* **36**: 856-859.

Sagawa Y, Shoji T and Shoji T (1966). Clonal propagation of Cymbidiums through shoot meristem culture. *Am. Orchid Soc. Bull.* **35**: 118-122.

Sagawa Y, Shoji T and Shoji T (1966). Clonal propagation of cymbidiums through shoot meristem culture. *Am. Orchid Soc. Bull.* **35**: 118-122.

Sagwa Y (1961). Vegetative propagation of Phalaenopsis - a pictorial review. *Fl. Orchid Soc. Bull.* **30**: 808-809.

Sahay NS and Varma A (2000). A biological approach towards icreasing the rates of survival of micropropagated plants. *Current Science.* **78**(**2**): 126-129.

Sakai K, Osuga M, Yonemura K, Higuchi H (1982). Effects of fertilizer application on growth and flowering in *Dendrobium* app. I. Application time and amount of fertilizer. *Research Bulletin of the Aichi-Ken Agricultural Research Center.* **14**: 187-192.

Sakamoto W, Kindle KL and Stern DB (1993). *In-vivo* analysis of *Chlamydomonas* chloroplast *pet*D gene expression using stable transformation of β-glucuronidase translational fusions, *Proc. Natl. Acad. Sci., USA.* **90**: 497-501.

Sanchez MI (1988). Micropropagation of *Cytropodium* (Orchidaceae) through root-tip culture. *Lindleyena*. **3**: 93-96.

Sanford JC (1990a). Biolistic plant transformation, *Physiol. Plant*. **79**: 206-209.

Sanford JC, Devit MJ, Russell JA, Smith FD, Harpending PR, Roy MK and Johnston SA (1991). An improved, helium-driven biolistic device, *Technique*. **3**: 3-16.

Sanford JC, Klein TM, Wolf ED and Allen N (1987). Delivery of substances into cells and tissues using a particle bombardment process, *Particulate Sci. Technol.* **5**: 27-37.

Sanford JC, Smith FD and Russell JA (1993). Optimizing the biolistic process for different biological applications, *Meth. Enzymol.* **217**: 483-509.

Sanford JC (1990b). The biolistic process — an emerging tool for research and clinical applications, from "Biomedical Engineering: Opening New Doors," *Proceedings of the 1990 Annual Meeting of the BioMedical Engineering Society*, D. C. Milulecky and A.M. Clarke (eds.), New York University Press, New York.

Sanford JC (1988). The biolistic process, *TIBTECH*. **6**: 299-302.

Sangawa Y (1990a). Biotechnology in Orchids. **pp.** 638-653 Amirato, P.V., Evans, D.A., Sharp., W.R., and Bajaj, Y.P.S (ed). Handbook of plant cell culture vol. 5. McGraw-Hill Inc. New York.

Scully RM (1966). Stem propagation of Phalenopsis JOT. *Orchid Soc. Bull.* **35**: 40-42.

Scully RM (1967). Aspects of meristem culture in the *Cattleya* alliance. *Am. Orchid. Soc. Bull.* **36**: 103-108.

Seeni S and Latha PG (1992). Foliar regeneration of the endangered Red Vanda, *Renenthera Imschootiana Rolfe* (Orchidaceae). Plant, Cell, Tissue and Organ Culture. **29**: 167-172.

Seeni S (1988). Micropropagation of blue vanda using leaf bases. **pp.** 22 S.P. Vij and S.P. Khullar [ed.], *Current research trends in Indian orchids*. Orchid Society of India, Dept. of Botany, Panjab University, Chandigarh, India.

Seeni S and Abraham A (1986). Screening of wild species and hybrid orchids for protoplast isolation, **pp.** 23-27 A.N. Rao [ed.], *Proc. 5th Asean Orchid Congr. Sent. Singapore*(1984). Parks and Recreation Dept., Ministry of National Development, Singapore.

Seki M, Komeda Y, Iida A, Yamada Y and Morikawa H (1991b). Transient espression of β-glucuronidase in *Arabidopsis thaliana* leaves and roots and *Brassica napus* stems using a pneumatic particle gun, *Plant Molec. Biol.* **17**: 259-263.

Seki M, Shigemoto N, Komeda Y, Imamura J, Yamada Y and Morikawa H (1991b). Transgenic *Arabidopsis thaliana* plants obtained by particle-bombardment-mediated transformation, *Appl. Microbiol. Biotechnol.* **36**: 228-230.

Serres R, Stang E, McCabe D, Russell D, Mahr D and McCown B (1992). Gene transfer using electric discharge particle bombardment and recovery of transformed cranberry plants, *J. Amer. Soc. Hort. Sci.* **117**: 174-180.

Shark KB, Smith FD, Harpending PR, Rasmussen JL and Sanford JC (1991). Biolistic transformation of a procaryote, *Bacillus megaterium, Appl. Environ. Microbiol.* **57**: 480-485.

Sharma A (1994). *In vitro* conservation through clonal propagation of a rare and endangered orchid *Renanthera inshootiana*. **pp.** 72. *Proceedings of VIII International Congress of Plant Tissue and Cell Culture,* Firenze, Italy.

Sharma A and Tandon P (1992). *In vitro* culture of *Dendrobium wardianum* warner: morphogenetic effects of some nitrogenous adjuvants. Indian *J. of Plant Physiology.* **35**(**1**): 80-85.

Sharma A, Tandon P and Kumar A (1992). Regeneration of *Dendrobium Wardianum* Warner (Orchidaceae) from synthetic seeds. Indian *J. of Experimental Biology,* **30**(**8**): 747-748.

Sharma AK and Chatturvedi HC (1988). Clonal multiplication and germplasm preservation of monopodial orchids through leaf and root-tip culture. **pp.** 23-24 S.P. Vij and S.P. Khullar (ed.), *National Seminar on Current Research Trends in Indian Orchids* with a *Special Reference to Tissue culture Technology,* Nehu, Shilong, India.

Sharma BD, Singh NP, Raghavan RS and Deshpande VR (1984). *Flora of Karnataka. An analysis,* BSI, Howrah.

Sharma BD, Singh NP, Raghavan RS and Deshpande VR (1984). *Flora of Karnataka. An analysis.* BSI, Howrah.

Sharma SK and Tandon P (1987). Axenic germination of some epiphytic ordchids of Meghalaya, India. *J. OrchidSoc. of India.* **1**(**1-2**): 85-90.

Sharma SK and Tandon P (1990). Asymbiotic germination and seedling growth of *Cymbidium elegans.* Lindl. And *Coelogyne punctulata* Lindl. as influenced by different carbon sources. *J. of the orchid soc. of India.* **4** (**1-2**): 149-159.

Sharon M and Vasundhara G (1990). Micropropagation of *Dendrobium* Foannie *Ostenhault. J. of the Orchid. Soc. of India* **4** (**1-2**): 145-148.

Sheehan TJ (1958). Orchid potting media. **pp**. 226-230 *Proc. Second World Orchid Conf.*

Shimaraki K and Uemoto S (1987). Studies on the micropropagation of Japanese *Calanthe* species. *Sci. Bull, of the Faculty of Agriculture.* **42**(**1-2**): 9-14.

Shimasaki K and Uemota S (1991). Rhizome induction and plantlet regeneration of *Cymbidium goeringii* from flower bud cultures *in vitro. Plant cell, tissue and organ culture.* **25**: 49 - 52.

Sigma Plant Cell culture (1991-1992). St. Louis Sigma Chemical Company.

Simard SW, Perry DA, Jones MD, Myrold DD, Durall DM and Molina R (1997). Net transfer of carbon between ectomycorrhizal tree species in the field. *Nature.* **388**: 579-582.

Simon S (1908). Experimentelle Untersuchungen uber die Differenzierungvorgange Kallusgewebe von Holzgewachsen. *Jahrb.* Wiss. Bot. **45**: 351-478.

Singh F and Prakash D (1984). *In vitro* propagation of *Thunia alba* through flower stalk cuttings. *Scientia Horticulture.* **24** (**3/4**):3 85-390.

Singh F and Prakash D (1985). Suspension culture technique for the culture of orchid embryos. *Garten bauwissens chaft.* **50** (**5**): 236-238.

Singh F (1986). Orchids. **pp.** 127-153. Chadha K.L. and Chowdhary B. (ed). *Ornamental Horticulture in India,* ICAR, India.

Singh F (1986). Two new orchid hybrids. *Indian Horticulture*. **31**(**3**): 14.

Singh F (1991). Enlapsulation of *Spathoglottis Plicata* Protocorms. *Lindleyana*. **6** (**2**): 61-63.

Singh F (1993). *In vitro* orchid seed germination and cloning of orchids. A success story. **pp.** 85-109 Prakash, J.P and Pierick, R.L.M. Plant biotech, Commercial Prospects and Problems. Oxford and IBH publishing Co. Pvt. Ltd., Calcutta.

Singh F and Prakash D (1984). *In vitro* propagation of *Thunia alba* (Wall.) Reichb. F. through flower stalk cuttings. *Sci. Hortic.* (Amesterdam). **24**: 385-390.

Singh F, Prakash J (ed.) and Pierik RLM (1993). *In vitro* orchid seed germination and cloning of orchids - a success story. **pp.** 85-109 *Plant biotechnology: commercial prospects and problems*. 85-109.

Singh H (1976). Meristem tissue culture of *Dendrobium Ng Eng Chow. Card. Bull. (Singapore)*. **28**: 259-267.

Singh H and Sagawa Y (1972). Vegetative propagation *of Dendrobium* by flower stalk cuttings. *Hawaii Orchid J.* **1**(**4**): 19 (pagination in each issue of this volume is separate).

Singh H and Sagawa Y (1972). Vegetative propagation of *Dendrobiumby* fower stalk cuttings. *Na Okika of Hawaii*. **1**: 19.

Singh M and Krikorian AD (1980). Chelated iron in culture media. *Ann. Bot.* (Lond.). **46**: 807-809.

Skoog F and Miller CO (1957). Chemical regulation of growth and organ formation in plant tissues cultivated *in vitro.* In: Biological Action of Growth Substances, *11th Symp. Soc. Exp. Biol.* **11**: 118-131.

Smith SE (1995). Discoveries, discussions and directions in mycorrhizal research. In: Verma A. and Hock B. (eds.) Mycorrhiza. Springer-Verlag Berlin. pp. 3-24.

Smith SW and Read DJ (1997). *Mycorrhizal symbiosis*. Academic Press, San Diego.

Smith FD (1992). Harpending P.R. and Sanford J.C., Biolistic transformation of prokaryotes- factors that affect transformation of very small cells, *J. Gen. Microbiol.* **138**: 239-248.

Smith HS (2000). *Plant Tissue culture techniques and experiments*. San Fransico, New York, Boston, London, Sydney, Tokyo.

Soediono N (1983). A method for saving contaminated tissue culture of *Dendrobium. Orchid Rev.* **91**: 22.

Soediono N (1983a). A method of saving contaminated tissue cultures *of Dendrobium. Orchid Review.* **91**: 1071:22.

Soediono N (1983b). Use of coconut water, NAA, 2,4-D and vitamins in shoot tip cultures of *Dendrobium* Jaquieline Thomas 'White'. *Orchd Rev.* **91**: 86-87.

Somers DA, Rines H W, Gu W, Kaeppler HF and Bushnell WR (1993). Fertile, transgenic oat plants, *Bio/Technol.* **10:** 1589-1594.

Sood A and Vij SP (1986). *In vitro* root segment culture of *Rhynchostylis retusa*. **pp.** 463-468 S.P. Vij (ed), Biology, conservation, and culture of orchids. Affiliated East-West Press, New Delhi.

Sounder Rajan T and Lokeshwari TS (1994). Multiplication of orchid protocorms in liquid medium. *of the orchid soc. of India*. **8**(**1**/**2**): 47-51.

Stanley MR, Koide RT and Shumway DL (1993). Mycorrhizal symbiosis increases growth, reproduction and recruitment of Abutilon theophrasti Medic. in the field. *Oecologia* **94**: 30-35.

Stasinopoulos TC and Hangarter RP (1990). Preventing photochemistry in culture media by long-pass light filters alters growth of cultured tissues. *Plant Physiol.* **93**: 1365-1369.

Steel RGD and Terrie JH (1980). *Principles and procedures of statistics.* McGraw-Hill, New York.

Stenberg ML and Kane ME (0000) *In Vivtro* seed germination and green house cultivation of *Encyclia boothiana* var. *crythronioides,* an endangered Florida Orchid. *Lindleyana.* **13(2)**: 101-112.

Stern DB and Kindle KL (1993). 3′ end maturation of the *Chlamydomonas reinhardtii* chloroplast *atp*B mRNA is a two-step process, *Mol. Cell. Biol.,* **13**, 2277-2285.

Steward FC and Mapes MO (1971b). Morphogenesis in aseptic cell cultures of *Cymbidium. Bot. Gaz.* **132(1)**: 65-70.

Steward FC, Mapes MO and Mears K (1958). Growth and organized development of cultured cells. II. Orgaixation in culstures grown from freely suspended cells. *Am. J. Botany.* **45**: 705-708.

Steward J (1978). Development of callus and plantlets from *Epidendrum* root tips cultured *in vitro. Am. Orchid Soc. Bull.* **47**: 607-612.

Steward J and Button J (1975). Tissue culture studies in *Paphiopedilum. Am. Orchid Soc. Bull.* **44**: 591-599.

Steward J and ButtonJ (1976a). Rapid vegetative multiplication of Epidendrum O'brienianum *in vitro* and in the greenhouse. *Am. Orchid Soc. Bull.* **45**: 922-930.

Steward J and Button J (1976b). Tissue culture studies in Paphiopedilum. Pages 372-378 in K. Senghas (ed.), *Proc. 8th World Orchid Conf.* Frankfurt (1975). German Orchid Society, Frankfurt.

Steward J and Button J (1978). Development of callus and plantlets from Epidendrum root tips cultured *in vitro* Am. Orchid Soc. Bull. **47**: 607-612.

Steward J and Button J (1979). Tissue culture of Paphiopedilum. S. Afr. Orchid J. **7**: 10-12.

Steweard FC and Mapes MO (197la). Morphogenesis and plant propogation in aseptic cultures of *Asparagus Bot. Gaz.* **132**: 70-79.

Stokes M, Thomas JE and Holdgate (1975). Notes on the aseptic propagation of members of the tribe Cypripedieae. *Orchid Rev.* **83**: 136-137.

Stokes MJ, Thomas E and Holdgate DP (1975). Notes on the aseptic propagation of members of the tribe Cypripedieae. *Orchid Rev.* **83**: 136-137.

Stomp AM, Weissinger A and Sederoff RR (1991). Transient expression from microprojectile-mediated DNA transfer in *Pinus taeda, Plant Cell Rep.* **10**: 187-190.

Street HE (1969). Growth in organized and unorganized systems. Knowledge gained by culture of organs and tissue explants. **pp.** 3-224. Stewad, F.C (ed). *In plant physiology: A treatise,* (Vol. VB). Academic Press. New York

Street HE (1996). The nutrition and metobolism of plant tissue and organ cultures.

pp. 533-630 Willmer, E.N. (ed). Cell and Tissue Culture (Vol-3). Academic press. New York.

Suner TG (1995). *In vitro* cultivation of orchids. *Horticulture Revista de Hortalizas, Flores plantas ornamentales* (spain). **108**: 99-100.

Svab Z and Maliga P (1993). High-frequency plastid transformation in tobacco by selection for a chimeric *aad*A gene, *Proc. Natl. Acad. Sci. USA*. **90**: 913-917.

Svab Z (1990). Hajdukiewicz, P., and Maliga, P., Stable transformation of plastids in higher plants, *Proc. Natl. Acad. Sci. USA*. **87**: 8526-8530.

Swamy BGL (1949a). Embroylogical studies in the orchidaceae II Embryogency. *A.M. Midland Naturalist*. **41**: 202-232.

Swamy BGL (I949b). Embryological studies in the orchidaceae-I. Gametophytes. *Am. Midland Naturalist*. **41**:184-201.

Takano T, Oyamada T and Hirai T (1990). Improvement in nutrient composition of culture media for the growth and multiplication of PLB in *Cymbidium*. **pp.** 95-101 Proceedings of the Nagoya International Orchid Society. Nagoya, Japan.

Takano T, Oyamada T and Hirai J (1990). Improvement in nutrient composition of culture media for the growth and multiplication of PLB in *Cymbidium*. **pp.** 95-100. *The proceedings of Nagoya International orchid Show, Nagoya,* Japan.

Takashi O, Taikichi T (1990). Clonal multiplication of *Dandrobium* Formidible through tissue culture. **pp.** 180-185 the proceedings of Nagoya International orchid Show, Nagoya.

Tanaka M (1990). Micropropagation of Phalaenopsis through leafsequent culture. **pp.** 113-118. *Proceedings of Nagoya international orchid Show,* Nagoya, Japan.

Tanaka M and Sakanishi Y (1977). Clonal propagation *of Phalaenopsis* by leaf tissue culture. *Amer. Orchid Soc. Bull.* (August). 733-737.

Tanaka M (1987). Studies on the clonal propagation of *Phalaenopsis* by *in vitro* culture. *Memoirs of the faculty of Agriculture,* Kagawa University. **49**: 85.

Tanaka M and Sakanishi Y (1978). Factors affecting the growth of *in vitro* cultured lateral buds from *Phalaenopsis* flower stalks. *Sci. Hortic.* (Amsterdam). **8**: 169-178.

Tanaka M and Sakanishi Y (1980). Clonal propagation *of Phalaenopsis* through tissue culture. **pp.** 215-221 R. Tanaka and K. Saito (ed), *Proc. 9th World 'Orchid Conf.* Tokyo.

Tanaka M and Sakanishi Y (1985). Regenerative capacity of *in vitro* cultured leaf segments excised from mature *Phalaenopsis* plants. *Bull. Of Osaka Prefectural University.* **37**: 1-4.

Tanaka M and Sakanishi Y (1976). Plantlet formation by root-tip culture. *Am. Orchid Soc. Bull.* **46**: 733-737.

Tanaka M, Goi M and Higashiura T (1988). A novel disposable culture vessel made of fluorocarbon polymer films for micropropagation. *Acta Horticulture*. **226(11)**: 663-670.

Tanaka M, Ikeda M, Fukai S, Goi M, Hayashi M (ed.), Kano A (ed.) and Goto E (1992). Effect of different films used for film culture vessels on plantlet development of Phalaenopsis and Cymbidium. International symposium. on transplant production systems. Biological, engineering and socioeconomic aspects, Yokohama, Japan, 21-26 July 1992. Acta-Horticulturae. **319**: 225-230.

Tanaka M, Kumura M and Goi M (1998). Optimal conditions for shoot production from *Phalaenopsis* flower cuttings cultured *in vitro. Scientia Horticultural.* **35**(**1-2**): 117-126.

Tanaka M, Senda Y and Hasegawa A (1976). Plantlet formation by root-tip culture in *Phalaenopsis Am. Orchid Soc. Bull.* **45**: 1022-1024.

Tang DC, Devit M, Johnston SA (1992). Genetic immunization is a simple method for eliciting an immune response, *Nature*. **356**: 152-154.

Taylor LD and Burns TD (1997). Independent, specialized invasion of ectomycorrhizal mutualism by two non-photosynthetic orchids. Proceedings of the National academy of Sciences, USA. **94**: 4510-4515.

Taylor LD and Burns TD (1999). Population, habitat and genetic correlates of mycorrhizal specialization in the "cheating orchids" *Corallorhiza maculata* and *C. mertensiana*. *Molecular Ecology.* **8**: 1719-1732.

Taylor MG and Vasil IK (1991). Histology of, and physical factors affecting, transient GUS expression in pearl millet (*Pennisetum glaucum* (L.) R Br) embryos following microprojectile bombardment, *Plant Cell Rep.* **10**: 120-125.

Teng W, Nicolson L and Yu LYY (1996). Clonal propagation of *Spathoglottis plicata* from young plants. *Acta Horticulturae.* **447**: 193-197.

Teo CKH (1978a). Clonal propagation of Haemaria discolor by tissue culture. *Am. Orchid Soc. Bull.* **47**: 1028-1030.

Teo CKH and Teo PS (1974). Clonal propagation of orchids by tissue culture. *Planter.* **50**: 213-216.

Teo CKH and Neumann KH (1977). The isolation and hybridisation of protoplasts from orchids. *Orchid Rev.* **86**: 186-189.

Teo CKH and Neumann KH (1978a). Gewinnung, Kultur and Fusion von Orchideenprotoplasten. *Die Orchidee.* **29**: 90-92.

Teo CKH and Neumann KH (1978b). The culture of protoplasts isolated from Renantanda Rosalind Cheok. *Orchid Rev.* **86**: 156-158.

Teo CKH, Kunisaki JT and Sagawa Y (1973). Clonal propagation of strap-leated Vanda by shoot-tip culture. *Am. Orchid. Soc. Bull.* **42**: 402-405.

Teuscher H (1977). *Phalaenopsis, Doritis* and *Kingidium*. Amer. Orchid Soc. Bull. (March). pp. 216-217.

Thomale H (1957). *Die Orchideen,* (*2* ed.) Verlag. Eugen Ulmer Stuttgart.

Thomas J (1996). Indian floriculture is poised to bloom. *Host. Science.* **31**(**4**): 651.

Thorpe TA (2000). History of plant cell culture in plant tissue culture. Academic Press. San Diego. pp. 1-132.

Thorsness PE and Fox TD (1990). Escape of DNA from mitochondria to the nucleus in *Saccharomyces cerevisiae, Nature*. **346**: 376-379.

Titomirov AV and Zelenin AV (1989). New methods of transfection of mammalian cells: a minireview, *Molec. Biol*. **359**: 1155-1159.

Toffaletti DL, Rude TH and Johnston SA (1993). Gene transfer in *Cryptococcus neoformans* by use of Biolistic delivery of DNA, *J. Bacteriol.* **175**: 1405-141.

Tokuhara K and Mii M (1993). Micropropagation of *Phalaenopsis* and *Doritaenopsis* by culturing shoot tips of flower stalk buds. *Plant cell Reports.* **13**: 7-11.

Tokuhara K (1992). Somaclonal variation of flowers in micropropagated plants through flower stalk bud culture. **pp.** 317-319, Ichihashi, S and Nagata, H (ed). Proceedings of Nagoya International Orchid Show, Nagoya, Japan.

Tomes DT, Weissinger AK, Ross M, Higgins R, Drummond BJ, Schaaf S, Malone-Schoneberg J, Staebell M, Flynn P, Anderson J and Howard J (1990). Transgenic tobacco plants and their progeny derived by microprojectile bombardment of tobacco leaves, *Plant Molec. Biol.* **14**: 261-268.

Tran Thanh and Van M (1974). Methods of acceleration of growth and flowring in a few species of orchids. *Am. Orchid Soc. Bull.* **43**: 699-707.

Trappe JM (1987). Phylogenetic and ecologic aspects of mycotrophy in the angiosperms from an evolutionary standpoint. In: Safir G.R. (ed.) Ecophysiology of VA Mycorrhizal Plants. CRC Press, Boca Raton, Florida. pp. 5-25

Tse AT, Smith RJ and Hackett WP (1971). Adventitious shoot formation on *Phalaenopsis nodes. Am. Orchid Soc. Bull.* **40**: 807-810.

Tsukamoto Y, Kano K and Katsuurat (1963). Instant media for orchid seed germination. Am. Orchid. Soc. Bull. **32**: 354-355.

Twell D, Klein TM, Fromm ME and McCormick S (1989). Transient expression of chimeric genes delivered into pollen by microprojectile bombardment, *Plant Physiol.* **91**: 1270-1274.

Ueda H and Torikata H (1972). Effects of light and culture medium on adventitious root formation by Cymbidium in aseptic culture. *Am. Orchid Soc. Bull.* **41**: 322-327.

Ueda H and Torikata H (1972). Effects of light and culture medium on adventitoius root formation by Cymbidiums in aseptic culture. *Am. Orchid Soc. Bull.* **41**: 322-327.

Uesato K, Yagi N and Odo S (1987). Effects of nitrogen and phosphate on the growth of *Ceratobium Phalaenanthe* type *Dendrobium. Sci. Bull, of the College of Agriculture,* University of the Ryukyus, Okinawa, **34**: 11-19.

Vacin E and Went FW (1949). Some pH changes in nutrient solutions. *Bot. Gaz.* **110**: 605-613.

Vajrabhaya M (1978). Tissue culture of dormant buds from *Cattleya* backbulbs. *Orchid Rev.* **86**: 256-257.

Vajrabhaya M and Vajrabhaya T (1970). Tissue culture of *Rhynchostylis gigantea,* a monopodial orchid. *Am. Orchid Soc. Bull.* **39**: 907-910.

Vajrabhaya M and Vajrabhaya T (1974), Variation in *Dendrobium* arising in meristem. **pp.** 231-242 *Proc. 7th World Orchid Conf.* Modellin, Colombia. (1972).

Vajrabhaya T and Vajrabhaya M (1976). The study of various organs of orchids *in vitro* I. Stems, leaves and buds. *Rep. Sci. Res. Fac. Sci., Chulaungkorn Univ.* **1**: 105-115.

Valmayor HL, Pimentel ML and Martinez MT (1986). Callus formation and plantlet morphogenesis in Vanda. Malay. *Orchid Rev.* (*Singapore*). **20**: 22-30.

Van Overbeek J, Conklin ME and Blakeslee AF (1941). Factors in coconut milk essential for growth and development of very young *Datura* embryos. *Science.* **94**: 350-351.

Van Overbeek J, Conklin MEand Blakeslee AF, 1942. Cultivation *in vitro* of small Datura embryos. *Am. J. Bot.* **29**: 472-477.

Van Rensburg JGJ and Veelar BM, 1984. Tissue culture of *Cymbidium. S. Afr. Orchid J.* **15**: 68-70.

Van Waes J and De Geest S, 1983. Disa uniflora Berg: Une Orchidee terrestre on multiplication *in vitro, Rev. Agric. (Bruss.)* **36**: 1416-1433.

Vasil V, Brown SM, Re D, Fromm ME, and Vasil IK, 1991. Stably transformed callus lines from microprojectile bombardment of cell suspension cultures of wheat, *Bio/ Technol.* **9**: 743-747.

Vasil V, Castillo AM, Fromm ME, and Vasil IK, 1992. Herbicide resistant fertile transgenic wheat plants obtained by microprojectile bombardment of regenerable embryogenic callus, *Bio/Technol.* **10**: 667- 674.

Vij SP, 1994a. Micropropagation of Orchids through root segments. **pp.** 72-80 *Proceedings of Nagoya International Orchid Conference,* Nagoya, Japan.

Vij SP, 1994b. Dermal peels and orchid propagation. **pp.** 133-137 Proceedings of Nagoya International Orchid Conferene, Nagoya, Japan.

Vij SP, 1998. The possibilities of commercialization of Indian Orchids. Application of tissue culture techniques. **pp.** 41. Proceedings of IX International Congress on Plant Tissue and Cell Culture Jeruselum, Israel.

Vij SP and Dhiman A, 1997. Regenerative competence of *Bletilla striata* pseudobulb segments: A study *in vitro. J. Orchid Soc. India.* **11**(**1-2**): 93-97.

Vij SP, Sood A and Plaha KK, 1986. *In vitro* breakdown of apical dormancy and development of vegetative shoots from inflorescence segments in *Saccolabium* calccolare Lindl. **pp.** 473-478 S.P. Vij (ed), Biology, conservation, and culture of orchids. Affiliated East-West Press, New Delhi.

Vij SP, Sood A, Sharma A and Shekhar N, 1983. *In vitro* tube culture of *Pachystonna senile*—a growing orchid. *Trap. Plant Sci. Res.* **1**: 211-213.

Vij SP and Pathak P, 1989. Micropropagation of *Dendrobium chrysanthum* Wall. Through pseudobulb segments *J. OrchidSoc. India.* **3**: 25-28.

Vij SP, Kondo K and Pathak P 1994. Regeneration potential *of Cymbidium Pendulum.* Nodal explants—a study *in vitro, of the Orchid Soc. of India.* **8**(**1/2**): 19-23.

Vij SP, Pathak P and Kher A, 1997. Regeneration response of *Rhynchostylis gigantia* inflorescence segments: A study *in vitro. J. Orchid Soc. India.* **11**(**1-2**): 75-78.

Villalobos AV, Mejia-Munoz J and Sosa-Moss C, 1994. Tissue culture of orchids; *Cattleya, Encyclia, Oncidium* and *Stanhopea. National Congress on ornamental horticulture,* Chapingo, Mexico. pp. 20-25.

Vitamic MJ and Arditti J, 1984. Effects of anticantaniants, *Cymbidium* shoot tip cultures. *Orchid Review.* **92**(**1086**): 118-121.

Waes JV, 1987. Effect of activated charcoal on *in vitro* propagation of Western European orchids. *Acta-Horticulturae.* **212**(**1**): 131-138.

Wainwright M, 1988. Metabolic diversity of fungi in relation to growth and mineral cycling in soil—a review. Transactions of the British

Walker C (1995). AM or VAM: what's in a word? In: Verma A. and Hock B. (eds.) Mycorrhiza. Springer-Verlag Berlin. pp. 25-26.

Wang HY (1989). Rapid clonal propation of *Phalaenopsis* by tissue culture. *Acta-Horticulturae-Sin.* **16(1):** 73-77.

Wang PJ and Huang LC (1976). Beneficial effects of activated charcoal on plnat tissue and organ culture. *In vitro.* **12**: 260-262.

Wang X, Chen JC, Liu GY, Gu MX and Bao CH (1981). Clonal propagation of orchids by means of tissue culture. *Acta. phytophysiol. Sin.* **10**: 391-396.

Wang X (1984). Studies on the orchid clonal propagation and floral bud differentiation by means of tissue culture. *Acta. Phytophysiol. Sin.* **10**: 391-396.

Wang X (1988). Tissue culture of *Cymbidium* plant and flower induction *in vitro. Lindleyana.* **3**(**4**): 184-189.

Wang YC, Cao J, Sanford JC and Wu R (1988). Transformation of rice, wheat and soybean by the microprojectile method, from *Symposium on the Molecular Basis of Plant Development, 17th Annual UCLA Symposia on Molecular and Cellular Biology,* March 26-April 2, 1988, *J. Cell Biochem. Suppl. 12,* Part C, p. 189.

Wang YC, Klein TM, Fromm M, Cao J, Sanford JC and Wu R (1988). Transient expression of foreign genes in rice, wheat and soybean cells following particle bombardment, *Plant Molec. Biol.* **11**: 433-439 (1988).

Weatherhead MA and Harberd DJ (1980). Micropropagation of *Pleione. Hortic. Res.* **20**: 87-89.

Weissinger A, Tomes D, Maddock S, Fromm M and Sanford J (1988). Maize transformation via microprojectile bombardment, in *Molecular Biology: Genetic Improvements of Agriculturally Important Crops*: *Progress and Issues Meeting*, pp. 21-25, Cold Spring Harbor Laboratory, Cold Spring Harbor, New York.

Wessler S and Hake S (1990). Maize harvest, *Plant Cell.* **2**: 495-499.

Wheilan T, Nicolson L and Yu LYY (1997). Clonal propagation of *Spathoglottis plicata* from young plants. **pp.** 193-197. Altaian, A and Ziv, M (ed). Proceedings of the 3rd latl. ISHA symposium on *in vitro* culture and horticultural breeding, Jerusetum, Israel.

White PR (1933). Results of preliminary experiments on the culturing of isolated stem tips *of Stellar ia media. Protoplasma.* **19**: 97-116.

White PR (1934). Potentially unlimited growth of excised tomato root tips in a liquid medium. *Plant Physiol.* (Bethesda). **9**: 585-600.

White PR (1963). The Cultivation of Animal and Plant Cells. Ronald Press.New York

Widiastoety D (1986). Experiments on various kinds of media and buds from different sources in the tissue culture of an orchid (*Aranthera* James Storie). *Bull. Penditian Horticultura.* **13**(**3**): 1-8.

Widiastoety D, Suseno H, Harran, D and Suseno R (1986). Modified media for shoot tip culture *of Aranda* cv. Christine 130. Bull *Penditian Horticulture.* **14** (**2**): 33-37.

Wilde HD, Meagher RB and Merkle SA (1989). Expression of foreign genes in transgenic yellowpoplar plants, *Plant Physiol.*, 98, 114-120 (1992). 105. Wu, R., Transformation of rice and generation of transgenic rice plants, in *Proceedings of Ag. Biotech.* **89**: 190-192 (1989).

Wilfret GJ (1966). Formation of protocorm-like bodies on excised *Cymbidium* shoot tips. *Am. Orchid Soc. Bull.* **35**: 823-827.

Wimber DD and Van Colt A (1966). Artificially induced polyploidy in Cymbidiums. In L DeGar MO. L.R. (ed). *Proc. 5th World Orchid Conf,* **pp**. 27-32. Long Beach, California.

Wimber DE (1963). Clonal multiplication of cymbidiums through tissue culture of the shoot meristem. *Am. Orchid Soc. Bull.* **32**: 105-107.

Wimber DE (1965). Additional observations on clonal multiplication of cymbidiums through culture of shoot meristems. *Cymbidium Soc. News.* **20**: 7-10.

Windholz M (1983). The Merck index: *An encyclopedia of chemicals and drugs* (10th Ed.) Merck. Rahway, N.J.

Winkler H (1908). Besprechung der Arbeit G. Haberlandt's Kultur Versuche mit isoliterten pflanzenzellen. *Bot. Gaz.* **60**: 262-264.

Withner CL (1955). Ovule culture and grwoth of *Vanilla* seedlings. *Am. Orchid. Soc Bull.* **24**: 380-392.

Withner CL (1959). The Orchids - A Scientific Survey, Ronald Press New York.

Wu HZ, Wang XY and Lin TB (1987). A study on shoot-tip culture for oriental orchids. *Acta-Horticulturae-Sinica,* **14**(**3**): 203-207.

Yam TW and Weatherhead MA (1991). Root tip cultures of several native orchids of Hong Kong. *Lindleyana* **6**(**3**): 151-153.

Yam TW (1989). *Conservation, ecology and propagation of the wild orchids of Hong Kong.* Ph.D thesis., Dept. of Botany, Univ. of Hong Kong.

Yam TW, Arditti J and Weatherhead MA (1989). The use of darkening agents in seed germination and tissue culture media of orchids: a review. *J. of the orchid soc. of India.* **3**(**1-2**): 35-39.

Yasugi S (1990). Protoplast culture and cell fusion in orchidaceal. **pp.** 120-124 *Proceedings of Nagoya International Orchid Show.* Nagoya, Japan.

Yasugi S, Kugimiya M and Katsura N (1986). Isolation of mesophyll protoplasts in orchids. *Bull Fac. Hortic. Minamikyushu Univ.* **16**: 91-99.

Ye GN, Daniell H and Sanford JC (1990). Optimization of delivery of foreign DNA into higher-plant chloroplasts, *Plant Molec. Biol.* **15**: 809-819.

Ye XL (1995). Histocytological study of *Dendrobium* hybrid propagation *in vitro. Acta-Horticulturae-Sinica.* **22**(**1**): 83-87.

Yoneda K and Momose H (1988). PLB and plantlet formation by root tip culture in *Phalaenopsis* Bull. *Coll. Agric. Vet. Med.* Nihon Univ. 191-196.

Yoneda K (1986). A study on the culture of the inflorescence tip of the *Phalaenopsis. Bull. Coll. Agri. and Vet. Med.*, Nihon Univ. **43**: 124-127.

Yoneda K, Sakamota T, and Sasaki H (1979). Studies on mericlone of orchids-I. Propagation of Cattleya protocorm by means of meristem culture. *Bull. Coll. Agric. Vet. Med.* Nibon Univ. **36**: 64-73.

Yoneda K, Sakamoto T and Sasaki H (1989). Studies on mericlone of orchids-II. Formation of buddings and plantlets of Cattleya hybrids *Bull. Coll Agric. Vet. Med. Nibon Univ.* **38**: 80-88.

Yu H, Yang SH and Goh CJ (2002). Spatial and temporal expression of the orchid floral homeotic gene DOMADS1 is mediated by its upstream regulatory regions. Plant Mol Biol. **49**(**2**): 225-37.

Zhang ZG, Liu H, Wang L, Cai ZG, Han XZ, Zhao LH and Luo ZJ (1992). Study on culture conditions of protocorm proliferation in white *Dendrobium (Dendrobium candicum). Zhongcaoyao—Chinese Traditional and Herbal Drugs*. **23(8):** 431-433.

Zimmer K and Pieper W (1978). Clonal propagation of *Phalaenopsis* by excise buds. *Orchids Rev.* **86**: 223-227.

Zimmer K and Pieper W (1979). Phalaenopsis - Zur vegetativen Vermehrung. *Gartebarse and Gartecuwelt.* **79**: 258-260.

Index

I

L

M

N

O

P

R

S